Communications in Computer and Information Science 940

Commenced Publication in 2007
Founding and Former Series Editors:
Phoebe Chen, Alfredo Cuzzocrea, Xiaoyong Du, Orhun Kara, Ting Liu,
Dominik Ślęzak, and Xiaokang Yang

More information about this series at http://www.springer.com/series/7899

Slobodan Kalajdziski · Nevena Ackovska (Eds.)

ICT Innovations 2018

Engineering and Life Sciences

10th International Conference, ICT Innovations 2018
Ohrid, Macedonia, September 17–19, 2018
Proceedings

Springer

Editors
Slobodan Kalajdziski (iD)
Faculty of Computer Science
  and Engineering
Saints Cyril and Methodius University
  of Skopje
Skopje
Macedonia

Nevena Ackovska (iD)
Faculty of Computer Science
  and Engineering
Saints Cyril and Methodius University
  of Skopje
Skopje
Macedonia

ISSN 1865-0929   ISSN 1865-0937 (electronic)
Communications in Computer and Information Science
ISBN 978-3-030-00824-6   ISBN 978-3-030-00825-3 (eBook)
https://doi.org/10.1007/978-3-030-00825-3

Library of Congress Control Number: 2018954659

This Springer imprint is published by the registered company Springer Nature Switzerland AG
The registered company address is: Gewerbestrasse 11, 6330 Cham, Switzerland

Preface

The ICT Innovations 2018 conference created and managed a vibrant environment, where participants shared the latest discoveries and best practices, and learned about the symbiosis between engineering and life sciences. The conference promoted the development of models, methods, and instruments of data science, as well as other aspects of computer sciences, and provided a unique environment for the presentation and discussion of new approaches and prototypes in the joint fields of engineering and life sciences.

ICT Innovations conferences are organized by the Association for Information and Communication Technologies (ICT-ACT), whose mission is the advancement of ICT technologies. The main co-organizer and supporter of the 10th International ICT Innovations conference was the Faculty of Computer Science and Engineering and Ss. Cyril and Methodius University in Skopje, Republic of Macedonia.

The ICT Innovations 2018 conference was held in Metropol Lake Resort - Ohrid, during September 17–19, 2018. The special conference topic was "Engineering and Life Sciences," and it celebrated a decade of successful ICT Conferences. Technological innovations have become an essential drive for modern life sciences development. With the advent of high-throughput techniques, life scientists are starting to grapple with massive data sets, encountering challenges with handling, processing and moving information that was once the domain of computer scientists and engineers. Today, cancer diagnostics introduces novel micro/nano-based technologies that can facilitate the detection of cancer biomarkers in early cancer phases, which are more amenable to treatment. Engineering has played a central role in the development of diagnostic and therapeutic instruments, such as prosthetic valves, pacemakers, implantable cardioverters/defibrillators (ICDs), and automated external defibrillators (AEDs). The non-invasive brain imaging techniques produce massive brain-related data that are a great basis for engineers and computer scientists to gather foundational knowledge of the brain and the nervous system, providing the basis for diagnosing and treating several neurological and mental illnesses. Robots have become part of life sciences as tools, models, and challenges. Robots are engaged in 3D printing, manipulating laboratory materials as well as surgery tools. They are emotional support of a class of patients and elderly. They are engaged in direct brain–robot communication using EEG signals. Genetic agents, on the other hand, can be modeled as nanorobots. The interleaving of engineering and the life sciences is becoming more imminent and present. Currently there is a great synergy between these two seemingly different areas. It is transforming the traditional ways of creation and of product assembly, the educational process, health care and other societal phenomena. However, in spite of the great synergy that exists, there are still many open issues and obstacles that need to be addressed and overcome in order to bridge the gap between life sciences and engineering.

ICT Innovations 2018 received 81 submissions from 169 authors coming from 20 different countries. All these submissions were peer reviewed by the ICT Innovations 2018 Program Committee consisting of 190 top researchers based in 47 different countries. In order to assure a high-quality and thorough review process, we assigned each paper to more than three reviewers, resulting in 3.91 reviewers per paper on average; at the end of the review process, there were an average of 3.68 reviews per paper. Based on the results of the reviews, 21 full papers were accepted, yielding a 25.9% acceptance rate.

We would like to express our sincere gratitude to the invited speakers for their inspirational talks, to the authors for submitting their work to this conference, and the reviewers for sharing their experience during the selection process. Special thanks to Ilinka Ivanoska, Bojana Koteska, Monika Simjanovska, Aleksandar Stojmenski, and Kostadin Mishev for their technical support during the conference and their help during the preparation of the conference proceedings.

September 2018   Slobodan Kalajdziski
                                                      Nevena Ackovska

Organization

ICT Innovations 2018 was organized by the Macedonian Society of Information and Communication Technologies (ICT-ACT).

Conference and Program Chairs

Slobodan Kalajdziski   Ss. Cyril and Methodius University,
                              Republic of Macedonia
Nevena Ackovska   Ss. Cyril and Methodius University,
                              Republic of Macedonia

Program Committee

Achkoski Jugoslav   Military Academy General Mihailo Apostolski,
                              Republic of Macedonia
Ackovska Nevena   Ss. Cyril and Methodius University,
                              Republic of Macedonia
Ahsan Syed   Technische Universität Graz, Austria
Aiello Marco   University of Groningen, The Netherlands
Akhtar Zahid   University of Udine, Italy
Aliu Azir   Southeastern European University of Macedonia,
                              Republic of Macedonia
Alor Hernandez Giner   Hernandez Instituto Tecnologico de Orizaba, Mexico
Alvarez Sabucedo Luis   Universidade de Vigo, Spain
Alzaid Hani   King Abdulaziz City for Science and Technology,
                              Saudi Arabia
Antovski Ljupcho   Ss. Cyril and Methodius University,
                              Republic of Macedonia
Armenski Goce   Ss. Cyril and Methodius University,
                              Republic of Macedonia
Astsatryan Hrachya   National Academy of Sciences of Armenia, Armenia
Baicheva Tsonka   Bulgarian Academy of Science, Bulgaria
Bakeva Verica   Ss. Cyril and Methodius University,
                              Republic of Macedonia
Balas Valentina Emilia   Aurel Vlaicu University of Arad, Romania
Balaz Antun   Institute of Physics Belgrade, Serbia
Basnarkov Lasko   Ss. Cyril and Methodius University,
                              Republic of Macedonia
Bojanic Slobodan   Universidad Politécnica de Madrid, Spain
Bozinovska Liljana   South Carolina State University, USA
Braun Torsten   University of Berne, Switzerland

Burmaoglu Serhat	Izmir Katip Celebi University, Turkey
Burrull Francesc	Universidad Politecnica de Cartagena, Spain
Calleja Neville	University of Malta, Malta
Chitkushev L. T.	Boston University, USA
Chorbev Ivan	Ss. Cyril and Methodius University, Republic of Macedonia
Chouvarda Ioanna	Aristotle University of Thessaloniki, Greece
Chung Ping-Tsai	Long Island University, USA
Cico Betim	EPOKA University, Albania
Conchon Emmanuel	Institut de Recherche en Informatique de Toulouse, France
Curado Marilia	University of Coimbra, Portugal
D'Elia Domenica	Institute for Biomedical Technologies, Italy
Damasevicius Robertas	Kaunas University of Technology, Lithuania
Davcev Danco	Ss. Cyril and Methodius University, Republic of Macedonia
De Nicola Antonio	ENEA, Italy
Delibašić Boris	University of Belgrade, Serbia
Dimitrievska Ristovska Vesna	Ss. Cyril and Methodius University, Republic of Macedonia
Dimitrova Vesna	Ss. Cyril and Methodius University, Republic of Macedonia
Dimitrovski Ivica	Ss. Cyril and Methodius University, Republic of Macedonia
Distefano Salvatore	University of Messina, Italy
Djukanovic Milena	University of Montenegro, Montenegro
Ellul Joshua	University of Malta, Malta
Fati Dr. Suliman Mohamed	INTI International University, Malaysia
Fetaji Majlinda	Southeastern European University of Macedonia, Republic of Macedonia
Filiposka Sonja	Ss. Cyril and Methodius University, Republic of Macedonia
Filipovikj Predrag	Mälardalen University, Sweden
Fischer Pedersen Christian	Aarhus University, Denmark
Gajin Slavko	University of Belgrade, Serbia
Ganchev Todor	Technical University Varna, Bulgaria
Ganchev Ivan	University of Limerick, Ireland
Gawanmeh Amjad	Khalifa University, United Arab Emirates
Georgievski Ilche	University of Groningen, The Netherlands
Gicheva Jana	Imperial College London, UK
Gievska Sonja	Ss. Cyril and Methodius University, Republic of Macedonia
Gjorgjevikj Dejan	Ss. Cyril and Methodius University, Republic of Macedonia
Gligoroski Danilo	Norwegian University of Science and Technology, Norway

Goleva Rossitza	Technical University of Sofia, Bulgaria
Gomes Abel	University of Beira Interior, Portugal
Grgurić Andrej	Ericsson Nikola Tesla d.d., Croatia
Gushev Marjan	Ss. Cyril and Methodius University, Republic of Macedonia
Haddad Yoram	Jerusalem College of Technology, Israel
Hadzieva Elena	St. Paul the Apostle University, Republic of Macedonia
Hao Tianyong	Guangdong University of Foreign Studies, China
Hoic-Bozic Natasa	University of Rijeka, Croatia
Hollmann Susanne	SB-Science Management, Germany
Hsieh Fu-Shiung	University of Technology, Taiwan
Huang Yin-Fu	University of Science and Technology, Taiwan
Huraj Ladislav	Ss. Cyril and Methodius University, Slovakia
Huynh Hieu Trung	Industrial University of Ho Chi Minh City, Vietnam
Ilarri Sergio	University of Zaragoza, Spain
Ilievska Natasha	Ss. Cyril and Methodius University, Republic of Macedonia
Ivanovic Mirjana	University of Novi Sad, Serbia
Jakimovski Boro	Ss. Cyril and Methodius University, Republic of Macedonia
Janeska-Sarkanjac Smilka	Ss. Cyril and Methodius University, Republic of Macedonia
Jovanov Mile	Ss. Cyril and Methodius University, Republic of Macedonia
Jovanovik Milos	Ss. Cyril and Methodius University, Republic of Macedonia
Jusas Vacius	Kaunas University of Technology, Lithuania
Kalajdziski Slobodan	Ss. Cyril and Methodius University, Republic of Macedonia
Kaloyanova Kalinka	University of Sofia - FMI, Bulgaria
Karaivanova Aneta	Bulgarian Academy of Sciences, Bulgaria
Karan Branko	Danube Robotics, Serbia
Kawamura Takahiro	The University of Electro-Communications, Japan
Kljajic Borstnar Mirjana	University of Maribor, Slovenia
Kocarev Ljupcho	Ss. Cyril and Methodius University, Republic of Macedonia
Koceska Natasa	Goce Delcev University, Republic of Macedonia
Koceski Saso	Goce Delcev University, Republic of Macedonia
Kon-Popovska Margita	Ss. Cyril and Methodius University, Republic of Macedonia
Kostoska Magdalena	Ss. Cyril and Methodius University, Republic of Macedonia
Kraljevski Ivan	VoiceINTERconnect GmbH, Germany
Kulakov Andrea	Ss. Cyril and Methodius University, Republic of Macedonia
Kundu Anirban	Netaji Subhash Engineering College, Singapore

Kurti Arianit	Linnaeus University, Sweden
Lameski Petre	Ss. Cyril and Methodius University, Republic of Macedonia
Lastovetsky Alexey	University College Dublin, Ireland
Lazarova-Molnar Sanja	University of Southern Denmark, Denmark
Lebedev Mikhail	Duke University, USA
Li Rita Yi Man	Hong Kong Shue Yan University, SAR China
Lim Hwee-San	Universiti Sains Malaysia, Malaysia
Loshkovska Suzana	Ss. Cyril and Methodius University, Republic of Macedonia
Machado Da Silva José	University of Porto, Portugal
Madevska Bogdanova Ana	Ss. Cyril and Methodius University, Republic of Macedonia
Madjarov Gjorgji	Ss. Cyril and Methodius University, Republic of Macedonia
Mancevska Sanja	Ss. Cyril and Methodius University, Republic of Macedonia
Mancevski Dejan	Center for Cardiovascular Diseases, Ohrid, Republic of Macedonia
Marina Ninoslav	St. Paul the Apostole University, Republic of Macedonia
Markovski Smile	Ss. Cyril and Methodius University, Republic of Macedonia
Martinovska Cveta	Goce Delcev University, Republic of Macedonia
Mastrogiovanni Fulvio	University of Genoa, Italy
Michalak Marcin	Silesian University of Technology, Poland
Mihova Marija	Ss. Cyril and Methodius University, Republic of Macedonia
Mileva Aleksandra	Goce Delcev University, Republic of Macedonia
Mileva Boshkoska Biljana	University of Novo Mesto, Slovenia
Mirceva Georgina	Ss. Cyril and Methodius University, Republic of Macedonia
Mirchev Miroslav	Ss. Cyril and Methodius University, Republic of Macedonia
Mishkovski Igor	Ss. Cyril and Methodius University, Republic of Macedonia
Mitreski Kosta	Ss. Cyril and Methodius University, Republic of Macedonia
Mitrevski Pece	St. Kliment Ohridski University, Republic of Macedonia
Mocanu Irina	University Politehnica of Bucharest, Romania
Mohammed Ammar	Cairo University, Egypt
Naumoski Andreja	Ss. Cyril and Methodius University, Republic of Macedonia
Nosović Novica	University of Sarajevo, Bosnia and Herzegovina
Ognjanović Ivana	Univerzitet Donja Gorica, Montenegro
Panov Pance	Jožef Stefan Institute, Slovenia

Paprzycki Marcin	Polish Academy of Sciences, Poland
Petcu Dana	West University of Timisoara, Romania
Petkovic Predrag	University of Niš, Serbia
Pinheiro Antonio	Universidade da Beira Interior, Portugal
Pleva Matus	Technical University of Košice, Slovakia
Pop Florin	University Politehnica of Bucharest, Romania
Pop-Jordanova Nada	Macedonian Academy of Sciences and Arts, Republic of Macedonia
Popeska Zaneta	Ss. Cyril and Methodius University, Republic of Macedonia
Popovska-Mitrovikj Aleksandra	Ss. Cyril and Methodius University, Republic of Macedonia
Popovski Petar	Aalborg University, Denmark
Porta Marco	University of Pavia, Italy
Poscic Patrizia	University of Rijeka, Croatia
Potolea Rodica	Technical University of Cluj-Napoca, Romania
Rechkoska-Shikoska Ustijana	St. Paul the Apostole University, Republic of Macedonia
Rege Manjeet	University of St. Thomas, USA
Reiner Miriam	Technion - Israel Institute of Technology, Israel
Ristevski Blagoj	St. Kliment Ohridski University, Republic of Macedonia
Ristov Sasko	Ss. Cyril and Methodius University, Republic of Macedonia
Rodic Aleksandar	Mihajlo Pupin Institute, University of Belgrade, Serbia
Roose Philippe	University of Pau, France
Rudnicki Witold	University of Białystok, Poland
Ruzic Jelena	Mediteranean Institute for Life Sciences, Croatia
Shafranek David	Masaryk University, Czech Republic
Saini Jatinderkumar	Narmada College of Computer Application, India
Samardjiska Simona	Ss. Cyril and Methodius University, Republic of Macedonia
Savovska Snezana	St. Kliment Ohridski University, Republic of Macedonia
Schwiebert Loren	Wayne State University, USA
Sezerman Osman Ugur	Acibadem University, Turkey
Siládi Vladimír	Matej Bel University, Slovakia
Silva Josep	Universitat Politècnica de València, Spain
Silva Manuel	Instituto Superior de Engenharia do Porto, Portugal
Singh Brajesh Kumar	RBS College, India
Sonntag Michael	Johannes Kepler University Linz, Austria
Spasov Dejan	Ss. Cyril and Methodius University, Republic of Macedonia
Spinsante Susanna	Università Politecnica delle Marche, Italy
Stojkoska Biljana	Ss. Cyril and Methodius University, Republic of Macedonia

Stulman Ariel   The Jerusalem College of Technology, Israel
Sun Chang-Ai   University of Science and Technology Beijing, China
Thiare Ousmane   Gaston Berger University, Senegal
Todorova Stela   University of Agriculture, Bulgaria
Tojtovska Biljana   Ss. Cyril and Methodius University,
                                 Republic of Macedonia
Trajanov Dimitar   Ss. Cyril and Methodius University,
                                 Republic of Macedonia
Trajkovic Ljiljana   Simon Fraser University, Canada
Trajkovik Vladimir   Ss. Cyril and Methodius University,
                                 Republic of Macedonia
Trcek Denis   University of Ljubljana, Slovenia
Trefois Christophe   University of Luxembourg, Luxembourg
Trivodaliev Kire   Ss. Cyril and Methodius University,
                                 Republic of Macedonia
Tudruj Marek   Polish Academy of Sciences, Poland
Valderrama Carlos   University of Mons, Belgium
Varbanov Zlatko   Veliko Tarnovo University, Bulgaria
Velinov Goran   Ss. Cyril and Methodius University,
                                 Republic of Macedonia
Vlahu-Gjorgievska Elena   University of Wollongong, Australia
Vrdoljak Boris   University of Zagreb, Croatia
Wac Katarzyna   University of Geneva, Switzerland
Wibowo Santoso   Central Queensland University, Australia
Xu Shuxiang   University of Tasmania, Australia
Xu Lai   Bournemouth University, UK
Yalcin Tolga   NXP Labs, UK
Yousef Malik   Zefat Academic College, Israel
Yue Wuyi   Konan University, Japan
Zdravev Zoran   Goce Delcev University, Republic of Macedonia
Zdravevski Eftim   Ss. Cyril and Methodius University,
                                 Republic of Macedonia
Zdravkova Katerina   Ss. Cyril and Methodius University,
                                 Republic of Macedonia
Zeng Xiangyan   Fort Valley State University, USA
Zucko Jurica   University of Zagreb, Croatia

Scientific Committee

Slobodan Kalajdziski   Ss. Cyril and Methodius University,
                                 Republic of Macedonia
Nevena Ackovska   Ss. Cyril and Methodius University,
                                 Republic of Macedonia
Danco Davcev   Ss. Cyril and Methodius University,
                                 Republic of Macedonia

Dejan Gjorgjevikj	Ss. Cyril and Methodius University, Republic of Macedonia
Boro Jakimovski	Ss. Cyril and Methodius University, Republic of Macedonia
Gjorgji Madzarov	Ss. Cyril and Methodius University, Republic of Macedonia

Organizing Committee

Gjorgji Madjarov	Ss. Cyril and Methodius University, Republic of Macedonia
Pece Mitrevski	St. Kliment Ohridski University, Republic of Macedonia
Cveta Martinovska	Goce Delcev University, Republic of Macedonia
Azir Aliu	Southeastern European University of Macedonia, Republic of Macedonia
Elena Hadzieva	St. Paul the Apostle University, Republic of Macedonia

Technical Committee

Ilinka Ivanoska	Ss. Cyril and Methodius University, Republic of Macedonia
Bojana Koteska	Ss. Cyril and Methodius University, Republic of Macedonia
Monika Simjanoska	Ss. Cyril and Methodius University, Republic of Macedonia
Kostadin Mishev	Ss. Cyril and Methodius University, Republic of Macedonia
Aleksandar Stojmenski	Ss. Cyril and Methodius University, Republic of Macedonia

Abstract of Keynotes

The Future of Brain Imaging

Vesna Prchkovska

COO and Co-Founder of QMenta, Spain
vesna@qmenta.com

Abstract. Neuroimaging has advanced rapidly in the last two decades. The MRI scanners are getting more powerful offering rich data that can provide detailed insights on the brain structure and function. New computational tools are being developed at a fast pace, and machine learning and big data are the new trends in brain imaging. In this talk I will address some of the most notable advancements in the brain imaging in the recent years from reconstruction to visualization and interaction techniques. Furthermore, I will talk about the future of brain imaging and patient care.

Keywords: Neuroimaging · Machine learning · Big data · Visualization

How Far Humans Are from the Time When Robots Will Become Superior?

Aleksandar Rodić

Head of Robotics Laboratory, Mihajlo Pupin Institute, Belgrade, Serbia
aleksandar.rodic@pupin.rs

Abstract. Human civilization in its long history passed through many delicate phases of its development, fighting for their biological survival, surviving brutal interethnic conflicts, natural disasters and large-scale epidemiological murders, experiencing their ups and downs through several techno-economic industrial revolutions globally stratifying divided into technologically developed and undeveloped communities (societies). Industrial revolutions in the 19th and 20th century contributed to the progress of humankind leaps and bounds, the rise of science and increasing the comfort and overall quality of life. Nowadays, on the scene is the so called 4th industrial revolution, whose main features are mass digitization, global communication and high automation and robotization of industry and society. On this wave of rapid development of mankind, Robotics as a highly interdisciplinary science, has built sophisticated machines for the first time in the history of human civilization that reached the level of skills, physical and intellectual, which can be comparable to the human skills. Are the robots created in the race for greater economic profit of rich industrialists or are they designed to help people in the times of major natural challenges (industrial pollution, climate change, risks from the Cosmos, etc.) to survive and prolong their biological type in the following centuries? Also a substantial analysis meaning will be exposed, can we expect (and when) the robots, as imitations of people (technology clones), to become superior to their biological models – human beings?

Keywords: Human beings · Humanoids · Degree of anthropomorphism Hyper realistic robots · Technology clones

Contents

Invited Keynote Papers

Reconstructing Gene Networks of Forest Trees from Gene Expression Data: Toward Higher-Resolution Approaches

Matt Zinkgraf[1], Andrew Groover[2], and Vladimir Filkov[3(✉)]

[1] Western Washington University, Bellingham, WA, USA
[2] United States Forest Service, Davis, USA
[3] University of California, Davis, CA, USA
filkov@cs.ucdavis.edu

Abstract. In two of our recent systems biology studies of forest trees we reconstructed gene networks active in wood tissue development for an undomesticated tree genus, *Populus*. In the first study, we used time series data to determine gene expression dynamics underlying wood formation in response to gravitational stimulus. In the second study, we integrated data from newly generated and publicly available transcriptome profiling, transcription factor binding, DNA accessibility and genome-wide association mapping experiments, to identify relationships among genes expressed during wood formation. We demonstrated that these approaches can be used for dissecting complex developmental responses in trees, and can reveal gene clusters and mechanisms influencing poorly understood developmental processes. Combining orthogonal approaches can yield better resolved gene networks, but the resulting network modules may contain large numbers of genes. This limitation reflects the difficulty in creating a variety of experimental conditions that can reveal expression and functional differences among genes within a module, thus imposing limits on the resolving power of network models in practice. To resolve networks at a finer level we are now adding a complementary approach to our work: using cross-species gene network inference. In this approach, transcriptome assemblies of two or more species are considered together to identify expression responses common to all species and also responses that are species specific. To that end here we present a new tool, fastOC, for identifying gene co-expression networks across multiple species. We provide initial evidence that the tool works effectively in calculating co-expression modules with minimal computing requirements, thus making cross-species gene network comparison practical.

Keywords: Bioinformatics · Gene networks · Tree genomics

1 Introduction

Transcription is a primary regulatory point in gene regulation. The expression level of a gene reflects in part the regulation from upstream genes encoding

© Springer Nature Switzerland AG 2018
S. Kalajdziski and N. Ackovska (Eds.): ICT 2018, CCIS 940, pp. 3–12, 2018.
https://doi.org/10.1007/978-3-030-00825-3_1

transcription factors (TFs), that bind to the regulatory cis-elements in the promoters of downstream gene targets [8,15]. The specific and combinatorial binding of transcription factors to target gene promoters contributes to spatial, and temporal regulation, as well as rates of transcription [8,11]. Additional levels of regulation include the stability and rates of degradation of gene mRNA transcripts. When RNA sequencing is used to measure mRNA transcripts, the combined influence of gene expression regulation and transcript stability properties is being observed.

Gene regulatory networks describe the links between the regulator genes and their targets. In aggregate these networks are extremely complex and in practice incompletely modeled, even for model species [11]. However, it appears that basic properties of these networks are shared by all organisms surveyed to date. For example, similar to what has been described in animals, yeast and bacteria, a gene regulatory network model in *Arabidopsis*was characterized by hierarchical relationships among regulators, with top-level master regulator TFs controlling expression of lower order TFs that in turn regulate expression of structural genes encoding proteins involved in building new cell components [5,24]. Additional regulatory features such as *feed forward loops* are also shared between the *Arabidopsis*and other gene regulatory networks [24]. Together these results suggest that concepts and approaches developed for modeling gene-based regulatory networks in model eukaryotic and prokaryotic species can be extended to plants [19]. We are exploring how to further extend gene regulatory network concepts and approaches further still, to undomesticated tree species.

Developmental processes in trees are complex, and are controlled through the interactions of thousands of genes [23]. Network modeling concepts and approaches thus hold promise to transcend simple gene-gene interactions to more comprehensively and realistically describe functional processes. Advances in network science and network analytics allow the understanding of gene networks in terms of their modularity [1,13], stability [6,26], controllability [18], and other emergent properties [2,3], some of which can be directly mapped onto phenotypic traits [9,21]. However, the difficulty has been in learning those networks in forest trees and other non-model organisms. Indeed for forest trees, there are thousands of species of ecological or economic importance, presenting the additional challenge of understanding how traits of interest vary across many species.

Advances in sequencing technologies now make gene expression and co-expression analyses tractable in non-model organisms. Projects such as oneKP are generating transcriptome assemblies for 1,000 undomesticated plants across the plant kingdom [20]. A major challenge is how to now leverage these types of data sets from individual and even multiple species to enable new analysis that provide insight into the regulation of complex traits both within and across species.

Co-expression networks provide a framework for integrating data types from multiple experiments [22,25]. Conceptually, gene transcript levels assayed across multiple conditions and developmental tissues can be clustered into groups of genes (i.e. modules) that show highly correlated expression. Overlaying such modules with functional annotations (e.g. Gene Ontology [7]) and correlations with phenotypes can provide insight into the biological pathways influencing developmental processes and the dissection of complex traits [29].

Calculating co-expression networks across species within a phylogenetic framework presents additional opportunities and challenges. On the one hand, the analysis of co-expression modules that show specific phylogenetic relationship can be extremely informative. For example, co-expression modules that are conserved across wide phylogenetic distance could represent signal from evolutionarily ancient regulatory mechanisms. On the other hand, these analyses are conceptually and computationally demanding. For example, the homologous relationships among genes must be established across all species being analyzed, and can include one to one, one to many, or many to many relationships among orthologs. A very practical challenge is thus to establish both orthologous relationships as well as co-expression relationships across multiple species in a computationally tractable way.

2 Our Recent Work on Trees

Plants modify their growth and development in response to external stimuli. As trees grow, they integrate environmental and developmental signals using complex but poorly defined transcriptional gene networks, allowing trees to produce woody tissues appropriate to diverse environmental conditions. Here, we summarize two of our recent studies that illustrate the different approaches we have undertaken to elucidate gene networks of forest trees [28, 29].

2.1 Recent Study #1: Time Series of Gene Expression

Plants respond to gravity to produce new growth that is properly oriented in space. For example, gravistimulation of leaning stems in angiosperm trees such as *Populus* results in modifications of wood development, to produce tension wood that pulls leaning stems upright (see Fig. 1, top row) [12]. This response provides an experimental system to perturb gene expression, and can be temporally calibrated against tissue development and stem movements.

In our recent work, we used gravistimulation and tension wood response to dissect the temporal changes in gene expression underlying wood formation in *Populus* stems [28].

Using time series analysis of transcriptome sequences at seven time points over a 14-day experiment, we identified 8,919 genes that were differentially expressed between tension wood (upper) and opposite wood (lower) sides of leaning stems. Clustering of differentially expressed genes showed four major transcriptional responses, including gene clusters whose transcript levels were associated with two types of tissue-specific impulse responses that peaked at about 24 to 48 h, and gene clusters with sustained changes in transcript levels that persisted until the end of the 14-day experiment. Our approach is illustrated in Fig. 1.

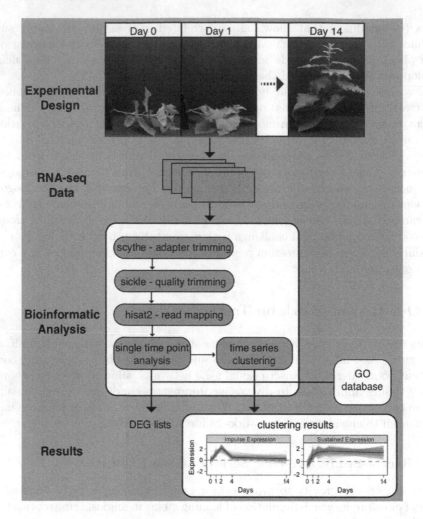

Fig. 1. The steps in our experimental and computational analysis of gene expression of different types of woody tissue in poplar trees, over 14 days [28]. The results include lists of differentially expressed genes (DEGs) over time, clusters of genes coexpressed over time, and functional annotation of those clusters from the GO database [7].

Functional enrichment analysis of those clusters suggested they reflect temporal changes in pathways associated with hormone regulation, protein localization, cell wall biosynthesis and epigenetic processes. Time series analysis of gene expression is an underutilized approach for dissecting complex developmental responses in plants, and can reveal gene clusters and mechanisms influencing development.

2.2   Recent Study #2: Integrating Data from Different Sources

Trees modify wood formation through integration of environmental and developmental signals in complex but poorly defined transcriptional networks, allowing trees to produce woody tissues appropriate to diverse environmental conditions. Basic, conceptual questions include whether multiple environmental inputs impinge upon common regulatory mechanisms, or if individual environmental inputs connect directly to independent regulatory mechanisms controlling growth.

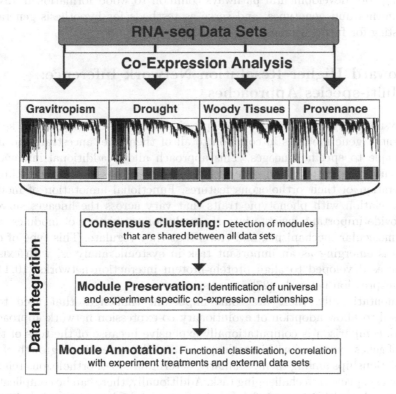

Fig. 2. Flow chart depicting the experimental approach for modeling co-expression networks underlying wood development using data integration and consensus clustering. Original figure appeared in New Phytologist [29].

In order to identify relationships among genes expressed during wood formation, in another recent study [29] we integrated data from our own and publicly available data sets in *Populus*. These data sets were generated from woody tissues and include transcriptome profiling, transcription factor binding, DNA accessibility and genome-wide association mapping experiments. Co-expression modules were calculated, each of which contains genes showing similar expression patterns across experimental conditions, genotypes and treatments. Conserved gene co-expression modules (four modules totaling 8,398 genes) were identified that were highly preserved across diverse environmental conditions and genetic backgrounds. Figure 2 illustrates our approach.

Functional annotations as well as correlations with specific experimental treatments associated individual conserved modules with distinct biological processes underlying wood formation, such as cell-wall biosynthesis, meristem development and epigenetic pathways. Module genes were also enriched for DNase I hypersensitivity footprints and binding from four transcription factors associated with wood formation. The conserved modules are excellent candidates for modeling core developmental pathways common to wood formation in diverse environments and genotypes, and serve as testbeds for hypothesis generation and testing for future studies.

3   Toward Higher Resolution Network Inference: Multi-species Approaches

More systemically, comparison of co-expression networks across multiple species can identify gene modules in common to all of them (i.e. ancestral) and modules unique to specific lineages. This approach allows additional inference of the evolutionary history and other features of co-expression modules through consideration of their orthologous features. Functional annotation of modules and association with phenotypic traits that vary across the lineages surveyed can provide important insights into the evolutionary history of modules, associated molecular mechanisms, and traits that they regulate. This type of comparison is emerging as an important task in systemic analyses, with existing approaches developed to align protein-protein interaction networks [10,14,17] and co-expression networks [27] across species.

Computationally, there are some important challenges that need to be addressed to allow adoption of evolutionary co-expression network approaches. Network comparison is computationally expensive because of the tens of thousands of genes expressed within each species and the super-linear growth of possible relationships among them. Similarly, determination of orthologous relationships across species is a challenging task. Additionally, there can be complications working with multiple non-model species, for which high quality transcriptomes must be established without the aid of guiding genomic sequence.

An existing state-of-the art approach, OrthoClust [27], e.g., is prohibitively slow on mammalian or plant size genomes, and can only work with two species at a time. Here we present a modified approach, fastOC, that eliminates those limitations and enables the construction and comparison of co-expression networks of multiple tree species, in tractable time and on typical hardware.

3.1   Results

We implemented fastOC, an extension of OrthoClust v1.0 [27], which works on two or more species, and is orders of magnitude faster. Both packages use a multi-layer network approach to compute co-expression networks across multiple species. These methods cluster genes based on the correlation of expression patterns within species and align networks across species using orthologous gene relationships. The within species co-expression relationships are defined by each gene and the top-N gene neighbors based on Pearson correlations.

The interactions of genes across species are defined as an orthologous weight that accounts for complex orthologous relationships, such as one-to-one, one-to-many and many-to-many relationships. The main difference between fastOC and OrthoClust v1.0, is that fastOC performs clustering of the entire multi-layer network using the Louvain community detection algorithm [4], instead of the original approach of simulated annealing. The Louvain algorithm is a heuristic approach that assigns genes (nodes) to communities to optimize community modularity. Using many Louvain runs, we calculate how often genes co-appear in the same Louvain communities, and identify both the gene modules conserved across species and those unique to specific lineages. Gene modules represent groups of genes that display high co-appearance in Louvain communities. We provide functions for module detection using dynamic tree cutting of hierarchical dendrograms [16]. Figure 3 illustrates the main steps in the fastOC program.

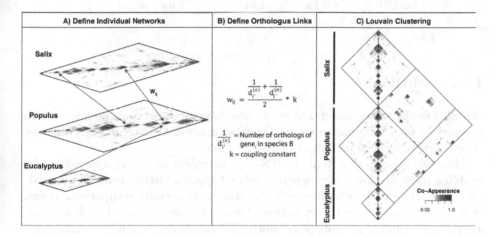

Fig. 3. The three steps in the cross-species co-expression analysis in fastOC.

We implemented fastOC as an R package and the source code is available at https://github.com/mzinkgraf/fastOC under a GPL-3 license. The package includes functions for (1) generating co-expression networks from expression data, (2) calculating orthologous weights, (3) summarizing expression of module genes, and (4) visualization of results. In addition, we developed functions to include parallel processing options to leverage the multi-core functionality of many modern computers. For example, the Louvain clustering function allows for parallel processing using the foreach and doParallel packages in R, to further increase computational efficiency and decrease run times.

Table 1 shows the comparison of running times between our tool, fastOC, and the existing OrthoClust using a single thread on typical hardware. In addition to being between two and three orders of magnitude faster, our examination of the results showed that fastOC finds similar modules in the worm-fly data set as those described in Yan et al. [27]. Specifically, the modules found by fastOC had statistically significant overlap with those found by OrthoClusts ($\chi^2 = 711.8$; $df = 462$; p-value = 6.244×10^{-13}).

Table 1. Comparison of run times for OrthoClust and fastOC. Run times for each method were calculated using 20 runs on a single CPU thread on MacBook Pro with a 3.0 GHz Intel Core i7, and 16 GB RAM. The Simulated and Worm-Fly data sets were obtained from https://github.com/gersteinlab/OrthoClust/. The 3 Woody Species data set is available at https://github.com/mzinkgraf/fastOC/.

	Simulated	Worm vs. Fly	3 Woody Species
# Genes	800	34K	83K
# Edges	13K	147K	625K
OrthoClust v.1	6.5 s	2,308.3 s	Not implemented
fastOC	0.1 s	1.8 s	7.6 s

4 Conclusion

Our past work demonstrated that combining different data types can help resolve gene networks, as can time series gene expression experiments. However, the sizes of resulting functional modules can still be large.

Here we described a tool, fastOC, to complement those approaches by enabling co-expression analyses across related species. fastOC is an almost real-time interactive tool for multi-species co-expression network comparison. It can make possible studies of the evolution of genomic function based on how genes interact in complex regulatory networks. Furthermore, fastOC can efficiently work with more than two species. As of this writing, we are using this tool on 13 tree species containing 291,375 genes and 4,642,738 edges in a multi-layer network.

References

1. Barabasi, A.L., Oltvai, Z.N.: Network biology: understanding the cell's functional organization. Nat. Rev. Genet. **5**(2), 101 (2004)
2. Bergman, A., Siegal, M.L.: Evolutionary capacitance as a general feature of complex gene networks. Nature **424**(6948), 549 (2003)
3. Bhalla, U.S., Iyengar, R.: Emergent properties of networks of biological signaling pathways. Science **283**(5400), 381–387 (1999)

4. Blondel, V.D., Guillaume, J.L., Lambiotte, R., Lefebvre, E.: Fast unfolding of communities in large networks. J. Stat. Mech.: Theory Exp. **2008**(10), P10008 (2008)
5. Brady, S.M., et al.: A stele-enriched gene regulatory network in the Arabidopsis root. Mol. Syst. Biol. **7**(1), 459 (2011)
6. Ciliberti, S., Martin, O.C., Wagner, A.: Robustness can evolve gradually in complex regulatory gene networks with varying topology. PLoS Comput. Biol. **3**(2), e15 (2007)
7. Gene Ontology Consortium: The Gene Ontology (GO) database and informatics resource. Nucleic Acids Res. **32**(Suppl_1), D258–D261 (2004)
8. Djebali, S., et al.: Landscape of transcription in human cells. Nature **489**(7414), 101 (2012)
9. Ellis, T., Wang, X., Collins, J.J.: Diversity-based, model-guided construction of synthetic gene networks with predicted functions. Nat. Biotechnol. **27**(5), 465 (2009)
10. Flannick, J., Novak, A., Srinivasan, B.S., McAdams, H.H., Batzoglou, S.: Graemlin: general and robust alignment of multiple large interaction networks. Genome Res. **16**(9), 1169–1181 (2006)
11. Gerstein, M.B., et al.: Architecture of the human regulatory network derived from ENCODE data. Nature **489**(7414), 91 (2012)
12. Gerttula, S., et al.: Transcriptional and hormonal regulation of gravitropism of woody stems in Populus. Plant Cell **27**, 2800–2813 (2015). pp. tpc-15
13. Han, J.D.J., et al.: Evidence for dynamically organized modularity in the yeast protein–protein interaction network. Nature **430**(6995), 88 (2004)
14. Kalaev, M., Bafna, V., Sharan, R.: Fast and accurate alignment of multiple protein networks. In: Vingron, M., Wong, L. (eds.) RECOMB 2008. LNCS, vol. 4955, pp. 246–256. Springer, Heidelberg (2008). https://doi.org/10.1007/978-3-540-78839-3_21
15. Kellis, M., et al.: Defining functional DNA elements in the human genome. Proc. Natl. Acad. Sci. **111**(17), 6131–6138 (2014)
16. Langfelder, P., Zhang, B., Horvath, S.: Defining clusters from a hierarchical cluster tree: the Dynamic Tree Cut package for R. Bioinformatics **24**(5), 719–720 (2007)
17. Liao, C.S., Lu, K., Baym, M., Singh, R., Berger, B.: IsoRankN: spectral methods for global alignment of multiple protein networks. Bioinformatics **25**(12), i253–i258 (2009)
18. Liu, Y.Y., Slotine, J.J., Barabási, A.L.: Controllability of complex networks. Nature **473**(7346), 167 (2011)
19. Long, T.A., Brady, S.M., Benfey, P.N.: Systems approaches to identifying gene regulatory networks in plants. Annu. Rev. Cell Dev. Biol. **24**, 81–103 (2008)
20. Matasci, N., et al.: Data access for the 1,000 plants (1KP) project. GigaScience **3**(1), 17 (2014)
21. Schadt, E.E.: Molecular networks as sensors and drivers of common human diseases. Nature **461**(7261), 218 (2009)
22. Serin, E.A., Nijveen, H., Hilhorst, H.W., Ligterink, W.: Learning from coexpression networks: possibilities and challenges. Front. Plant Sci. **7**, 444 (2016)
23. Shinozaki, K., Yamaguchi-Shinozaki, K.: Gene networks involved in drought stress response and tolerance. J. Exp. Bot. **58**(2), 221–227 (2007)
24. Taylor-Teeples, M., et al.: An Arabidopsis gene regulatory network for secondary cell wall synthesis. Nature **517**(7536), 571 (2015)
25. Usadel, B., et al.: Co-expression tools for plant biology: opportunities for hypothesis generation and caveats. Plant Cell Environ. **32**(12), 1633–1651 (2009)

26. Von Dassow, G., Meir, E., Munro, E.M., Odell, G.M.: The segment polarity network is a robust developmental module. Nature **406**(6792), 188 (2000)
27. Yan, K.K., Wang, D., Rozowsky, J., Zheng, H., Cheng, C., Gerstein, M.: Ortho-Clust: an orthology-based network framework for clustering data across multiple species. Genome Biol. **15**(8), R100 (2014)
28. Zinkgraf, M., Gerttula, S., Zhao, S., Filkov, V., Groover, A.: Transcriptional and temporal response of Populus stems to gravi-stimulation. J. Integr. Plant Biol. (2018). https://doi.org/10.1111/jipb.12645
29. Zinkgraf, M., Liu, L., Groover, A., Filkov, V.: Identifying gene coexpression networks underlying the dynamic regulation of wood-forming tissues in Populus under diverse environmental conditions. New Phytol. **214**(4), 1464–1478 (2017)

Standardization and Quality Assurance in Life-Science Research - Crucially Needed or Unnecessary and Annoying Regulation?

Susanne Hollmann[1,2(✉)], Teresa K. Attwood[4],
Erik Bongcam-Rudloff[9], Deborah Duca[8], Domenica D'Elia[5],
Christoph Endrullat[3], Marcus Frohme[7], Katrin Messerschmidt[6],
and Babette Regierer[2]

[1] Potsdam University, 14476 Potsdam, Germany
susanne.hollmann@uni-potsdam.de
[2] SB Science Management UG (haftungsbeschränkt), 12163 Berlin, Germany
regierer@sb-sciencemanagement.com
[3] Formerly at Technical University of Applied Sciences Wildau,
Wildau, Germany
cen@katece.de
[4] University of Manchester, Manchester, UK
attwood@bioinf.man.ac.uk
[5] CNR - Institute for Biomedical Technologies, 70126 Bari, Italy
domenica.delia@ba.itb.cnr.it
[6] Department of Synthetic Biosystems, Potsdam University,
14476 Potsdam, Germany
katrin.messerschmidt@uni-potsdam.de
[7] Molecular Biotechnology and Functional Genome Analysis,
Technical University of Applied Sciences Wildau, Wildau, Germany
mfrohme@th-wildau.de
[8] Research Support Services, University of Malta, Msida, Malta
deborah.duca@um.edu.mt
[9] SLU - Global Bioinformatics Centre, Department of Animal Breeding
and Genetics, Swedish University of Agricultural Sciences, Uppsala, Sweden
Erik.Bongcam@slu.se

Abstract. Open Science describes the ongoing transitions in the way research is performed, i.e. researchers collaborate, knowledge is shared, and science is organized. It is driven by digital technologies and by the enormous growth of data, globalization, enlargement of the scientific community and the need to address societal challenges [23]. It has now widely been recognized that making research results more accessible to all societal actors contributes to better and more efficient science, as well as to innovation in the public and private sectors [1, 17]. However, the reuse of research results can only be achieved reliably and efficiently, if these data are valorized in a specific manner. Data are to be generated, formatted and stored according to Standard Operating Procedures (SOPs) and according to sophisticated Data Management Plans [23]. Hence, to generate accurate and reproducible data sets, to allow interlaboratory comparisons as well as further and future use of research data it is mandatory to work in line with good laboratory practices and well-defined and validated

S. Kalajdziski and N. Ackovska (Eds.): ICT 2018, CCIS 940, pp. 13–20, 2018.
https://doi.org/10.1007/978-3-030-00825-3_2

methodologies. Within this article, members of the Cost Action CHARME [10] will discuss aspects of quality management and standardization in context with Open Access (OA) efforts. We will address the question: Are Standardization and Quality Management measures in life-science research crucially needed or introduce further unwanted means of regulation?

Keywords: Open Access (OA) · Standardization
Validity and reliability of data · Standard operating procedures (SOPs)
Quality management (QM) · Quality control (QC) · Quality assurance (QA)
Seal of science · Education

1   Introduction into the Problem

The root word for "science" is the Latin word "scientia" which means "knowledge", with research being the tool towards obtaining this knowledge [21]. Posing research questions and designing experiments to answer those questions have enabled the scientific community and the society in general to gain a deep understanding of the world around us [4]. Access to scientific knowledge is essential for any research activity. Over hundreds of years researchers have been focused on generating data and gaining knowledge to answer particular research questions and to expand the insight to their field of interest. Results and conclusions were published but not open to everybody, and thereby shared with only a limited number of the peer researchers. This resulted in recognition, intellectual merit, and building a common knowledge base but the potential of research results obtained was not fully exploited. In contrast, unimpeded flow of knowledge is important for the implementation of research results in innovations and as a source of inspiration for new ideas [7, 23]. The easier research results are findable and accessible to anyone, the better they can be the basis of further research and innovation. Open Access allows for quick and easy access to relevant scientific content by making scientific information openly available. Besides availability, reusability of research results is crucial for the success of Open Access. This demands a high quality of the data and information to be shared. This also benefits society as a whole: anyone interested can find research results and scientific publications on the Internet, download, read and share them [14, 22].

With the advent of the new Framework Program Horizon 2020 the Open Access policy is broadly implemented throughout the funding program, and many national funders across Europe follow this example. The sharing of research results in Open Access format publications is no longer an option, but a mandatory task for all publicly funded research to ensure accessibility and reusability of research results [23]. With this new policy of openness, we face a new challenge of ensuring quality of research across all scientific disciplines and actors resulting in a need to share and implement standards, SOPs and Good Scientific Practice among all these groups involved.

General repositories of data are not sufficient; rather descriptions and detailed annotations are required to provide open science in order to accelerate the innovation processes. This includes the background behind the generation of certain data formats as well as the potency for interoperability and transferability between different data formats. Definition of quality benchmarks for data are also important in order to define metrics which are applicable and reasonable for building a framework around good data quality [14]. The data quality is directly linked to the quality of the biological samples and procedure/protocols used. Hence, high-quality data can only be obtained by the respective use of high-quality samples [6].

Those aspects face questions about who is responsible for reviewing the data quality and how the handling of low quality data should be performed. Proficiency testing methods for data generation using well-known and high-quality reference data sets are an additional point.

The reproducibility debate in the life sciences revealed that scientific results are not only suffering from lack of reusability but have been demonstrated by some examples that the data seem to a large extent to be irreproducible [8]. The impact is enormous, not only affecting the scientific progress but also limiting the translation of research results into application and increasing the costs of research. Furthermore the perception of the "truth of science" in the public may be deteriorated. In knowledge-based societies this is unacceptable and the recent discussions about fake and true news (also from science) started by political leaders for economic interests opens a Pandoras Box of misinformation with global impact.

The reasons for this "reproducibility crisis" could be manifold, among them lack of a good study design, controls or insufficient documentation, but also non-scientific reasons might contribute to this like pressure to publish, lack of funding to replicate experiments thoroughly or simply the exclusion of negative results. Therefore, a proper implementation of both, standards and standard operating procedures (SOPs) is crucial if we want to overcome the problem. Many initiatives have emerged in recent years to provide standards and tools to their scientific communities ensuring reproducibility through a common framework. The basis to achieve such a common framework demands the agreement for a common language that could allow a successful interaction and cooperation. Besides the commonly used scientific jargon, it needs an agreement on a harmonized terminology and ontology for a successful implementation of a standards framework. Well known examples of existing standardization in today's life-science research are the usage of SI units in all publications as well as defined formats for data of nucleic acids (DNA and RNA) and protein sequences (FASTA format), and for three-dimensional (3D) structure of biological molecules such as the Protein Data Bank (PDB) format which provides a standard representation for macromolecular structure data derived from X-ray diffraction and NMR studies [2].

Hence, standards are the key for addressing and neutralizing the majority of problems related to the management and reuse of big data. There is a strong need for consensus agreements on measurements and stringent performance criteria when dealing with process, data and differences in definitions/terminology.

Data are the most valuable resource for investigating biological systems, but their value is null if their formats do not allow sharing and integration from different sources. Interoperability is a must for omics disciplines. To answer to this pressing need, in 2005 the Research Data Management (RDM) initiative introduced a new set of principles for data management services. Following their principles, data should be FAIR – Findable, Accessible, Interoperable and Reusable [19, 23]. Applying the FAIR data principles mainly addresses the metadata levels in research and does not necessarily take into account the quality of the source datasets. Hence, even if the datasets are published following the FAIR data principles, the quality of the data might be unsatisfactory. As a result, downstream calculations, analyses and proceedings based on such data might be questionable [8, 14].

High-throughput technologies, such as Next Generation Sequencing (NGS), have turned the Life Sciences into a data-intensive discipline and require that the analysis of data is performed using high-performance computing resources. Researchers are now using informatics tools and computational models to decipher the biological information and predict the functioning of cells, organs and whole organisms. These approaches require the integration of data from different types of sources and of different levels of biological information. To achieve interoperability is therefore mandatory and the only way to make data and resources available for their easy exchange and integration is the use of standards. Working across scales and (biological) systems demands now also the harmonization of existing standards - including a common language - between particular fields or analytical technologies. But "...the lack of full standardization hinders effective integration of results from cross-disciplinary collaboration studies" [5, 8]. This implies also to carry out data management and analysis tasks on large scale. One way to standardize such data analysis is by the use of bioinformatics Workflow systems that simplify and automatize the construction of analysis pipelines. Well documented and deposited on suitable databases these workflows can then support reproducibility and provide measures for fault-tolerance. Workflow systems for data analysis need therefore to be part of the documentation connected to any deposited data [13, 20].

Harmonization and interfacing on the level of data formats and structures, descriptors and metadata represent just one side of the coin. The quality of the data provided is an issue of fundamental importance which, as we have described above, has not yet been resolved satisfactorily. The diversity of data sources precludes any straightforward and coherent strategy for maintaining and documenting the quality of the data. Data quality implies not only the fit for use of the data but also metrological traceability, repeatability, reproducibility, consistency and comparability. At best some but not all of the requirements are met by the prevailing data standards. More work is required to ensure that the experimental (including clinical) data underlying the computational models are not only appropriate for the context in which it is being used but is also of sufficient quality for that purpose [22].

For raw data to be of value and of use, they must be both reliable and valid. Reliability refers to the repeatability of findings. Reliability also applies to individual measures starting with the experiments in every wet lab. To test the validity of instruments, procedures, or experiments, research may replicate elements of prior projects or the project as a whole.

A good starting point in standardization measures would be the introduction of quality documentation of experiments which is frequently an obvious lack. Thus, it is crucial to develop and establish procedure-, operating- and inspection instructions as well as quality records. Furthermore, verification documents, particularly for providing a string of documents for the verifiable origin of data is an essential point. Especially the quality records could act as a certificate for potential users (customers) and the general documentation would improve the traceability and transparency with the aim to prove the reliability of results. Another important parameter in quality management

(QM) considerations is the quality assurance (QA). A QA program should contain predetermined quality control (QC) checkpoints for monitoring QA and an extensive documentation including, among others, used devices, reagent lot numbers and any deviation from standard procedures [9, 18]. Moreover, for sequencing data, the QA program should contain QC methods for contamination identification at several stages within the sequencing workflow. These stages comprise the initial sample evaluation, the fragmentation step, the final library assessment, the monitoring of error rates during the sequencing process and the raw data analysis with a focus on reads quality [6, 18].

2 Synopsis

To maximize the impact of research it is not sufficient to ask for OA alone. To this end, the availability of data via portal, platforms and repositories is indeed not enough. The principle of OA can only enter in force successfully, if the actors know and understand clearly what needs to be done at all stages towards them. Here, a huge backlog exists for measures in context with education and training at all connected levels. Validated research allows reproducible performing of experiments based on different techniques and technologies and this would serve in parallel as an indicator and a quality seal for reliable generated data. The idea behind education is that it is dynamic and continuous with questioning itself, its means and content. Education in life science gets stuck without providing validated and shareable research. Without, there is danger of mis-interpretation, misuse and a drift towards fundamentalist dullness. Validation could become realized by trainers and "train the trainer" workshops which should be offered to all research data generators. Those trainings could be set as mandatory for certifi-cation issues, where the EU and EC should be responsible for encouraging and even enforcing standardization and quality issues in research data generated within EU Member Country Institutions.

A huge obstacle nowadays to OA is antiquated mindsets regarding the claim of ownership of results. Although is necessary to deal with the three O (Open Access, Open Science and Open Innovation) in research projects, the idea of making results and data available for further use is still looked upon with distrust by many researchers and university leaders. Here several factors are contributing. There is a delicate balance to be struck between the freedom to preserve the autonomy of scientists on the one hand and the requirements of economic use on the other [16, 23].

Opinions differ on this question by different stakeholders. The researcher or research group often claims the ownership of generated results because they were generated during their projects and are therefore intellectual property. On the other hand, the public claims free accessibility due to the nature of public funding. Fur-thermore, open data are subject to conflict of interest issues as far as the generated data are confidential and therefore linked to financial aspects.

In addition to the Open Access policy the European Commission also promotes Open Science as a new strategy. "Open Science represents a new approach to the scientific process based on cooperative work and new ways of diffusing knowledge by using digital technologies and new collaborative tools. The idea captures a systemic change to the way science and research have been carried out for the last fifty years:

shifting from the standard practices of publishing research results in scientific publications towards sharing and using all available knowledge at an earlier stage in the research process" [3]. This new policy of the European Commission now acknowledges that science and innovation are not restricted to the academic world and the industry, but also takes place in other societal groups. The European Commission intends to embrace these groups typically excluded from the research process and promotes their inclusion. Prerequisite is the openness of research data, results and scientific communication [11, 15].

With this new policy of openness, we face a new challenge of ensuring quality of research across all scientific disciplines and actors. There will be a need to share and implement standards, SOPs and Good Scientific Practice among all these groups involved.

3 Outlook/Perspective

The **initiative CHARME aims to harmonize** standardization strategies to increase efficiency and competitiveness of European life-science research. The members of CHARME welcome the implementation of OA and the resulting possibility of sharing data and results. However, a general uploading of data is not sufficient; rather descriptions and detailed annotations are required to provide open science in order to accelerate the desired innovation processes. This includes the background behind the generation of data as well as the potency for interoperability and transferability between different data formats. At the metadata level huge efforts have been made in the past (Data Management Plan, FAIR Principle, etc.), whereas in context with the raw data hardly any efforts have been made.

We think, that there is a strong need for mechanisms of control for the quality of data which are openly accessible. This data check must be upstream of the open access.

A "seal of quality" similar to a DMP with clear definition of quality benchmarks for data is needed in order to define metrics which are applicable and reasonable for building a framework around good data quality [14] which than are unthinkingly usable for further proceeding by everyone.

This seal of quality should be supported by incentives by funder and publishers.

Incentives should also consider another important aspect, that is education to acceptance of QA plans. There are many examples of resistance of researchers to accept rules that QA plans impose and of how these resistances can be easily overcome by education [12]. To enlarge as much as possible the possibility for courses and implementation of these courses as part of university curricula is a crucial step forward the universalization of a safe and reliable way to make research.

We encourage interested stakeholder to join our discussion and to contribute to enabling the credibility of data and publications which are available from OA.

Acknowledgement. This publication is based upon work from COST Action CHARME (CA15110) supported by COST (European Cooperation in Science and Technology). www.cost.eu.

References

1. Editorial: Benefits of sharing. Nature **530**(7589), 129 (2016). https://doi.org/10.1038/530129a
2. Berman, H., Henrick, K., Nakamura, H.: Announcing the worldwide protein data bank. Nat. Struct. Mol. Biol. **10**(12), 980 (2003). https://doi.org/10.1038/nsb1203-980
3. Directorate-General for Research and Innovation. ISBN 978-92-79-65567-8, https://doi.org/10.2777/380389
4. Understanding Science. University of California, Museum of Paleontology, 3 January 2018. http://www.understandingscience.org
5. Ellingsen, G., Monteiro, E.: Seamless integration: standardisation across multiple local settings. Comput. Support. Coop. Work CSCW **15**(5–6), 443–466 (2006). https://doi.org/10.1007/s10606-006-9033-0
6. Endrullat, C., Glöckler, J., Franke, P., Frohme, M.: Standardization and quality management in next-generation sequencing. Appl. Transl. Genom. **10**, 2–9 (2016). https://doi.org/10.1016/j.atg.2016.06.001
7. European Commission EUR 22836 - Improving knowledge transfer between research institutions and industry across Europe: embracing open innovation (2007). ISBN 978-92-79-05521, http://ec.europa.eu/invest-in-research/pdf/download_en/knowledge_transfe_07.pdf
8. Freedman, L.P., Venugopalan, G., Wisman, R.: Reproducibility 2020: progress and priorities. F1000Research **6**, 604 (2017). https://doi.org/10.12688/f1000research.11334.1. [version 1; referees: 2 approved]
9. Gargis, A.S., et al.: Assuring the quality of next-generation sequencing in clinical laboratory practice. Nat. Biotechnol. **30**(11), 1033–1036 (2012). https://doi.org/10.1038/nbt.2403
10. Harmonising standardisation strategies to increase efficiency and competitiveness of European life-science research (CHARME) Cost Action CA15110. https://www.cost-charme.eu/the-action/about-charme
11. Impact of Emerging Technologies on the Biological Sciences, Report of a Workshop, 26–27 June 1995. https://nsf.gov/bio/pubs/reports/stctechn/stcmain.htm
12. Monya, B.: How quality control could save your science. Nat. News **529**, 456–458 (2016)
13. Wilkinson, M.D., et al.: The FAIR guiding principles for scientific data management andstewardship. Sci. Data **3**, Article no. 160018 (2016). Winkler-News. http://www.ariadne.ac.uk/issue64/datacite-2010-rpt
14. Nickerson, D., et al.: The Human Physiome: how standards, software and innovative service infrastructures are providing the building blocks to make it achievable. Interface Focus **6**(2), 20150103 (2016). https://doi.org/10.1098/rsfs.2015.0103
15. OECD: Frascati Manual 2015: Guidelines for Collecting and Reporting Data on Research and Experimental Development. The Measurement of Scientific, Technological and Innovation Activities. OECD Publishing, Paris (2015). Accessed 27 Oct 2017
16. OECD Principles and guidelines for access to research data from public funding – OECD (2007). http://www.oecd.org/sti/sci-tech/38500813.pdf
17. Open Access in Deutschland. https://www.bmbf.de/pub/Open_Access_in_Deutschland.pdf
18. Rehm, H.L., et al.: ACMG clinical laboratory standards for next-generation sequencing. Genet. Med. **15**(11), 733–747 (2013). https://doi.org/10.1038/gim.2013.92. Working Group of the American College of Medical Genetics and Genomics Laboratory Quality Assurance Committee
19. Sansone, S., et al.: FAIRsharing: working with and for the community to describe and link data standards, repositories and policies. bioRxiv No. 1, 0–9 (2018)

20. Smalter Hall, A., Shan, Y., Lushington, G., Visvanathan, M.: An overview of computational life science databases & exchange formats of relevance to chemical biology research. Comb. Chem. High Throughput Screen. **16**(3), 189–198 (2013)
21. Source: Definition of 'science'. https://www.collinsdictionary.com/dictionary/english/science
22. Stuart, D., et al.: Practical challenges for researchers in data sharing. Springer Nature (2018). https://doi.org/10.6084/m9.figshare.5975011.v1
23. The transition towards an Open Science system - Council conclusions, 8791/16 RECH 133 TELECOM 74. http://data.consilium.europa.eu/doc/document/ST-9526-2016-INIT/en/pdf. Accessed 27 May 2016

Foresight as a Tool for Increasing Creativity in the Age of Technology-Enhanced Learning

Derek Woodgate(✉)

University of Agder, Grimstad, Norway
derek.woodgate@uia.no

Abstract. The inclusion of established methodologies from the science of foresight within an ICT course structure can lead to increased levels of student creativity when these processes are learned through a multimedia enhanced learning environment.

In 2014, I was asked by the ICT Department of the Faculty of Engineering and Science at the University of Agder to design and teach a master's course on the future of mobile learning. After examining the student cohort, department need, technology resources, etc. I decided to undertake a three-pillar approach to the course design, using the science of foresight as an overarching framework and to build out the curriculum around this future-focused platform.

This framework leveraged the unique characteristics of the science of foresight as it seeks to determine and create future opportunities in any specific domain. This provided an opportunity to involve each student in ultimately solving potential future real-life problems by learning the methodologies and techniques for alternative thinking in a plausible future landscape full of unknowns and to develop novel multimedia approaches and tools for the future of mobile learning. This placed considerable emphasis on the development of creative skills, both cognitive and physical and allowed them to experience present and emerging mobile learning challenges for themselves. Accordingly, the final project for the students involved creating a fully structured project thesis that demonstrated how emerging multimedia could deliver future content of a mobile learning course curriculum that he or she had designed and deemed would be salient in 2030. It required each student to create a future course based upon the needs of a plausible, future workforce landscape and potential future jobs incorporating critical 21st Century skills, with increased creativity as a central outcome.

Subsequent research around the question of whether or how the future of mobile learning course delivered increased creativity, established that the science of foresight elements, such as having to place oneself and think in an alternative future landscape, creating novel future scenarios and solving potential future problems of discontinuity were as important or more important for the students' increased levels of creativity as the emerging multimedia learning approaches and tools they used or developed.

Keywords: Creativity · Foresight · Future learning · Multimedia

© Springer Nature Switzerland AG 2018
S. Kalajdziski and N. Ackovska (Eds.): ICT 2018, CCIS 940, pp. 21–35, 2018.
https://doi.org/10.1007/978-3-030-00825-3_3

1   Introduction

In 2014, prior to my joining the University of Agder (UiA) Faculty, the ICT Department of the Faculty of Engineering and Science at UiA established a two-years Master's level, Multimedia and Educational Technologies Program. Whilst the program essentially emphasizes knowledge of existing and emerging technology, new tools and methods for dissemination of knowledge using modern educational technology, I was invited to create and teach a 7.5 ECTS course on the future of mobile learning. I was selected by the then Department Dean, now Rector, Frank Reichert, because of the work I had undertaken as a professional futurist and President of The Futures Lab, Inc. on the future of learning, as well as the fact that I was at the time teaching a PhD class in the Learning Technologies Department of Georgia State University's College of Education and Human Development. I was given substantial freedom in every aspect of the course structure and curriculum development.

In the initial phase, my design architecture was founded on eight key objectives:

1. To apply constructivism-based blended learning as a pedagogical platform [1];
2. To use the science of foresight as a practical framework for studying and creating the future of the domain, namely mobile learning;
3. To apply the principles of opportunity-oriented problem-based learning theory [19];
4. To make increased engagement and creativity pivotal to the outcome and assessment criteria (competency-based);
5. To include key 21st Century learning skills based upon potential future workforce needs in the field [23, 26];
6. To deliver a "multimedia technology in – multimedia technology out" approach for accelerating learning of the technologies and creating future learning technologies;
7. To emphasize alternative thinking techniques to better deal with the complexity of discontinuous change, which is central to creating the future;
8. To create new spatial narratives [4] as a critical platform for delivering student-centered, self-directed learning and the continuing development of each learner's individual competencies and interests.

Given the frequent debate around the scope and precise description of the science of foresight as an academic methodology, I use the term foresight here as a synonym for futures studies, namely, a set of competencies designed to shape or create the future, while taking account of potential discontinuities and disruptors in terms of societal, technological, environmental, economic and political change.

The science of foresight has at its core the development of alternative futures involving a complex framework and toolset of alternative thinking techniques, computational forecasting and modeling, environmental scanning, scenario-building and strategic implementation, commonly based upon horizons longer than a decade. It is practiced by professional futurists. The practice of foresight typically follows a six-stage process [28], which delivers coherent visions with measurable outcomes. Although linear by structure, it is extremely fluid, and non-linear in practice. While it involves the comprehensive application of systems modeling and computational thinking, it equally engages high levels of emotional intelligence, intuition, transdisciplinarity and nomadic thinking. Foresight operates way beyond strategic planning, conventional R&D or innovation practices. It is based upon the creation of potential and opportunities within a future landscape based upon a future baseline of unknowns and uncertainties rather than an extrapolation, projection or predictive modeling of the past or present. Foresight deals with weak signals and emerging issues and not trends, which tend to be already framed as more, or less of something that already exists. To work with discontinuity, it is essential to hijack unexpected signals and break though ideas that help explore and discover novel worlds and paradigms within future contexts that reflect social, cultural and human change, technological inventions, our relationships to the environment, economic transformation and new standards, political structures and legislative platforms.

For the past 50 years or so, foresight has found a growing role in the understanding of plausible and probable futures in the political, military, space and economic arenas, such as the projection of China's role in the world in 2050, the economic growth of Brazil or the peak levels of energy and oil or the likely military strategy of Iran over the next 80 years. However, over the past couple of decades, we have witnessed its accelerated application in industry and business (smart infrastructure, 3D printed excavator, adaptive driving systems, etc.) as well as the advancement of technologies and fields such as developing future learning systems [29], or reshaping the future of medicine [21], or projecting advances in connecting AI and brain science for the development of cognitive capacities and mental well-being [25].

Prior to developing the pedagogical structure and syllabus, I examined: (a) the student cohort: twelve M/F master's students, mainly with multimedia background and development experience, but with no knowledge or experience of deep theory or studies of the future (foresight); (b) the department need, specifically to integrate the Future of Mobile Learning course with the other new courses, namely: Interaction Design, Communication, Cooperation and Research Methods, eLearning and Games, eTeaching, eCourse development and Education, Visualization; (c) technology resources relevant to the future of mobile learning: augmented reality, virtual reality and 3D/4D worlds, holograms, simulation, new devices technologies, frameworks and platforms, structures, materials, batteries, and interfaces, xAPI, avatars and learning agents, apps, Web 4.0. and 5.0., GPS, LMS, etc. These were augmented with instruction on areas such as machine learning & deep learning, neural networks, cognitive feedback, natural interfaces, quantum computing, implants, claytronics, etc.

Furthermore, I wanted the students to master:

- How to learn in alternative spatial narratives such as 3D worlds;
- How to create augmented reality tools;
- How to consider holograms for teaching and learning potential;
- How to build a learning avatar/agent;
- How to develop learning apps, distributed component-based architecture for student adaptive eLearning, and much more.

Accordingly, I decided to build a three-unit approach to the course design, using the science of foresight as an overarching framework:

(1) Unit One: Exploring the future of mobile learning;
(2) Unit Two: Creating the future of mobile learning;
(3) Unit Three: Creating a design brief for future mLearning course to be taught in 2030. This involves incorporating the future of the subject domain, the future student in terms of human change, learning climate and environment, learning tools, pedagogical advances.

2   The Pedagogical Framework

The choice of a constructivist-blended pedagogical framework was grounded in the need to combine face-to-face and eLearning/online and distance learning, due to both my being primarily located in the USA at the time, also the decision to integrate a kinetic/experiential learning aspect to the course, which required face-to-face group collaboration. This approach also facilitated an in-depth study of the learner and course performance, enabling continuous adaptation for the purpose of learner, mentor, content, and delivery optimization. In addition, I built in elements of kinetic/experiential learning. Blending integrates a variety of event-based learning activities, including face-to-face class room, live e-learning, and a week-long (40 h) experiential learning module. Essentially, the course emphasized student-centered learning, and a high level of customization, based upon the learner having the possibility of creating her/his own story, making the course their own, by contributing to its continued development, whilst extending and enhancing her/his incoming skillset and having the freedom to deliver in any format and try new approaches and novel ideas. Constructivism provides a robust platform for learners to become more active in developing and creating knowledge, both collaboratively and individually, based upon their experiences, perspectives and interpretations.

In building off the eight objectives described in the introduction, I had the freedom to create a new learning system or integrated structure, which I named the Living Learning System (LLS) (Fig. 1). It translated the objectives into a pedagogical framework and incorporated the key aspects of constructivist-blended learning, with the needs for 21st Century learning skills, and primary approaches for increasing engagement and creativity such as multimodal delivery of immersive experiences.

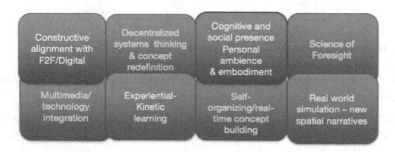

Fig. 1. The pillars of the Living Learning System [28]

The LLS builds on ten underlying learner pillars, namely: (1) each student learns her/his own strategy (personal relevance); (2) alternative thinking techniques are critical; (3) technology enhanced immersive exposure and interaction; (4) optimized ZPD (zone of proximal development) transitioning; (5) open communication; (6) opportunities for constant testing and enhancement of individual competencies; (7) technology supported learning; (8) increased creativity (input and output); (9) freestyle delivery of assignments; (10) contribution to the course design and progression [28]. The Living Learning System was designed to create a learning climate, structure and pedagogical platform that provides a sustainable future for learning in line with the plausible future workforce needs and the specific objectives and requirement of the Future of Mobile Learning course.

In terms of pedagogical approach, instead of using Problem Based Learning (PBL), which is where an initial problem serves as a catalyst for subsequent learning [8] and is an important principle of Engagement Theory [17], I opted for Opportunity Oriented, Problem Based Learning [19]. This is particularly relevant as the course is based upon creating and delivering a plausible future landscape and scenarios, where the problems are rarely identified until the process is completed. Although I designed in elements of experiential learning throughout the course, the week-long workshop in the middle of the course, has proven ideal for bringing everybody up to speed on the course content, as well as enabling students to develop the future-focused multimedia enhanced scenarios, which simulate a plausible future reality.

The Practical Enquiry Model below (Fig. 2) reflects how the key course elements of my Science of Foresight course are integrated. This is especially relevant to blended learning, where a community of enquiry is required to create greater cognitive presence

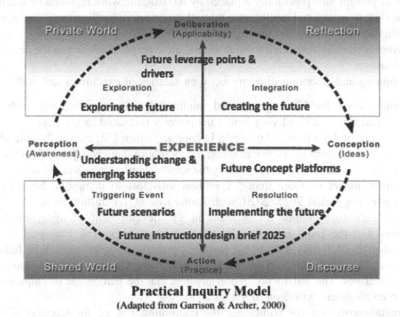

Practical Inquiry Model
(Adapted from Garrison & Archer, 2000)

Fig. 2. Integration of the science of foresight into the practical enquiry model [31]

for higher order thinking processes. Alternative critical thinking techniques are key aspects, of the science of foresight, especially when pedagogical objects include creativity, problem solving, intuition, and insight [10].

3   The Growing Need for Creatives

The focus on engagement and creativity reflects the changing economic structure, driven by emergence of the "gig economy" or alternative workforce and specifically the growing need for creatives (i.e. the growing creative class) [7]. In the "gig economy", temporary positions are common and organizations contract with independent workers for short-term engagements. A study by Intuit [13] predicted that by 2020, 40% of American workers would be independent contractors. This is being driven predominantly by digitalization, in the sense that the workforce is increasingly mobile and work can increasingly be done from anywhere, so that job and location are decoupled. The research by Rand found that 94% of net employment growth from 2005–2015 in US was from alternative work arrangements, and that 25% of workers aged 55–74 have alternative work [18].

Whilst, there is considerable debate over the accuracy of recent predictions regarding global economic growth and particularly growth within specific areas of the economy, an analysis of recent reports from the OECD/RAND/OXFORD/TFL provides the following indicators:

- 6.8 million creative new jobs in US by 2020;
- Employment in arts, design, and media rising by 12%;
- 50% of present jobs potentially replaced by AI. Routine work replaced by machines – the shift from assisted intelligence through augmented intelligence to autonomous intelligence and connected intelligences;
- Unemployment up (expected to be at around an average of 8% across the developed economies);
- Continuing and widening disparity between salaries of executives and staff.

Potential need for GMI (guaranteed minimum income) particularly with 43.1 million Americans (12.7%) living below the poverty threshold in US [24].

The Future of Jobs Report by World Economic Forum [32] claims that the global workforce is expected to experience significant churn between jobs, families and functions. Across the countries covered by the Report, current trends could lead to a net employment impact of more than 5.1 million jobs lost to disruptive labor market changes over the period 2015–2020, with a total loss of 7.1 million jobs.

These indicators assume an increase in the development of agile, intelligent enterprises and the rapid reduction in full-time employees. The Futures Lab, Inc. projects that by 2020 only 40% of the corporate American workforce will be full-time employees. The remainder will be made up of contingent workers, consultants and external creatives. The 2016 PWC report predicts that the percentage of employees in full-time employment will drop to 9% in 2030.

Simultaneously we are witnessing the continuing rise of the Creative Class – changing workforce structure [7]. Between 1980 and 2015, we have witnessed a significant shift in the make-up of the global workforce, particularly in developed western countries. This shift has been driven primarily by the emergence of digital technologies, globalization, expansion of the service industry in line with considerable social and lifestyle changes, the migration from rural areas to cities (especially mega cities), growth in alternative economies, growth in non-routine cognitive jobs, women in the workplace, increase in gained bachelor and advanced degrees, etc. This shift has resulted in a major percentage growth in the share of the creative class in the global workforce structure with growth in the USA over the 35-year period from 8%–36% and the creative class now representing 42% of the workforce in Norway [3]. A further 6.8 million creatives will be required in the USA by 2020, while the current US education system is projecting a considerable shortfall. The power of the growing creative class is such that there are currently over 50 million workers in the USA and a similar number in Europe.

Another major metric is R&D investment and this element is a critical contributor to future economic growth in the western economies. Globally, the spending on R&D has shown a consistent growth, which more than trebled between 1996 and 2015, reaching $1.7 trillion. There is also a projected growth in the spend as a percentage of GDP with the forerunners being the US, which is planning to invest 2% of GDP in R&D; and EU, which is targeting 3%. To sustain this level of growth in research investment, a pipeline of innovations is required, together with a growing need for creativity.

4 Incorporating 21st Century Skills

The necessity of interpreting the 21st Century skills into curriculum design approach and content was critical. Expanding the 4Cs and the work of Tittenberger and Siemens [26] and Rotherham and Willingham [23], I reworked the 11 skills into the emerging skill repertoire:

(a) Sense making
(b) Cognitive interaction
(c) Domain expertise
(d) Social/emotional intelligence
(e) Cross-cultural competency
(f) Virtual collaboration
(g) Transmedia literacy
(h) Computational thinking
(i) Innovation & design thinking
(j) Social-motivated creativity
(k) Novel and adaptive thinking
(l) Transdisciplinarity.

If one were to build a future-focused course based upon traditional thinking it would become redundant in the immediate future. Also, it would make it difficult for the learner to project him/herself into the future, which is an essential quality when studying the science of foresight or the future of a designated domain.

In order to incorporate the 21st Century skills into the course design, I ensured that each of the aspects of the emerging skill repertoire above were woven into the course design in line with the LLS pillars. This is the outline I used for building the learning strategy:

(a) Constructive alignment • Curriculum, teaching methods, learning environment an assessment procedures constructivist/learner centric • Blended – community of inquiry • Multiple learning environments incl. experiential • Bring it what they know and experience (multimedia skills) • Learners to build course in situ • Future-focused • Competency-based assessment

(b) Decentralized systems thinking/concept redefinition • Rethink all concepts such as "learning, mobility", etc. (transformative concept development) • Science of Foresight methods – change theory • Systems thinking, Causal Layered Analysis, Opportunity mapping, etc. • Rhizomatic thinking skills for scenario development • Think like a DJ • Amorphoscapes, data knitting, jouissance (Free Play) techniques

(c) Cognitive & social presence • Student selection of integrated multimodal academic activities and cadence • Contribute to building the course in real-time • Deliver key assignments in any format • Create a context of personalized, relevant learner modeling • Social learning • Multi-disciplinary teams • Student contribution to development of learning climate

(d) New spatial narratives • F2F 3 times per semester • "Creating the future" workshop (1 week; interaction and simulation) • Student telepresence system • Skype – open mentoring • Testing mobile learning strengths • Interactive video • "Living the future" – testing and creating VR and AR tools for learning and 3D environments

(e) Technology enhanced accelerated learning • Development of AI learning agents • Collaboration with Neuroscience Department • On-line interactive communications and support – live reading • Use of multimedia learning tools through combination of Media Lab and Learning Lab • Develop UI & platform knowledge and skills • Gamification and transmedia development

(f) Experiential–kinetic learning • Week-long workshops to apply theory • Building of future learning environments • Programming skills for new learning tools • Applying those tools to existing instructional design to understand gaps • Maker lab • Research in regional schools and institutions

(g) Self-organizing, real-time course building • Creating own story and relevance and approach to course • Accumulated acquisition of capabilities • Constant updating of competence and contribution • Rather than mastery of a single body of knowledge – integrate multiple disciplines i.e. multimedia and learning, neuroscience, etc. • Changing perspectives – shifting directions

(h) Real-world simulation • Development of current status reports and strategic vision • Building of "live" future scenarios • Development of AR prototypes • Development of VR worlds for learning • Test various LMS systems and support tech • Final project – building future-based curricula.

5   Foresight as a Tool for Increasing Engagement and Creativity

There are multiple studies identifying the intrinsic affordances that multimedia and ICT have the potential to transform learning in a way that our basic conceptions of knowledge will be rewired (e.g. Lankshear [15] or Cigman and Davis [6]). Equally, technology-rich learning environments or learning spaces designed accordingly can enable higher levels of motivation, participation, engagement and learner performance [27]. However, there is little research demonstrating how the science of foresight as a course framework can act as a credible driver for increased engagement and creativity, whilst there is research demonstrating the power of multimedia and design tools and strategies as a foresight tool [9].

By making the science of foresight, the framework (Fig. 3) for the future of the domain, in this case The Future of Mobile Learning, there is an opportunity to find an alternative track towards the immersive experience, which can be studied through a very different lens than the immersion created by augmented multimedia.

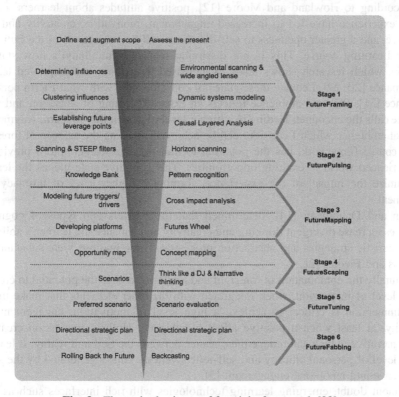

Fig. 3. The revised science of foresight framework [28]

Three key aspects of the science of foresight stand out as potential generators of immersion, namely: having to project yourself into a future landscape of unknowns; opportunity to fearlessly develop novel ideas; the requirement to create future scenarios. In this particular course, I emphasized six key elements in the course design that would potentially enhance engagement and creativity: self-direction (active control), opportunity to enhance existing skills and deliver assignments in any/multiple formats (freedom), ability to contribute to the course development and the future of learning in general (pride), having to think and work in a future landscape (excitement), learning, applying and building multimedia (self-esteem), creating and delivering the unexpected, novel ideas (inspiration). Accordingly, the science of foresight played an important role in creating an immersive learning narrative for the learner.

This learning narrative leveraged the spatial narrative [4] that merges physical, virtual and cognitive space, which are all part of the constructivist-blended learning experience. This narrative structure has the potential to deliver new perspectives, metaphors and visions for the learner's personalized, externalized world, leading to increased engagement, creativity and learning. The objective is to transport the learner from participant to immersant [16].

According to Howland and Moore [12], positive attitudes about learners' online course experience have led to increased engagement, motivation, creativity and performance, and a greater openness to self-direction – all key objectives of the Future of Mobile Learning course. This is not dissimilar to Csikszentmihalyi's flow state of pleasure, which research shows leads to increased motivation and improved learner performance [20, 22]. I equate these states of consciousness with what I term personal ambience [31]. Personal ambience reflects an extended feeling of sensation and what Deleuze calls the encounter leading to a higher level of desirability to learn. Augmented personal ambience which I consider can be achieved by using the science of foresight as the course framework, has the potential to transport the learner into a previously unexperienced world of exciting opportunities. This cognitive state incites the learner to optimize the impact of the experience, forming a higher level of intimacy and attachment.

Isen and Daubmann [14], determined that these levels and types of cognitive processes increase divergent thinking and creativity and enhance the learner's ability to apply heuristic strategies and abstraction processes. These findings were substantiated by Bless and Fiedler [2].

Naturally the combination of foresight and multimedia have the potential to create a higher level of immersion, when integrated with the other elements that make up the LLS. Immersion is critical to achieve a high level of motivation, engagement and creativity, at least when a positive affect is achieved. Holistic immersion creates a spatial narrative, which can reduce the distance between the Self and reality. It leads to a high level of personal efficacy and self-reflectiveness. This is generated by the sense of presence and emotional engagement [5].

Without doubt, emerging learning technologies with rich interfaces such as augmented reality, virtual reality, simulation, etc. embedded into intelligent learning environments; and the power of personalized AI learning agents to create interactive simulation and representation can expand human imagination beyond real-world knowledge. It is precisely this imagination when inspired and augmented by the demands of the science of foresight that is able to tap into higher levels of immersion, cognitive transformation, engagement and creativity. Creating imaginative worlds through the process of potential future scenarios can change the learner's ability to extend his or her capability for expressing multiples dimensions of perception beyond an evolution of their current projections. The visions created with these future scenarios allow the learners to expand what they see, by creating new narratives and iterations of the original scenarios. The process is seen to produce positive affect which according to a recent Swedish study [11] changes mindfulness and enables students to take an increasingly positive attitude to future potential. The same research suggests that mindfulness training is a key attribute to achieving higher levels of positive affect. Such practices dovetail with the science of foresight alternative thinking approaches that are a key part of the Future of Mobile Learning course. These include the ability to develop new perspectives through approaches such as rhizomatic/nomadic thinking, and the remixing of creative imagination ("Think like a DJ") [30]. This latter process involves a bundle of alternative thinking methods aimed at subverting assumptions, peeling away the surface, revisiting values and signifiers, determining aspects of fracture, critical impact points and disruptors, reconstructing dystopian realities, paradoxes and hybrids and adding potential events and wildcards. The bundle includes techniques such as: convergent disconnects, amorphoscapes and random worlds, parallel realities, collision – asymmetries, biomimicry – bionomics, parallel realities, Body > Data > Space and "magical and alchemical". Given a deep sense of freedom and reducing the fear of failure, by understanding that there is not one, but many plausible futures, none of which are absolute, when created a decade ahead, allows students to apply these in multiple combinations or singularly to challenge more conventional future forecasting and modeling methods that use an extension of the present as their baseline.

6 Research

After five semesters of teaching The Future of Mobile Learning course at the University of Agder, I conducted some qualitative and quantitative research to understand whether the LLS-directed design had met the goals of immersing students in a way that would lead to increased engagement, creativity and overall performance. The quantitative research delivered 33 respondents (from the 45 contacted). All 33 are current or have been graduate students and have taken The Future of Mobile Learning Course within the past four years. A part of the results from the quantitative research, which was conducted in the form of a questionnaire under the guidance of NSD are shown below. This was followed up with personal interviews with six of the respondents. Here is a summary of the research.

Responses about engagement:

- 57% felt developing futures scenarios (future focus subject matter) the most engaging element, through excitement and inspiration (the joy of learning).
- 67% considered learning, experiencing and creating future multimedia technologies enhanced their level of engagement.
- 47% considered the experiential learning aspects to be critical.

- 47% expressed that future thinking techniques were critical to engagement.
- 47% considered freedom of expression and the ability to deliver in any format important for engagement.

Responses about creativity:

- 75% stated that having to think about the future and place themselves into the future landscape increased their creative skills.
- 70% strongly agreed or agreed that the course helped increase their personal creativity.
- Increased personal creativity was due to: (a) ability to develop unique, novel ideas (60%); (b) deliver in any format (56%); (c) learn in and with, and to create multimedia (60%); (d) apply alternative thinking techniques (43%); (e) creative thinking (50%).
- 50% claimed that the course was better than any other parallel courses at increasing personal creative skills.

Responses about the Living Learning System:

- 80% found the course to be very interesting and novel.
- 50% considered applying futures thinking was critical to the success of the course.
- 40% felt that working in the future was a critical point of the course.
- 67% claimed that learning and using multimedia technologies were crucial for the their success throughout the course.
- 70% stated that the new learning modes experienced throughout this course led to higher engagement and creativity.
- 63% said that the integrated elements within the LLS together increased their creativity levels.
- 67% stated that the LLS approach significantly increased their overall well-being and attitude towards learning.
- They felt that the fact that the LLS allowed them to be more creative, which made it easier to be creative.

Respondents were asked to elaborate of why they did feel that LLS approach had significantly increased their overall well-being and attitude towards learning:

- "By going deeper into theoretical design of future learning, it made me more engaged."
- "The new learning modes made it easier to step out from the comfort zone."
- "Thinking outside the box and engaging interesting technologies, helped me with my motivation and engagement."
- "The ability to choose approaches and tools that I felt comfortable with or interested in, vastly increased my ability for self-direction/self-determination."
- "I felt confident, because the course was really personalized."
- "I felt like in was in complete control of my learning, and how the information was given to me. This feeling of control of my own ideas and execution allowed to feel that I wanted to go on, rather than I was instructed to."
- "Interesting to see how futurists believe the world will change regarding learning and how students will interact through digital tools and devices."

– "The future thinking processes made us feel more free and engaged than other courses and allowed us to express ourselves in a creative manner."

7 Conclusion

The introduction of new industries, business and economic, social and political structures, will continue to seek higher levels of employee creativity. Therefore, it is essential that future learning systems and approaches should be designed to deliver increased levels of creative students with a repertoire of competencies that are optimized for the emerging future. Consequently, incorporating 21st Century skills into the holistic framework of future curricula design is crucial. This will require a future-focused platform that enables students to project themselves into the future to explore and discover unexpected potential opportunities. Our research shows that a course, such as the Future of Mobile Learning master's course that I designed, which requires the student to think and create in a future landscape, create novel future scenarios and apply alternative thinking approaches significantly increases learner engagement and creativity. The dynamics of these levels can be further heightened and accelerated, when augmented multimedia is integrated into the learning approach. I created a new learning system (LLS) specifically designed to accommodate the 21st Century skills, a constructivist-blended learning approach based upon opportunity-oriented problem-based learning theory and with a clear future focus in terms of input and output based upon a 10 to 15-year horizon. By making the science of foresight the basic platform and framework, I created a new learning narrative that maximized the benefits of real, virtual, and moreover cognitive transformation, including mindfulness. The framework focused on a high level of personal efficacy and self- reflectiveness generated by the sense of presence, motivation and emotional engagement. Which focused on the reduction the distance between the learner Self and reality. With immersion at its core and increased personal ambience as its desired outcome, the use of the science of foresight, with its necessary alternative thinking approaches, proved to be a very effective tool for increasing learner engagement and creativity. While the LLS demonstrated a powerful potential new approach to the future of learning for advanced level students.

References

1. Al-Huneidi, A., Schreurs, J.: Constructivism based blended learning in higher education. Int. J. Emerg. Technol. Learn. (iJET) **7**(1), 4–9 (2012)
2. Bless, H., Fiedler, K.: Mood and the regulation of information processing and behavior. In: Forgas, J.P., Williams, K.D., van Hippel, W. (eds.) Hearts and Minds: Affective Influences on Social Cognition and Behavior, pp. 65–84. Psychology Press, New York (2006)
3. Bock, M.: Life worlds and information habitus. Salzburg Vis. Commun. **3**(1), 281–293 (2009)
4. Bodenhamer, D., Corrigan, J., Harris, T.: Deep Maps and Spatial Narratives. Indiana University Press, Bloomington (2015)

5. Chirico, A., Yaden, D., Riva, G., Gaggioli, A.: The potential of virtual reality for the investigation of awe. Front. Psychol. **7**, 1766 (2016)
6. Cigman, R., Davis, A.: New Philosophies of Learning. Wiley-Blackwell, Malden (2009)
7. Adler, P., Florida, R., King, K.E., Mellander, C.: Creative economies. Martin Prosperity Institute Report (2015)
8. Fogarty, R.: Problem-Based Learning and Other Curriculum Models for the Multiple Intelligences Classroom. IRI/Skylight Training and Publishing, Arlington Heights (1997)
9. Gabrielli, S., Zoels, J.C.: Creating imaginable futures: using human-centered design strategies as a foresight tool. In: Proceedings of the 2003 Conference on Designing for User Experiences, DUX 2003, San Francisco, California, 06–07 June 2003, pp. 1–14 (2003)
10. Garrison, D.R., Anderson, T., Archer, W.: Critical inquiry in a text-based environment: Computer conferencing in higher education. Internet High. Educ. **2**(2–3), 1–19 (2000)
11. Hansen, E., Lundh, L., Homman, A., Wångby-Lundh, M.: Measuring mindfulness: pilot studies with the Swedish versions of the mindful attention awareness scale and the Kentucky inventory of mindfulness skills. Cogn. Behav. Therapy **38**(1), 2–15 (2009)
12. Howland, J., Moore, J.: Student perceptions as distance learners in internet-based courses. Distance Educ. **23**(2), 183–195 (2002)
13. Intuit Inc.: Intuit 2020 Report: Twenty Trends That Will Shape The Next Decade. Intuit Inc. (2010). http://http-download.intuit.com/http.intuit/CMO/intuit/futureofsmallbusiness/intuit_2020_report.pdf. Accessed 24 May 2017
14. Isen, A.M., Daubman, K.A.: The influence of affect on categorization. J. Pers. Soc. Psychol. **47**, 1206–1217 (1984)
15. Lankshear, C., Knobel, M.: Digital literacy and digital literacies: policy, pedagogy and research considerations for education. Nord. J. Dig. Lit. **9**(4), 8–20 (2015)
16. McRobert, L.: Char Davies' Immersive Virtual Art and the Essence of Spatiality. University of Toronto Press, Toronto (2007)
17. Miliszewska, I., Horwood, J.: Engagement theory. ACM SIGCSE Bull. **38**(1), 158 (2006)
18. Nber.org: December 2016 NBER Digest (2018). http://www.nber.org/digest/dec16/w22667.html. Accessed 25 July 2018
19. Oganisjana, K., Laizans, T.: Opportunity–oriented problem–based learning for enhancing entrepreneurship of university students. In: Proceedings of Procedia - Social and Behavioral Sciences, vol. 213, pp. 135–141 (2018)
20. Pearce, J., Ainley, M., Howard, S.: The ebb and flow of online learning. Comput. Hum. Behav. **21**(5), 745–771 (2005)
21. Rose, N., Aicardi, C., Reinsborough, M.: Foresight report on future neuroscience and the human brain project (2015)
22. Rossin, D., Ro, Y.K., Klein, B.D., Guo, Y.: The effects of flow on learning outcomes in an online information management course. J. Inf. Syst. Educ. **20**, 87–98 (2009)
23. Rotherham, A.J., Willingham, D.T.: "21st-century" skills. Am. Educ. **34**, 17–20 (2010)
24. Semega, J.L., Fontenot, K.R., Kollar, M.A.: Income and Poverty in the United States: 2016. US Census Bureau, US Department of Commerce, 12 September 2017. https://www.census.gov/library/publications/2017/demo/p60-259.html
25. The Government Office for Science: Foresight Mental Capital and Wellbeing Project. Final Project report, pp. 20–21. The Government Office for Science, London (2008). https://assets.publishing.service.gov.uk/government/uploads/system/uploads/attachment_data/file/292450/mental-capital-wellbeing-report.pdf. Accessed 3 June 2018
26. Tittenberger, P., Siemens, G.: Handbook of Emerging Technologies for Learning. University of Manitoba, Winnipeg (2009)
27. Wankel, L., Blessinger, P.: Increasing Student Engagement and Retention Using Multimedia Technologies. Emerald, Bingley (2013)

28. Woodgate, D., Isabwe, G.: Developing future vision landscape and models of technology enhanced learning. In: INTED 2018 Proceedings, pp. 6824–6835 (2018)
29. Woodgate, D.: The "sense event" – from sensation to imagination. Comput. Entertain. **10**, 3:1–3:33 (2012)
30. Woodgate, D.: Future Frequencies. Fringecore Publishing, Austin (2004)
31. Woodgate, D.: Optimizing student interaction, engagement and collaboration through blended learning (keynote presentation). In: Adila Conference, Kristiansand, 8–9 December 2016 (2016)
32. World Economic Forum: The Future of Jobs Employment, Skills and Workforce Strategy for the Fourth Industrial Revolution, 1st edn. World Economic Forum (2016). http://www3.weforum.org/docs/WEF_Future_of_Jobs.pdf. Accessed 22 May 2017

Proceeding Papers

Electrophysiological and Psychological Parameters of Learning in Medical Students with High Trait Anxiety

Sanja Mancevska[1]($\boxtimes$), Adrijan Božinovski[2]($\boxtimes$), and Jasmina Pluncevic-Gligoroska[1]

[1] Department of Physiology, Faculty of Medicine,
University "Ss. Cyril and Methodius", Skopje, Republic of Macedonia
sanjamancevska@gmail.com, jasnapg65@yahoo.com
[2] School of Computer Science and Information Technology,
University American College Skopje, Skopje, Republic of Macedonia
bozinovski@uacs.edu.mk

Abstract. The aim of the study was to evaluate the learning process in 30 subjects with high trait anxiety (mean TMAS score = 33.9 ± 6.7) and in 30 subjects with low trait anxiety (mean TMAS score = 7.5 ± 2.9) aged 19 to 22 years, using psycho-physiological and psychological tests. The electroexpectogram (EXG) paradigm and a computerized psychological test of pattern recognition (P-R) were used. The Taylor Manifest Anxiety Scale (TMAS) was used for the evaluation of the levels of trait anxiety. The EXG paradigm is an experimental design which is a modified and an expanded auditory CNV paradigm. Based on a biofeedback design, the occurrence of the S2 tone in the EXG paradigm depends on the amplitude of the CNV potential recorded from Cz. If CNV reaches a predefined threshold level, the S2 tone turns off, which causes an extinction of the CNV potential after several consecutive trials. Electrophysiological parameters of the associative learning were: duration of the acquisition and the extinction of the conditioned response, mean amplitude of the contingent negative variation (CNV) during the blocks of acquisition and the extinction of the conditioned response and the speed of the motor response. Psychological parameters included: number of mistakes, number of trials, and time necessary for the P-R test to be completed, as well as the learning efficiency index. The results of the study suggest that subjects with high trait level show significantly slower and less efficient learning process during more complex cognitive tasks compared to subjects with low levels of trait anxiety.

Keywords: CNV · Learning · Anxiety

1 Introduction

Many studies have recognized higher rates of anxiety and depression among university students worldwide compared to general population [1, 24]. During this period of life, there are many stressors, including low socioeconomic status, uncompleted processes of separation and individualization of the young person, lack of social relations, high

© Springer Nature Switzerland AG 2018
S. Kalajdziski and N. Ackovska (Eds.): ICT 2018, CCIS 940, pp. 39–50, 2018.
https://doi.org/10.1007/978-3-030-00825-3_4

everyday workload, worries about academic achievement and future career goals. This can lead to an adoption and maintenance of maladaptive behaviors, such as anxiety and depressive symptoms, and poor academic achievement. Greater rates of high anxiety, depression and other symptoms of psychological distress have been reported for medical students in general [10, 16, 22, 23].

As a personality trait, anxiety is a relatively stable feature, which determines a person's motivational and emotional response when coping with different situations. During stressful situations, individuals with high levels of trait anxiety show strong emotional responses, which do not correspond with the objective reality, and an anticipation of failure and/or threats to self-esteem. These, in turn, can have negative prolonged effects on the efficiency of their cognitive performances and can enhance further development of clinically manifested anxiety disorders [17]. In contrast, people with low anxiety in stressful conditions show more efficient behaviors.

Effective cognitive performances, especially a student's learning ability as well as his/her motivation, are important for their academic achievement [9, 20]. However, many reports, employing different sets of psychological and electrophysiological tools for assessment, suggest that high levels of anxiety have negative impact on attention, memory, learning and problem solving, while low and moderate anxiety levels are related to better cognitive performance [2, 13, 14, 18, 25, 36].

The contingent negative variation (CNV potential) is an endogenous anticipatory event related brain potential. It is elicited during experimental paradigms employing a cognitive task. A taxonomy of brain potentials including anticipatory potentials was first proposed in 1992 [4]. CNV has been investigated for more than fifty years with different experimental paradigms since 1964, when Walter et al. recorded it from vertex and called it "expectancy wave" [38]. It is an objective electrophysiological parameter, which reflects anticipation, attention and the underlying mental processes (working memory and learning). Reports from studies on the effects of high anxiety on the recorded CNV potential suggest that subjects with high anxiety show reduced CNV amplitudes and are more prone to distraction [3, 26, 28–34, 37].

A forewarned reaction time task (S1-S2-MR paradigm, where S1 is a warning, S2 is an imperative stimulus which requires a motor reaction - MR) is often used as a standard paradigm which is known to elicit the CNV potential [3, 26, 28–34, 37]. The establishment of clear association (contingency) between the warning stimulus and the imperative one and the existence of focused attention during the preparation for the response towards S2 are important attributes of this paradigm. They are in the essence of conditioning. It leads to generation of cognitive expectancies (acquisition of conditioned response) which enable an adequate and easy answer towards predictable events [31, 34, 37]. However, when unpredictable events happen during conditioning, earlier established cognitive expectancies degrade (inhibition of the conditioned response) and new are created. Their strength is as big as the surprise from the new event [31]. During an unpredictable shift of the paradigm shape S1-S2-MR, known as "go condition", towards the shape S1 - (in which S2 is omitted and there is no need for MR), known as "no-go condition", the expectancy towards S2 (represented by the CNV amplitude) gradually reduces during several consecutive trials in healthy subjects. As a result, the dynamics of the recorded CNV amplitude oscillations can be monitored during a certain period of time. These oscillations were implemented in a control paradigm named the CNV Flip-Flop paradigm [6], which was later used to control a single robotic arm [7], as well as two robotic arms at the same time [8]. Later this concept was generalized as an EEG emulated flip-flop control circuit [5].

The electroexpectogram (EXG) paradigm used in this study is a modified and an expanded CNV paradigm which employs a biofeedback design. It was proposed by Bozinovski in 1984 and initial experiments were done using the IBM Series/1 computer, with further development using a PC [4]. The dynamics of expectancy and attention represented by the CNV amplitude is observed during 100 trials. The subject is placed in a "game with the paradigm", based on a biofeedback. The appearance of "go condition" and "no-go condition" exclusively depends on the amplitude of the late component of the CNV potential, recorded from the vertex in the current and recent trials. As a result, an oscillatory curve (electroexpectogram) can be recorded. The electroexpectogram curves in healthy subjects show features of damped oscillations. In subjects with different neuropsychiatric conditions (such as epilepsy and chronic schizophrenia) they tend to be non-oscillatory [27].

The aim of the present study was to estimate the parameters of the learning process in subjects with high trait anxiety, at the early stage of education with two different tools (an electrophysiological and a psychological).

2 Materials and Methods

2.1 Participants

Sixty students, volunteers from both genders, aged 19–22 years, enrolled in the second year of the Medical Faculty in Skopje were tested with the EXG paradigm and with a computerized pattern recognition test of memory and learning (P-R). Thirty subjects had high scores (HA - high anxiety) and other 30 students had low TMAS scores (LA - low anxiety group). Previously, the Taylor Manifest Anxiety Scale was given to a cohort of 176 students. They anonymously answered up to 50 questions of the scale, related to the existence of cognitive, somatic, emotional and behavioral symptoms of anxiety during a longer period of time. The main value of the obtained TMAS scores of the cohort was 19.8 and the standard deviation was 8.3. The TMAS scores of the HA group were between 25–40 with a mean value of 33.9 and standard deviation of 6.7, while the TMAS scores of the LA group were from 1 to 11, with a mean value of 7.55 and standard deviation of 2.9.

The subjects undertook the EXG test and the computerized Pattern Recognition (P-R) test at the Institute of Physiology, Medical Faculty, UKIM, Skopje.

2.2 Stimuli

The electroexpectogram (EXG) paradigm started as a classical CNV paradigm - "go condition", in which the warning tone (75 ms duration, 1000 Hz), was followed 2000 ms later by a tone with a frequency of 1500 Hz and a possible duration of 4000 ms, unless interrupted by a motor reaction. The second tone was an imperative one, requiring MR. There was a variable interval (3–10 s, randomly selected) until the next warning stimulus. The probability of the imperative tone appearance was less than 100% and it depended on the amplitude of the measured CNV potential in recent trials. All stimuli were presented through headphones at 60 dB SPL on both ears. Subjects were instructed to press as quickly as possible to the longer tone. All subjects responded with their dominant hand thumb.

The time measured from the occurrence of the second tone until its interruption signaled by a pressure of the subject's dominant hand thumb on a taster knob, was the reaction time and was measured automatically by the computer.

2.3 Procedure

The participants were familiarized with the testing procedure and the laboratory. After the recording electrodes were fitted, subjects lay on a bed in a sound-proof room. They were encouraged to keep as still as possible throughout the task and to keep the eye movements to a minimum using a central fixation spot on the wall, or to keep their eyes closed.

2.4 Electrophysiological Recording

Three Ag/AgCl standard disk electrodes were applied on Cz, Fz, and right mastoid according to the 10–20 system. The electrode on Cz was active and was referenced to the mastoid. The impedance was lower than 5 kΩ. The subject was grounded by the Fz electrode. EEG was recorded with a computerized device. Standard parameters of the two channel amplifier were TC = 3 s, high frequency band 15 kHz, and gain 20000 (50 μV/V).

2.5 Electrophysiological Data Extraction

The EEG signal was recorded during 7000 ms in every trial: 1000 ms before the warning stimulus, 2000 ms during the inter-stimulus interval and 4000 ms after the beginning of imperative stimulus. The ERP was extracted by an optimal averaging algorithm which enables reduction of the influence of earlier measurements proportionally to the time interval until the actual measurement.

The EXG paradigm consisted of 100 trials. A biofeedback design was employed. Its leading parameter was the maximal amplitude of the late component of the CNV potential (mean amplitude of the recorded signal during 1000 ms before the S2 occurrence). Its value of 10 μV was the threshold line for the biofeedback-based automatic turning-off and turning-on of the S2 tone during the paradigm.

If the CNV amplitude, measured during three repeated trials, reaches the value of 10 μV, the computer automatically turns off S2 tone in the paradigm. So, the initial "go-condition" of the classical CNV paradigm, turns into a "no-go condition" during which there is no S2 tone in the paradigm and no MR request for the subject. It happens exclusively because the value of the leading parameter, the maximal CNV amplitude on Cz, reaches the threshold line level. Subsequently, the maximal amplitude of the recorded CNV in the next trials reduces (a well known phenomenon in the CNV studies [31]) towards the values below the threshold line, as a result of the no-go condition of the paradigm.

When the below-threshold value of the recorded CNV is confirmed by the computer, the S2 tone is automatically turned on again and the paradigm appears in its "go" condition again. During 100 trials, depending on the subject's mental capacity, the CNV paradigm modifies from a classical CNV paradigm to a dynamic one. During the EXG paradigm, an oscillatory curve, containing the values of the maximal CNV amplitude during each trial, can be recorded. We call it the electroexpectogram. It shows the oscillatory dynamics of the value of the recorded CNV potential around the threshold line (Fig. 1).

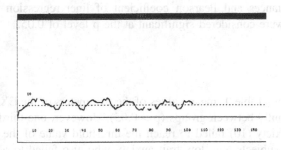

Fig. 1. Oscillatory electroexpectogram curve obtained from a healthy subject.

2.6   Data Analysis

One of the analyzed parameters was the motor speed, represented by the minimal reaction time (ms) towards the S2 tone. The mean reaction time throughout the whole experiment was also analyzed. The subject's attention was analyzed through the mean values of the maximal and minimal amplitude of the CNV potential during the whole test. The maximal CNV amplitude was the mean value of the signal during 1000 ms before the imperative tone occurrence, referenced to the base line (the mean value of the signal during 1000 ms before the S1 tone). The minimal CNV amplitude was the lowest value of the recorded maximal CNV amplitude during the test. The number of the oscillations of the CNV amplitude around the threshold level, during the electroexpectogram, was analyzed as a parameter of the adaptive dynamics of the attention and expectancy. The duration (number of trials) of the consecutive blocks of acquisition (GO) and the extinction of the conditioned response (NOGO), and the mean values of the CNV amplitude (μV) during them were the electrophysiological parameters of learning during the EXG paradigm.

The computerized pattern recognition (P-R) test of memory and learning consisted of 8 test objects - patterns. They were represented by different combinations of binary numbers 1 and 0, divided into two classes (A and B). Each pattern was shown on the computer monitor during the time interval of 5 s. The subject had to memorize all of them for correct classification of patterns during a training period. While testing, each pattern of binary numbers was shown on the monitor along with two keys labeled with the letters A and B. The subjects had to choose the correct answer within a period of

5 s. If the answer was correct, the sign "correct" appeared; if not, the sign "not correct" appeared. The appearance of the patterns was random. The test was completed when all eight test patterns were correctly recognized and classified by the subject. The total number of false recognitions (total number of mistakes - TM), number of trials necessary to complete the tests (TT), total duration of the test (TD) and the learning index (LI) for the P-R test were the analyzed parameters of cognitive performance.

2.7   Statistical Analysis

The results are presented as mean values and their standard deviations.

Standard statistical methods: descriptive statistics, Student's t-test for two samples with unequal variances and Pearson coefficient of liner regression were used. All statistical values were considered significant at the p level of 0.05.

3   Results

As can be seen from Table 1, mean values of the parameters of the EXG paradigm did not differ significantly between groups (p > 0.05). However, one third of the subjects with high trait anxiety (10) failed to reach the threshold value of the CNV potential, compared to four subjects with low trait anxiety who also failed to obtain oscillatory EXG curves (chi square = 14.4 df = 1, p = 0.04). The minimal reaction time and the average reaction time during all trials were insignificantly faster in subjects with high trait anxiety compared to subjects with low trait anxiety (p = 0.09).

Mean values of the electrophysiological parameters of learning during the EXG paradigm are shown on Tables 2, 3 and 4. As can be seen from Table 2, subjects from

Table 1. Mean values of psychological, electrophysiological and behavioral parameters in both groups of subjects.

Parameters/subjects	Subjects with low trait anxiety (LA) N = 30	Subjects with high trait anxiety (HA) N = 30
Age (years)	19.8 ± 0.7	19.9 ± 0.8
TMAS (scores)	7.5 ± 2.9	33.9 ± 6.7
Maximal CNV amplitude (μV)	12.1 ± 2.9	12.2 ± 3.7
Average CNV amplitude (μV)	7.9 ± 3.2	7.7 ± 3.7
Minimal CNV amplitude (μV)	0.9 ± 3.6	0.6 ± 1.8
Number of consecutive NOGO blocks	2.1 ± 2.3	2.8 ± 2.0
Minimal reaction time (ms) during GO blocks	153.8 ± 37.5	146.5 ± 37.4
Average reaction time (ms) during GO blocks	248.4 ± 126.5	245.7 ± 135.9

both groups, who succeeded to reach the threshold value of 10 μV for the recorded CNV potential obtained five consecutive GO blocks and four consecutive NOGO blocks.

Table 2. Mean values of the duration (number of trials) of the consecutive blocks of acquisition (GO) and the extinction of the conditioned response (NOGO) in both groups of subjects.

Subjects/blocks	1GO	2GO	3GO	4GO	5GO
LA	52.1 ± 31.6	12.8 ± 8.1	11.1 ± 7.8	12.4 ± 12.4	6.4 ± 5.0
HA	41.3 ± 36.2	31.0 ± 33.2	10.3 ± 6.0	13.0 ± 9.3	8.3 ± 5.5
	1NOGO	2NOGO	3NOGO	4NOGO	/
LA	8.0 ± 7.6	7.9 ± 3.9	7.4 ± 7.3	4.2 ± 4.1	
HA	10.8 ± 9.6	15.2 ± 16.4	10.3 ± 9.9	4.0 ± 2.5	/

There was no significant difference between the duration of each of the consecutive GO and NOGO blocks between the two groups (p > 0.05). There was strong negative linear correlation between the duration of the consecutive blocks and their ordinal number for LA group (r = −0.776) and also for the HA group (r = −0.909) for the GO blocks and r = −0.852 and −0.708 respectively for the NOGO blocks. The linear regression models of the learning curves for the duration of the blocks of acquisition of the conditioned response were Y = −26.162 Ln(x) + 44.012 (p = 0.044) for the subjects with low trait anxiety, and Y = −21.86 Ln(x) + 41.717 (p = 0.0127) for the subjects with high trait anxiety. The linear regression models of the learning curves for the duration of the blocks of extinction of conditioned response were Y = −1.192x + 9.845 (p = 0.14) for the subjects with low trait anxiety, and Y = −2.547x + 16.475 (p = 0.29) for the subjects with high trait anxiety.

As can be seen from Table 3, there was no significant difference between the mean values of the CNV amplitude during the consecutive GO blocks between the two groups of subjects (p > 0.05). The obtained CNV amplitudes during the first, second and the third NOGO block in subjects with high anxiety levels were significantly higher than the ones in subjects with low trait anxiety. A strong positive linear correlation was obtained between the mean values of the recorded CNV amplitude during consecutive GO blocks and their ordinal number for LA group (r = 0.792) and also for the HA group (r = 0.821). There was no linear correlation between the mean amplitudes of CNV potential during consecutive NOGO blocks and their ordinal number in both groups of subjects (r = −0.153 and r = −0.02 respectively). Therefore, there was no significant reduction of the CNV amplitude during the consecutive NOGO blocks and there was a very weak tendency for approximation of the amplitude towards the threshold value. The linear regression models of the learning curves for the CNV amplitude during consecutive GO blocks was Y = 2.06 Ln(x) + 6.581 (p = 0.044) for the subjects with low trait anxiety, and Y = 2.608 Ln(x) + 5.675 (p = 0.028) for the subjects with high trait anxiety. The learning curves for the CNV amplitude during the NOGO were statistically insignificant for both groups.

As can be seen from Table 4, subjects from the HA group showed significantly faster motor reaction towards S2 during the second and the third GO block compared to

Table 3. Mean values of the CNV amplitude (μV) during the consecutive blocks of acquisition (GO) and the extinction of the conditioned response (NOGO) in both groups of subjects.

Subjects/blocks	1GO	2GO	3GO	4GO	5GO
LA	6.1 ± 3.1	8.7 ± 1.9	9.5 ± 1.1	8.6 ± 1.2	9.8 ± 0.6
HA	5.5 ± 2.9	7.3 ± 2.4	9.5 ± 1.3	9.7 ± 1.1	8.9 ± 1.4
	1NOGO	2NOGO	3NOGO	4NOGO	/
LA	10.9 ± 1.1	12.3 ± 1.9	10.9 ± 0.8	11.1 ± 1.1	/
HA	11.6 ± 1.3	13.3 ± 2.6	13.1 ± 1.7	11.6 ± 1.5	/
Student t-test (p)	0.0004	0.0031	<0.0001	0.306	

Table 4. Mean values of the average reaction times towards S2 tone (milliseconds) during the consecutive blocks of acquisition (GO) of the conditioned response (NOGO) in both groups of subjects.

Subjects/blocks	1GO	2GO	3GO	4GO	5GO
LA	251.3 ± 144.8	232.1 ± 74.6	254.3 ± 97.7	250.7 ± 89.3	230.3 ± 60.4
HA	260.8 ± 163.4	222 ± 78.2	227.8 ± 67.6	227.4 ± 99	214.4 ± 79.1
Student t-test (p)	0.228	**<0.0001**	**0.048**	0.12	0.396

Table 5. Mean values and standard deviations of analyzed parameters of learning during P-R test in medical students with high anxiety.

Parameters of the P-R test	LA	HA	Student t-test (p)
Learning index	91.4 ± 8.4	83.3 ± 11.8	**0.002**
Total number of mistakes	11.4 ± 15	26.1 ± 22.5	**0.004**
Total number of trials	71.32 ± 55.7	108 ± 57.6	**0.012**
Total duration of the test (seconds)	177.6 ± 142.9	291.5 ± 163.4	**0.005**

subjects from LA group. There was a weak negative correlation between the speed of the MR and the ordinal number of the consecutive GO block in subjects with low trait anxiety (r = −0.323) and a strong negative correlation in subjects with high trait anxiety (r = −0.777). However, the obtained learning curves for the speed of motor response for both groups were not significant.

As can be seen from Table 5, subjects with high trait anxiety showed significantly worse results on all analyzed parameters during the pattern recognition test, compared to subjects with low trait anxiety.

4 Discussion

The results from the electrophysiological investigation with the EXG paradigm, with a threshold value of 10 μV, show that a learning process with an aim of minimization of the surprise and successful adaptation to variable environment with differential efficacy was established in both groups. These results are in accordance with the reports that a learning process takes place during the CNV paradigm [15, 31, 34]. Recorded individual EXG curves for every subject and the derived learning curves for both groups are graphical displays of the realized learning process. Learning curves for the CNV amplitude during both GO and NOGO blocks of the EXG paradigm, the duration of the consecutive GO and NOGO blocks, and the minimal reaction time during GO blocks were derived.

One third of the subjects with high trait anxiety compared to 13% of the LA group did not succeed to reach the threshold level of 10 μV for the recorded CNV potential, as an indicator of successful acquisition of the conditioned response. Other subjects who managed to reach the threshold value and to obtain oscillatory EXG curves, from both groups did not differ regarding the number of EXG oscillations and the maximal amplitude of the recorded CNV potential.

The derived learning curves for the speed of acquisition of the conditioned response show significant tendency for reduction of the time necessary for every consecutive acquisition of the conditioned response. A significant increment of the anticipation of S2 tone and approximation to the threshold value in every consecutive acquisition block (GO block) in both groups is apparent. The average reaction time in the EXG paradigm did not differ in both groups of subjects. Nevertheless, subjects with high trait anxiety showed shorter motor responses towards S2 during the later GO blocks. Reports suggest that a motor learning process also happens during the CNV paradigm as a function of the adaptation of the preparatory set [15]. Both groups showed a successful learning during the blocks of acquisition of the conditioned response.

The derived learning curves for the extinction of previously established association i.e., the inhibition of the conditioned response (during NOGO blocks) did not show strong tendency towards reduction of the duration of the trials and towards decrement of the maximal CNV amplitude and approximation to the threshold value during every consecutive extinction of the conditioned response. High uncertainty during NOGO blocks prevents from degradation of previously established association between the two stimuli, which is shown by the increment of the maximal CNV amplitude during the second and the third NOGO block in both groups. The analysis of the electrophysiological parameters of associative learning during the auditory modified CNV paradigm did not show any significant differences between subjects with high and low trait anxiety.

Subjects with high trait anxiety performed significantly worse than the subjects with low trait anxiety on the computerized pattern recognition test, which is a psychological test of memory and learning. All analyzed behavioral parameters of cognitive performance (total number of false pattern recognitions, total duration of the test and the learning efficacy index) showed reduction in the efficacy of the learning of the patterns. Subjects with high trait anxiety learned more complicated patterns significantly less efficiently compared to subjects with low trait anxiety. This result is in accordance with the reports of many authors, who documented reduction in the efficiency of learning, memory and problem solving in subjects with high anxiety, especially in patients with clinically manifest anxiety disorders. According to Eysenck et al. [12], the effects of high anxiety are such that the processing capabilities of the brain are always engaged by the excessive arousal. This limits the informational intake and the efficiency during the task [15, 17]. Therefore, subjects with high anxiety need increased activation of the brain systems, like the prefrontal dorsolateral cortex, which supports the cognitive control in order to maintain performance equivalent to the one in subjects with low anxiety. It has negative impact on the efficacy of the behavioral performance and especially on the quality of the information processing [11–15, 19, 21, 35].

A small sample size in this study and its cross-sectional design might limit the generalization of practical application of the findings. Further studies with larger samples, which include subjects with clinically manifest anxiety disorders could provide more information on the impact of severe anxiety on the dynamics of focused and sustained attention during variable paradigms. The results from our study provide initial information on the effect of high trait anxiety in medical students in the early stage of their education on their cognitive performance. It is a contribution to cognitive science. It is also very important in the context of the length of medical undergraduate studies and the possible cumulative negative effects of prolonged academic stress on the mental health of medical students. It will be beneficial for the practitioners (student counselors and curriculum developers) to recognize this information and to take measures in order to reduce the academic stress in medical schools.

5   Conclusion

This paper presents a study of subjects with both high and low trait anxiety involved in psycho-physiological and physiological tests, consisting of the electroexpectogram (EXG) and a computerized pattern recognition (P-R) test, whereas the Taylor Manifest Anxiety Scale (TMAS) was used to determine the levels of anxiety. An explanation of the P-R test and the EXG paradigm were presented, alongside the results obtained by both groups of subjects. The results suggest that subjects with high trait anxiety show more efficient learning process during less demanding cognitive tasks (such as an auditory paradigm) compared to more demanding and more complex cognitive tasks which involve simultaneous engagement of several cognitive systems (visuospatial orientation, engagement of the semantic and working memory, decision making) to achieve the designated goal.

References

1. Andrade, L., Caraveo-Anduaga, J., Berglund, P.: Cross-national comparisons of the prevalences and correlates of mental disorders. WHO International Consortium in Psychiatric Epidemiology. Bull. World Health Org. **78**, 413–426 (2000)
2. Aronen, E.T., Vuontela, V., Steenari, M.R., Salmi, J., Carlson, S.: Working memory, psychiatric symptoms and academic performance at school. Neurobiol. Learn. Mem. **83**(1), 33–42 (2005)

3. Boudarene, M., Timsit-Berthier, M.: Stress, anxiety and event-related potentials. L'Encéphale **23**(4), 237–250 (1997)
4. Bozinovska, L., Bozinovski, S., Stojanov, G.: Electroexpectogram: experimental design and algorithms. In: Proceedings of the IEEE International Biomedical Engineering Days, pp. 58–60 (1992)
5. Bozinovski, S., Bozinovski, A.: Mental states, EEG manifestations, and mentally emulated digital circuits for brain-robot interaction. IEEE Trans. Auton. Ment. Dev. **7**(1), 39–51 (2015)
6. Božinovski, A.: CNV flip-flop as a brain-computer interface paradigm. In: 7th Conference of the Croatian Association of Medical Informatics, pp. 149–154. Croatian Association of Medical Informatics, Rijeka (2005)
7. Božinovski, A., Božinovska, L.: Anticipatory brain potentials in a brain-computer interface paradigm. In: IEEE EMBS Conference on Neural Engineering, pp. 451–454 (2009)
8. Božinovski, A., Tonković, S., Išgum, V., Božinovska, L.: Robot control using anticipatory brain potentials. Automatika **52**(1), 20–30 (2011)
9. Chandavarkar, U., Azzam, A., Mathews, C.: Anxiety symptoms and perceived performance in medical students. Depress. Anxiety **24**, 103–111 (2007)
10. Dyrbye, L.N., Thomas, M.R., Shanafelt, T.D.: Systematic review of depression, anxiety and other indicators of psychological distress among U.S. and Canadian medical students. Acad. Med. **81**(4), 354–373 (2006)
11. Epstein, S.: The nature of anxiety with emphasis upon its relationships to expectancy. In: Spielberger, C.D. (ed.) Anxiety: Current trends in Theory and Research, vol. 2, pp. 292–334. Academic Press, New York (1972)
12. Eysenck, M.W., Santos, R., Derakshan, N., Calvo, M.G.: Anxiety and cognitive performance: attentional control theory. Emotion **7**(2), 336–353 (2007)
13. Falconer, E.M., et al.: Developing an integrated brain, behavior and biological response profile in posttraumatic stress disorder (PTSD). J. Integr. Neurosci. **7**(3), 439–456 (2008)
14. Fales, C.L., et al.: Anxiety and cognitive efficiency: differential modulation of transient and sustained neural activity during a working memory task. Cogn. Affect. Behav. Neurosci. **8**(3), 239–253 (2008)
15. Friston, K.J., Stephan, K.E.: Free-energy and the brain. Synthese **159**, 417–458 (2007)
16. Goldman, M.L., Shah, R.N., Berstein, C.A.: Depression and suicide among physician trainees: recommendations for a national response. JAMA Psychiatry **72**(5), 411–412 (2015)
17. Hardy, L.: Stress, anxiety and performance. J. Sci. Med. Sport **2**(3), 227–233 (1999)
18. Hayes, S., Hirsch, C., Mathews, A.: Restriction of working memory capacity during worry. J. Abnorm. Psychol. **117**(3), 712–717 (2008)
19. Henderson, R.K., Snyder, H.R., Gupta, T., Banich, M.T.: When does stress help or harm? The effects of stress controllability and subjective stress response on Stroop performance. Front. Psychol. **3**, 179 (2012). https://doi.org/10.3389/fpsyg.2012.00179
20. Kernan, W.D., Wheat, M.E., Lerner, B.A.: Linking learning and health: a pilot study of medical students' perceptions of the academic impact of various health issues. Acad. Psychiatry **32**(1), 61–64 (2008)
21. Kizilbash, A.H., Vanderploeg, R.D., Curtiss, G.: The effects of depression and anxiety on memory performance. Arch. Clin. Neuropsychol. **17**, 57–67 (2002)
22. Mancevska, S., Bozinovska, L., Tecce, J., Pluncevik-Gligoroska, J., Sivevska-Smilevska, E.: Depression, anxiety and substance use in medical students in the Republic of Macedonia. Bratisl. Med. J. **109**(12), 568–572 (2008)
23. Mancevska, S., Pluncevic, J., Todorovska, L., Dejanova, B., Tecce, J.: Substance use and perceived hassles among junior medical students with high anxiety levels in the Republic of Macedonia. Iran. J. Public Health **43**(10), 1451–1453 (2014)

24. Mikolajczyk, R.T., Maxwell, A.E., Naydenova, V., Meier, S., El Ansari, W.: Depressive symptoms and perceived burdens related to being a student: survey in three European countries. Clin. Pract. Epidemiol. Ment. Health 4(10), 19 (2008). https://doi.org/10.1186/1745-0179-4-19

25. Moritz, S., et al.: Extent, profile and specificity of visuospatial impairment in obsessive-compulsive disorder (OCD). J. Clin. Exp. Neuropsychol. 27(7), 795–814 (2005)

26. Picton, T.W.: The endogenous evoked potentials. In: Başar, E. (ed.) Dynamics of Sensory and Cognitive Processing by the Brain. Springer Series in Brain Dynamics, vol. 1, pp. 266–274. Springer, Heidelberg (1988). https://doi.org/10.1007/978-3-642-71531-0_18

27. Gligoroska, J.P., Bozinovska, L., Mancevska, S., Sivevska, E., Todorovska, L., Brezovska, J.: EXG parameters in healthy population: electrophysiological standards. Physioacta 2(2), 91–103 (2008)

28. Proulx, G.B., Picton, T.W.: The CNV during cognitive learning and extinction. In: Kornhuber, H.H., Deecke, L. (eds.) Motivation, Motor and Sensory Processes of the Brain. Electric Potential Behavior and Clinical Use, pp. 309–313. Elsevier (1980)

29. Proulx, G.B., Picton, T.W.: The effects of anxiety and expectancy on CNV. Ann. N. Y. Acad. Sci. 23, 617–623 (1984)

30. Rebert, C.S., Tecce, J.J.: The cerebral physiology of preparatory set: understanding the contingent negative variation. In: Ogura, C., Koga, Y., Shimokochi, M. (eds.) Recent Advances in Event-Related Brain Potential Research, pp. 863–873. Elsevier Science B.V. (1996)

31. Rescorla, R.A., Wagner, A.R.: A theory of Pavlovian conditioning: variations in the effectiveness of reinforcement and non-reinforcement. In: Black, A.H., Prokasy, W.F. (eds.) Classical Conditioning II: Current Research and Theory, pp. 165–189. Appleton-Century-Crofts (1972)

32. Rochstroh, B., Birbaumer, N., Lutzenberger, W.P.: Slow Brain Potentials and Behavior. Urban and Schwarzenberg, Baltimore (1982)

33. Rochstroh, B., Cohen, R., Berg, P., Klein, C.: The post-imperative negative variation following ambiguous matching of auditory stimuli. Int. J. Psychophysiol. 25(2), 155–167 (1997)

34. Rose, M., Verleger, R., Wascher, E.: ERP correlates of associative learning. Psychophysiology 38, 440–450 (2001)

35. Rose, E.J., Ebmeirer, K.P.: Pattern of impaired working memory during major depression. J. Affect. Disord. 90(2–3), 149–161 (2006)

36. Roth, R.M., Baribeau, J., Milovan, D.L., O'Connor, K.: Speed and accuracy on tests of executive function in obsessive-compulsive disorder. Brain Cogn. 54(3), 263–265 (2004)

37. Ritter, F.E., Schooler, L.J.: The learning curve. In: International Encyclopedia of the Social and Behavioral Sciences, Pergamon, pp. 8602–8605 (2002)

38. Walter, W.G., Cooper, R., Aldridge, V.S., McCallum, W.C., Winter, A.L.: Contingent negative variation: an electric sign of sensory-motor association and expectancy in the human brain. Nature 203, 380–384 (1964)

Group Decision Making for Selection
of Supplier Under Public Procurement

Dilian Korsemov[1], Daniela Borissova[1,2(✉)], and Ivan Mustakerov[1]

[1] Institute of Information and Communication Technology at Bulgarian
Academy of Sciences, 1113 Sofia, Bulgaria
dilian_korsemov@abv.bg,
{dborissova, mustakerov}@iit.bas.bg
[2] University of Library Studies and Information Technologies,
1784 Sofia, Bulgaria

Abstract. The article deals with problem of group decision making for selection of supplier under public procurement. For the goal, a generalized algorithm for multi-attributes group decision-making is proposed. The distinguish feature of the described algorithm is consideration of knowledge and experience of each expert from the group by using of weighted coefficients. Simple additive weighting and weighted product model are modified to cope with the differences in experts' knowledge and experience. The applicability of proposed group decision making algorithm is illustrated by using of new modified utility functions for simple additive weighting and weighted product model. The numerical testing considers a real-life problem for selection of the most preferable supplier of personal computers under a public procurement. The results demonstrate the flexibility of proposed approach when using a group of experts with different expertise.

Keywords: Multi-attribute group decision making · Simple additive weighting
Weighted product model · Optimization techniques · Pure integer linear model

1 Introduction

In the contemporary economics, proper supplier determination becomes a key strategic decision for success of different business activities. The process of selection takes into account two major factors: the presence of useful information for evaluation of parameters and corrupt behavior possibility [10]. The supplier selection problem is recognized as a complex problem consisting of both quantitative and qualitative criteria [4, 9]. In most cases, these evaluation criteria are in conflict. The availability of qualitative and quantitative criteria, require they to be evaluated simultaneously in decision making process [12]. To be more transparent, the selection process should involve a group of experts with different skills, experience and knowledge capable to evaluate all of the criteria [2, 8]. All of these considerations received relatively large amount of attention in both academia and industry by proposing of different approaches to tackle with problems of selection [5, 10]. This requires involving different analysis and techniques to support business decision making processes [2]. Using of business

© Springer Nature Switzerland AG 2018
S. Kalajdziski and N. Ackovska (Eds.): ICT 2018, CCIS 940, pp. 51–58, 2018.
https://doi.org/10.1007/978-3-030-00825-3_5

intelligence tools improve decision making and optimize business processes that contribute to the business competitiveness.

The supplier selection for the goal of public procurement is an important case of group decision making where spending of funds has to be public. An increased interest in public procurement is observed over the last few decades, as the purchase of goods by the public sector is increased [7]. The public procurements are to be transparent and in accordance to legislation, administrative regulations and should follow particular public procurement procedures. Different mathematical methods and models are proposed for supplier selection in procurement environment. Systematic reviews of literature for application of decision making techniques in supplier selection are given in [3, 11]. Compensatory strategies in decision making rely on rational decision choices based on multi-attribute utility models [6]. These utility models represent the preferences of the expert and very often are expressed as a sum of the utilities that each criterion determines [5]. Most commonly used method based on compensatory strategy is simple additive weighting (SAW). The SAW does not consider the different preferential levels and preferential ranks for each decision maker's assessment of alternatives in a decision group [1]. A very similar to SAW is weighted product model (WPM) where the main difference is that instead of addition in the model utility function multiplication is used [13]. The key idea of the both methods based on multi-attribute utility theory relies on construction of utility function used to evaluate given alternatives toward performance criteria.

In the article, the multi-attribute utility theory is used to formulate a combinatorial optimization decision making model in group environment. The proposed model is applied for supplier selection under a public procurement. The final choice relies on evaluations of group of experts capable to estimate given set of alternatives in respect of predefined quality and quantity indicators. The proposed in the article modeling approach takes into account the difference in knowledge and experience of expert within the group.

The rest of the article is structured as follows: Sect. 2 describes the problem for supplier selection. Section 3 is focused on generalized algorithm for group decision making considering the differences in experts' knowledge and expertise. Section 4 describes application of the proposed modification of SAW and WPM for group decision making while Sect. 5 presents obtained results, and conclusions are given in Sect. 6.

2   Problem Description

The considered decision making problem consists in selection of the most preferable supplier for delivery of personal computers (PCs) in accordance to a public procurement. There exist a number of suppliers with different PCs offers and the choice must be done by a group of different experts which are relevant to the problem. The experts in the group have expertise in different (but related to the problem) fields. This means that their evaluations are with different importance accordingly to their relevance to the problem. Each supplier (vendor) is considered as possible alternative that can be described by different parameters expressed as evaluations criteria. The most preferred alternative should be determined by considering all of the above considerations.

3 Generalized Algorithm for Group Decision Making

To support the business decision making processes by business intelligence, a generalized algorithm for group decision making considering the differences in experts' knowledge and expertise is proposed. This algorithm is composed of nine stages as shown in Fig. 1.

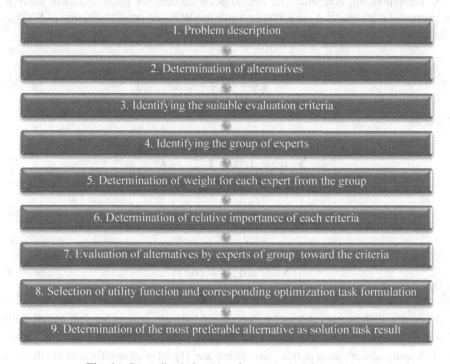

Fig. 1. Generalized algorithm for group decision making

The first stage concerns the description of the existing group decision making problem. The 2-nd stage considers the determination of a set of acceptable alternatives appropriate to cope with the goal of problem. Next stage is focused on determination of important criteria for evaluation of the alternatives. Stage 4 deals with determination of relevant group of competent experts in the area of the problem. On 5-th stage the corresponding weighted coefficients for the experts' knowledge and experience are to be set. Next stage requires determination of relative importance between criteria by assigning of corresponding weighted coefficients in accordance to the point of view of each expert. On the 7-th stage the evaluation of alternatives toward criteria from all experts are to be done. At this stage, normalization may be needed depending on the units of dimension for evaluation criteria. The applied normalization transforms experts' evaluations are in the range of 0 to 1 in dimensionless units.

The stage 8 concerns the choice of proper utility function and formulation and solving of corresponding optimization task. This task should incorporate the information from the above stages including: (1) weights for experts from the group, (2) performance of alternatives in accordance to the evaluation criteria, (3) weights for relative importance of evaluation criteria. In the last stage, the solution of the optimization task will give information for the best alternative performance considering different point of view of the experts.

In the article, two types of utility functions based on SAW and WPM are used to define corresponding combinatorial optimization models for group decision making. The proposed mathematical pure integer linear model based on modification of SAW is as follows:

$$\text{maximize} \left(\sum_{i=1}^{M} \sum_{k=1}^{K} \sum_{j=1}^{N} w_j^k a_{i,j}^k \lambda^k x_i \right) \tag{1}$$

subject to

$$\sum_{j=1}^{N} w_j^k = 1, \forall k = 1, 2, \ldots, K \tag{2}$$

$$\sum_{k=1}^{K} \lambda^k = 1 \tag{3}$$

$$\sum_{i=1}^{M} x_i = 1, \ x_i \in \{0, 1\} \tag{4}$$

where w_j^k are coefficients for relative importance between criteria for k-th expert point of view, $a_{i,j}^k$ represents the i-th alternative evaluation to j-th criterion from k-th expert. The following sets are used in this formulation are: (1) set of alternatives to get the final selection $\{i = 1, 2, \ldots, M\}$; (2) set of evaluation criteria $\{j = 1, 2, \ldots, N\}$; (3) set of experts $\{k = 1, 2, \ldots, K\}$; (4) set of decision variables $\{x_i\}$ assigned to each alternative; (5) set of weighted coefficients for experts $\{\lambda^k, k = 1, 2, 3, \ldots, K\}$ representing their level of expertise and knowledge.

The proposed modification of WPM for group decision making is as follow:

$$\text{maximize} \left(\sum_{i=1}^{M} \sum_{k=1}^{K} x_i \lambda^k \prod_{j=1}^{N} (a_{i,j}^k)^{w_j^k} \right) \tag{5}$$

subject to the same restrictions (2)–(4).

Here multiplication $\prod_{j=1}^{N} \left(a_{i,j}^{k} \right)^{w_j^k}$ expresses the performance of *i-th* alternative toward *j-th* evaluation criteria accordingly *k-th* expert' opinion.

The group decision making process is managed by a leader of higher management responsible to organize overall decision making process including: identification of possible alternatives and determination of essential criteria for evaluation of alternatives. He also determines the weighted coefficients that express the knowledge and experience of group members or uses the values provided by other competent authorities for the purpose of considered problem. The leader is authorized to make the final decision or to propose most appropriate alternative to higher management. The distribution of responsibilities between leader and group members in the group decision-making process are illustrated in Fig. 2.

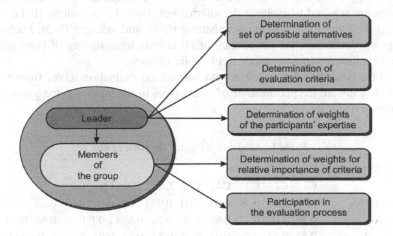

Fig. 2. Distribution of responsibilities between leader and group members

All of the described above data are used to formulate a combinatorial optimization task (on stage 8), which decision determines the most preferable alternative (on stage 9).

4 Numerical Testing

Three offers from different suppliers for delivery of PCs are considered as three possible alternatives (A-1, A-2, A-3). Each supplier is evaluated by 12 quality and numeral criteria: (1) technical performance; (2) bid price; (3) price breaks and quantity discounts; (4) payment terms – possibility of deferred payment; (5) warranty; (6) out-of-warranty service; (7) number of available repair shops; (8) availability of experienced staff; (9) certifications; (10) previous experience; (11) lead time; (12) customer recommendations.

The criterion for technical performance of PCs incorporates processor type, core and frequency, memory frequency, type and volume, graphics resolution and available ports, keyboard type and existence of installed operating system. The bid price criterion is related with the price for a single unit and the offered price for PCs delivery as a whole. The criteria for payment terms and price breaks and quantity discounts consider the possibility of different payment options – payment on delivery, in advance or deferred payment as well as the possibility for decreasing of cost per unit of goods for certain quantity. The criteria for warranty terms and out-of-warranty service conditions are used to guarantee as much as possible flawless working of purchased PCs. The criteria for number of repair shops and experienced staff examine how quick and qualitative will be handling of possible problems. Availability of certifications and previous experience are indicators for reliability of the supplier. Lead time is essential criterion as it determines the needed time between the initiation of contract and delivery. Last but not least, for evaluation criterion is information about satisfaction of other customers and their recommendations.

The group of five authorized experts relevant to the problem (excluding leader of the group) are selected to evaluate the alternatives: financial consultant (E-1), two IT specialists (E-2 and E-3), system administrator (E-4), and manager (E-5). Each expert of the group determines the importance of all criteria by assigning of corresponding weighted coefficients in accordance to his point of view.

All of the described data together with normalized evaluation scores from experts' point of view toward the performance of alternatives in accordance to the given criteria are shown in Table 1.

Table 1. Modified weighted decision matrix

Experts	Alternatives	Evaluation criteria											
		C1	C2	C3	C4	C5	C6	C7	C8	C9	C10	C11	C12
Criteria weights		0.08	0.10	0.09	0.09	0.08	0.07	0.08	0.08	0.09	0.08	0.07	0.10
E-1	A-1	0.76	1.00	0.72	0.82	0.76	0.56	0.69	1.00	0.68	0.66	0.66	0.56
	A-2	0.83	0.88	0.78	0.93	0.77	0.58	0.65	0.93	0.79	0.76	0.68	0.55
	A-3	0.81	0.90	0.76	1.00	0.72	0.66	0.72	0.92	0.74	0.75	0.70	0.63
Criteria weights		0.11	0.07	0.07	0.06	0.09	0.09	0.09	0.09	0.08	0.08	0.08	0.09
E-2	A-1	0.81	0.85	0.73	0.67	0.67	0.89	0.73	0.67	0.74	0.85	0.80	0.93
	A-2	0.84	0.78	0.76	0.65	0.77	0.91	0.75	0.69	0.78	0.82	0.81	0.85
	A-3	0.82	0.74	0.73	0.72	0.72	0.82	0.78	0.65	0.79	0.78	0.73	0.91
Criteria weights		0.15	0.06	0.06	0.06	0.11	0.10	0.09	0.09	0.06	0.06	0.07	0.09
E-3	A-1	0.86	0.81	0.78	0.77	0.95	0.82	0.84	0.82	0.81	0.76	0.96	0.79
	A-2	0.72	0.79	0.76	0.81	0.83	0.76	0.70	0.81	0.78	0.79	0.89	0.82
	A-3	0.81	0.78	0.79	0.69	1.00	0.80	0.79	0.88	0.83	0.72	1.00	0.76
Criteria weights		0.13	0.06	0.06	0.06	0.12	0.11	0.09	0.08	0.06	0.07	0.08	0.08
E-4	A-1	1.00	0.85	0.61	0.62	0.95	0.93	0.73	0.73	0.94	0.85	0.85	0.93
	A-2	0.88	0.74	0.66	0.66	0.96	1.00	0.75	0.79	0.88	1.00	0.92	0.88
	A-3	0.92	0.76	0.73	0.75	0.83	0.91	0.78	0.76	1.00	0.88	0.78	0.90
Criteria weights		0.10	0.10	0.10	0.10	0.10	0.06	0.06	0.07	0.06	0.09	0.10	0.06
E-5	A-1	0.92	0.79	0.88	0.73	0.83	0.78	1.00	0.73	0.72	0.76	0.63	0.84
	A-2	0.78	0.88	1.00	0.72	0.84	0.71	0.88	0.74	0.75	0.78	0.68	1.00
	A-3	0.84	0.76	0.82	0.68	0.88	0.73	0.91	0.79	0.81	0.82	0.69	0.86

5   Results and Discussion

The numerical testing is based on a real-life problem for selection of supplier for PCs under public procurement procedure. The flexibility of proposed approach is illustrated by using of different sets of weighted coefficients for importance of experts' opinions in the group. The data from Table 1 are used to formulate the corresponding optimization tasks. Their solution results are shown in Table 2.

Table 2. Solution results

Sets	Coefficients for expertise of the group members					Selected alternative by modified SAW	Selected alternative by modified WPM
	E-1	E-2	E-3	E-4	E-5		
Set-1	0.20	0.20	0.20	0.20	0.20	A1	A3
Set-2	0.25	0.10	0.15	0.25	0.25	A3	A3
Set-3	0.27	0.12	0.13	0.18	0.30	A2	A3

Using of different weighted coefficients for expertise of group members' influence essentially the final selection of the most preferable alternative. The consideration of equal expertise for all group members (set-1) leads to choice of alternative A-1 for modified SAW and A-3 for modified WPM. The weighting coefficients from set-2 put more weights on opinions of experts E-1, E-4 and E-5 followed by the importance of experts E-2 and E-3. In this case, the defined preferable alternative is A-3 for both models. When the major attention is paid on opinions of experts E-5 and E-1 and lees on experts E-2, E-2 and E-4, the most preferable alternative is A-2 for modified SAW and A-3 for modified WPM.

From Table 2 it is seen that the modified SAW is more sensitive than modified WPM in determination of different choices of alternatives when different weighted coefficients for expertise of the experts are defined. Additional experiments are to be done to prove this fact. Nevertheless, both proposed modifications of SAW and WPM could be used for group decision making to determination of the most preferable alternative.

6   Conclusions

Two of the most widely used methods for group decision making (SAW and WPM) are modified by introducing a weighting coefficients of group decision making experts' opinions. The key idea of the proposed approach relies on construction of utility functions used to evaluate potential alternatives toward performance criteria taking into account the differences in knowledge and experience of the group experts. The including of weighting coefficients for expertise of each group members' adds flexibility in group decision making process. Using of these coefficients allow determination of optimal alternative in accordance to the experts competency and conforms better to the organization goals.

The described modifications of SAW and WPM are implemented in a generalized algorithm for group decision-making. The practical usability of the described approach was confirmed by numerical testing on an example of real-life problem for PCs supplier selection under public procurement requirements.

Future studies need to be done to make a more robust comparison of the practical application of the modified SAW and WPM methods. It is also interesting how similar modifications of other group decision methods will perform in practice.

References

1. Abdullah, L., Adawiyah, C.W.R.: Simple additive weighting methods of multi criteria decision making and applications: a decade review. Int. J. Inf. Process. Manag. **5**(1), 39–49 (2014)
2. Borissova, D., Mustakerov, I., Korsemov, D.: Business intelligence system via group decision making. Cybern. Inf. Technol. **16**(3), 219–229 (2016)
3. Chai, J., Liu, J.N.K., Ngai, E.W.T.: Application of decision-making techniques in supplier selection: a systematic review of literature. Expert Syst. Appl. **40**(10), 3872–3885 (2013)
4. Danielson, M., Ekenberg, L., Göthe, M., Larsson, A.: A decision analytical perspective on public procurement processes. In: Papathanasiou, J., Ploskas, N., Linden, I. (eds.) Real-World Decision Support Systems. ISIS, vol. 37, pp. 125–150. Springer, Cham (2016). https://doi.org/10.1007/978-3-319-43916-7_6
5. Figueira, J., Greco, S., Ehrgott, M.: Multiple Criteria Decision Analysis: State of the Art Surveys. Springer, Heidelberg (2005). https://doi.org/10.1007/978-1-4939-3094-4
6. Lee, L., Anderson, R.: A comparison of compensatory and non-compensatory decision making strategies in IT project portfolio management. In: International Research Workshop on IT Project Management 2009.9. (2009). http://aisel.aisnet.org/irwitpm2009/9
7. Mimovic, P.: Application of multi-criteria analysis in the public procurement process optimization. Econ. Themes **54**(1), 103–128 (2016)
8. Mustakerov, I., Borissova, D.: A web application for group decision-making based on combinatorial optimization. In: International Conference on Information Systems and Technologies, Spain, pp. 46–56 (2014)
9. Sarkar, S., Pratihar, D.K., Sarkar, B.: An integrated fuzzy multiple criteria supplier selection approach and its application in a welding company. Journal of Manufacturing Systems **46**, 163–178 (2018)
10. Sciancalepore, F., Falagario, M., Costantino, N., Pietroforte, R.: Multi-criteria bid evaluation of public projects, In: Management and Innovation for a Sustainable Built Environment, Amsterdam, The Netherlands (2011)
11. Simic, D., Kovacevic, I., Svircevic, V., Simic, S.: 50 years of fuzzy set theory and models for supplier assessment and selection: a literature review. J. Appl. Log. **24**(Part A), 85–96 (2017)
12. Tavana, M., Fallahpour, A., Di Caprio, D., Santos-Arteaga, F.J.: A hybrid intelligent fuzzy predictive model with simulation for supplier evaluation and selection. Expert Syst. Appl. **61**, 129–144 (2016)
13. Webster, J.G. (ed.): Encyclopedia of Electrical and Electronics Engineering, vol. 15, pp. 175–186. Wiley, New York (1998)

Emotion-Aware Teaching Robot: Learning to Adjust to User's Emotional State

Frosina Stojanovska, Martina Toshevska(✉), Vesna Kirandziska, and Nevena Ackovska

Ss. Cyril and Methodius University, Faculty of Computer Science and Engineering, Skopje, Macedonia
stojanovska.frose@gmail.com, martina.tosevska.95@gmail.com, {vesna.kirandziska,nevena.ackovska}@finki.ukim.mk

Abstract. Robots today are taking more and more complex roles thus they are getting smarter and more human-like. One complex function, specific to social robots, is the role of robots in human-robot interaction. They are helpful in the process of social human-robot interaction while performing a specific task like teaching, assisting, entertaining, etc. The ability to recognize emotions has a significant role for social robots. A robot that can understand emotions could be able to interact according to that emotion. In this paper, we propose a model for robotic behavior adapting to the user's emotions. The humanoid robot Nao is used in the role of emotion-aware teacher for teaching math. Its main purpose is to teach and entertain the user while adapting its behavior to the user's emotional state derived from the facial expression. The robot uses reinforcement learning to learn which action to perform in a specific emotional state. It employs the Q-learning algorithm, maximizing the next action's award - a value that depends on the current emotional state of the user. An experimental study with a selected group of subjects is conducted to assess the proposed behavior. We evaluated the robot's ability to recognize emotions and the subjects' experience of interacting with the robot.

Keywords: Human-robot interaction · Emotional robot
Emotion-aware robot · Social robot · Teaching robot
Reinforcement learning · Emotion recognition · Face analysis
Decision making

1 Introduction

Human-robot interaction (HRI) is a study field that encompasses the study of communication between humans and robots, as well as the design and adaptation of robotic systems used by humans [8,19]. The social behavior aspect of human interaction leads the field into the development of social robots [15,18,20]. These robots are autonomous robots that interact with humans following rules originated in a social manner in order to construct a human-like interaction.

© Springer Nature Switzerland AG 2018
S. Kalajdziski and N. Ackovska (Eds.): ICT 2018, CCIS 940, pp. 59–74, 2018.
https://doi.org/10.1007/978-3-030-00825-3_6

Although we do not have a formal definition and full knowledge of the influence of emotions, studies show that emotions play a key role in many cognitive tasks of humans, such as decision-making, reasoning, memory, attention, learning, etc [35,45]. Therefore, the ability to recognize emotions has a significant role for social robots. A robot that can recognize emotion in users can be able to interact according to that emotion. There are already developed frameworks for emotion modeling for social robots [34]. Social robots that detect and adapt to the human emotions are able to perform an empathic behavior accompanied by compassion and feeling of warmth [17]. These are the key points in the social human interaction process [37]. Understanding the emotional state of the human can help in adjusting the robot's behavior to that state. This can affect the process of learning and problem solving, making it more pleasant and fun experience.

Today's methods used for recognizing emotions include emotion detection through text, where the data can be: posts, comments, messages, criticisms, certain blogs, e-commerce websites, movies, music, etc [10]. Another approach is the identification of emotions through an image, where mainly emotion is detected through certain facial expressions and body language [3,27]. A third method used to solve this problem is the emotion recognition through sound, i.e. human speech, sound features [25] and the intonation of the human voice [14]. Other methods incorporate processing of various types of physiological signals such as EEG (electroencephalogram) to detect electrical activity in the brain, ECG (electrocardiogram) to detect the electrical activity of the heart, skin conductance, changes in respiration, and so onwards [9]. It is a big challenge for a person to determine the emotional state of another person, especially because not everyone expresses their emotion in the same way and with the same intensity. Sometimes humans hide their emotions or express some emotion that is not the true emotion when interacting with other people, especially if those people are unknown to them. Knowing this, building smart solutions to enable machines and robots to determine the emotional state of the humans with greater accuracy than themselves, is a nontrivial problem that presents a big challenge [26].

Social robots are already designed and used for several tasks in different domains. Robots like Paro [48,49], Robovie [22,38], and Kaspar [12,21] are successfully adopted in the health-care and therapy domain. Keepon [29], Tega [28], iCat [7,30], Robotis OP2 [11] are utilized for educational purposes. The entertainment field includes several designed robots like Aibo [47].

Nao is an autonomous robot developed in 2006 by SoftBank Robotics[1] and continues to develop as a small humanoid robot. This robot has been used for many research and educational purposes in numerous academic institutions. Paper [40] introduces the Nao robot with the capability of mimicking the emotions of users and providing feedback based on sentiment apprehension. Authors in [2] include the emotion conveyed through touch in the HRI between human and Nao. This robot was also designed for therapy of autistic children with good response and results [39,42,43]. The same robot is used in [5] as a tool to offer

---

[1] https://www.ald.softbankrobotics.com/en, last accessed: May 2018.

a tour guide for informatics laboratory, while the study in [44] explores a model for adaptive emotion expression in child-robot interaction. Another emotion and memory model intended for the robot Nao is presented in [1]. The robot adapts its behavior based on memory accounts of the child's emotional events.

The purpose of this research is to build an emotionally conscious teacher with the help of the humanoid robot Nao. The robot Nao used in our experiments is shown in Fig. 1a. The main task of the robot is to try to teach and entertain the user adapting to its current emotional state. The emotional state is recognized using the facial expression while the behavior is determined using optimal policy learned with reinforcement learning. The prime target group are children from primary school. However, in order to initially test the model and build a safe environment for children, the experiments for determining the ability of robot's learning and emotion recognition are carried out with adults. Testing the proposed robot behavior with children remains our aim for future work.

(a) Nao robot.

(b) Nao robot in the experimental setup.

Fig. 1. Nao robot utilized in this study.

The rest of the paper is organized as follows. Section 2 describes the model of the robotic behavior. The details of the implementation are included in Sect. 3. Section 4 presents the experimental study and evaluation of the results. Suggestions for the future work are given in Sect. 5, and at the end, Sect. 6 concludes the study.

2 A Model for Robotic Behavior

A model for robotic behavior based on user emotions is proposed in this paper. The complete workflow of the robot's behavior is shown in Fig. 2. The model is formulated with a set of robotic actions and states. Robot's learning is implemented as reinforcement learning, where the emotional state of the user is taken as a reward. Each part of the proposed behavior is described in details in the following subsections.

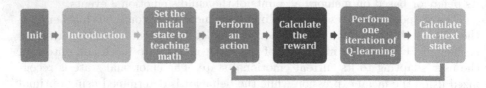

Fig. 2. Robot workflow.

2.1 Reinforcement Learning

The central element of robot control is the robot's learning. The robot should perform actions with fewer mistakes over time, utilizing the knowledge derived from previously gained experience. The learning of a robot can be categorized into two groups: offline and online learning. In this paper, we implemented reinforcement learning [41], which is an online learning method. The idea of this type of learning is to maximize the award that the robot receives for the actions it takes, that is, to build an optimal policy (behavior) with which the robot performs actions that give the best reward.

A mathematical framework for defining a solution for a reinforcement learning scenario is Markov Decision Processes, composed of:

- finite set of states S
- finite set of actions A
- function $f(s, s')$ for determining the reward r for transiting from state s to state s'
- policy π, which is a function that takes the current environment state to return an action
- value $V^\pi(s)$ - state-value function that returns the expected value from state s following policy π

One model of reinforcement learning is Q-learning [6,50]. The basis of Q-learning is learning the Q value that represents the value of a given action in a given state. The state-value function $V^\pi(s)$ is determined as

$$V^\pi(s) = \sum_{a \in A} \pi(a|s) * Q^\pi(s, a) \tag{1}$$

where s refers to the state, a is an action, A is the set of all possible actions, π represents the policy, and $Q^\pi(s, a)$ is the is the action-value function following the policy π in a given state s. First, the robot begins with initial values and then over time, it learns by exploring the space state-action and updates the values accordingly.

Q-learning is implemented with a simple version of the algorithm. The essential components are the defined states S, the possible actions A and the function $f(s, s')$ for determining the reward for mapping a given state and action. Having the current state s, after performing the action a the robot receives a reward r (the value of the function $f(s, s')$) and changes the state from s (previous state) to s' (current state). If the value of r is a penalty, then for the previous state s and the action a, the value Q is updated so that it is decreased. Therefore, the next time the robot finds itself in that state, the other actions will have a greater Q value and they will be selected. On the other hand, if r represents a reward, then the value for that combination of state-action increases and that action is favorable for the given state s.

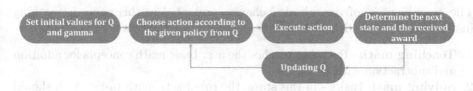

Fig. 3. Q-learning algorithm steps.

The equation for updating Q is given as

$$\mathbf{Q}(s, a) = \mathbf{Q}(s, a) + \alpha[r + \gamma \max_a \mathbf{Q}(s', a) - \mathbf{Q}(s, a)] \tag{2}$$

where s refers to the previous state, s' is the current state, $\mathbf{Q}$ is the matrix with state-action values, r represents the award from the environment, γ is a gamma factor $(0 < \gamma < 1)$ and α is the learning rate from the reinforcement learning. The goal is to learn an optimal policy which maps the states S into the robot's actions A. Selecting an action according to the values in $\mathbf{Q}$ is obtained according to the most beneficial action from the mappings in $\mathbf{Q}$ using ϵ-greedy policy (ϵ is the probability to select a random action), given the current state and possible actions. The algorithm is presented with the diagram on Fig. 3.

2.2   User's Emotional State

The ability to recognize user's emotion plays a significant role in the robotic behavior. Being aware of that emotional state the robot can choose its next action in accordance with it. When the robot is able to perceive the user's emotional state, it will determine how well the user is satisfied with the performed action. Positive emotion, if for example, the user is happy, signifies that the robot chooses an appropriate action. On the contrary, negative emotion indicates that the action should be changed in order to improve the emotional state of the interlocutor.

The emotion felt at a specific time by the user, i.e. its emotional state, represents a reward for the robot in the learning process described in the previous subsection. After every performed action, the user's emotional state is analyzed. Subsequently, the reward is calculated based on five emotions: happiness, neutral, surprise, sadness and anger. The probability of each emotion is calculated from facial expressions. The final reward is defined as a weighted sum of these probabilities as follows:

$$r = P(h) * 10 + P(n) * 5 + P(su) * 5 - P(sa) * 5 - P(a) * 10 \qquad (3)$$

where $P(h)$, $P(n)$, $P(su)$, $P(sa)$ and $P(a)$ are the probabilities for happiness, neutral, surprise, sadness and anger, respectively.

2.3   States and Actions

The main purpose of the robotic behavior presented in this paper is teaching math. Therefore, we define three different states for the robot:

1. **Teaching math** - the robot teaches the user basic math concepts for addition and subtraction with examples
2. **Solving math tasks** - in this state, the robot sets math tasks which should be solved by the user
3. **Playing/Taking rest** - the robot and the user play, i.e. they take rest

Schematic view of the states is displayed in Fig. 4. The robot takes actions and as a result it can change its state. We define two actions which are performed by the robot:

1. **Stay in the same state**
2. **Change state**

According to this, if the robot is, for example, in the state **Teaching math** it can perform one of the three actions:

- *Stay in the same state* - with this action the state is not changed. The new state is again **Teaching math**.
- *Change state into **Solving math tasks*** - with this action the state is changed from **Teaching math** to **Solving math tasks**.
- *Change state into **Playing/Taking rest*** - with this action the state is changed from **Teaching math** to **Playing/Taking rest**.

3   Implementation

The Nao robot[2] has a special operating system called NAOqi, which is Linux based. Also, the robot has a software package that includes a graphical programming tool called Choregraphe[3] [36], simulation software package and SDK.

---

[2] http://doc.aldebaran.com/2-1/home_nao.html, last accessed: May 2018.

[3] http://doc.aldebaran.com/2-1/software/choregraphe/index.html, last accessed: May 2018.

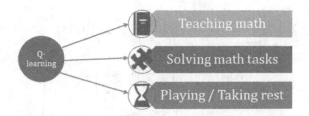

Fig. 4. Changing states with Q-learning.

Supported working languages are C++, Python, Java, MATLAB, Urbi, C, .Net. To implement the emotion-aware behavior of the robot, we used Choregraphe and the Python programming language. This tool enables simple application development using already implemented modules, as well as the ability to define custom modules. Figure 5 shows a visual representation of the implemented modules.

An initialization module is executed before the start of the interaction with the robot. This module sets predefined values for the parameters to be used. To begin interacting with the robot, the user needs to touch the robot at the front of the head. In this way, the interaction with the robot starts and the robot firstly introduces itself. Then, it performs tasks that are part of the first state - teaching math. The tasks for this state, as well as for other states, are implemented in the form of modules.

Within the state for teaching math, which is the initial state, the robot randomly chooses one of the two topics (addition or subtraction) to teach the user. Then, for the selected theme, the robot tells a brief explanation and illustrates it with an example. The robot interaction includes talking, which is done with the *AlTextToSpeech*[4] module which enables the robot to speak. Choosing the mathematical tasks for the user is part of the second possible state. In this state, in a similar way as in the previous one, the robot selects the theme for the task question. After selecting the topic (addition or subtraction), the task is randomly selected. Next, the robot waits for an answer. Information for the correctness of the answer is received by a third person in the following way: the touch of the front of the head indicates that the child answered correctly while the touching the back of the head indicates that the answer is wrong. After receiving information about the answer, the robot informs the child about the correctness of the answer. Within the third possible state - taking rest, the robot plays music that aims to entertain the child.

Next follows the estimation of the emotional state of the user. This is defined in a separate module that gains information about the five emotions defined previously and calculates the robot's reward. The module *ALFaceCharacteris-*

---

[4] http://doc.aldebaran.com/2-1/naoqi/audio/altexttospeech.html, last accessed: May 2018.

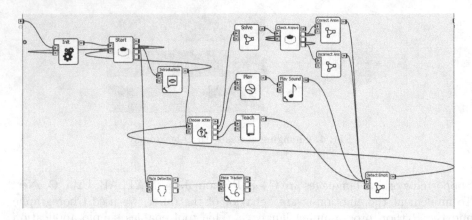

Fig. 5. The modules of the robot behavior in Choregraphe.

tics[5] is used to recognize and calculate the intensity of the emotional state of a person. This module analyses the person's characteristics for the people who are near the robot. *ExpressionProperties* represents one attribute of the conducted analysis. This attribute contains a list that holds the probability of a certain emotion for each of the five emotional states: neutral, happiness, surprise, anger, and sadness. It should be noted that the facial recognition and facial features analysis modules are sensitive to room illumination. This means that brightness can influence whether a person near the robot is detected and whether the values as features (probabilities of emotions) are obtained at all.

A specific module defines the Q-learning algorithm and the selection of next action. This module takes the previously calculated award and uses it to update the Q values. Depending on these values, the next action is selected. The parameters used in the multiple modules are stored in the robot's memory.

In parallel with these modules, the *ALFaceDetection*[6] and *ALFaceTracker*[7] modules are executed. These modules serve to detect and track people in the proximity of the robot. The memory of the robot contains the information about each of the "reachable" people.

4   Experimental Results

To evaluate the proposed behavior, we performed an experimental study with a group of 17 subjects from different gender and age. The distribution of age and gender is displayed in Fig. 6. We evaluated the robot's ability to recognize

---

[5] http://doc.aldebaran.com/2-1/naoqi/peopleperception/alfacecharacteristics.html, last accessed: May 2018.

[6] http://doc.aldebaran.com/2-1/naoqi/peopleperception/alfacedetection.html,   last accessed: May 2018.

[7] http://doc.aldebaran.com/2-1/naoqi/trackers/alfacetracker-api.html?highlight=tracker, last accessed: May 2018.

emotions and the experience of interacting with the robot. The results from these experiments are explained in details in the following subsections.

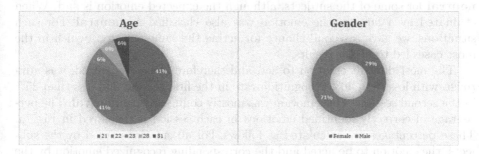

Fig. 6. Distribution of age and gender of the subject participating in the experimental study.

4.1   Experimental Procedure

The subjects were divided into two groups (of size eight and nine), and two sessions for interaction with the robot Nao were performed. Each subject had approximately five minutes for interacting with the robot during which he or she had to act all five defined emotions. The robot was placed on a desk in a sitting position which guarantees that the robot will stay safe during the experiments. The subjects were sitting in front of the robot at less than 1-meter distance. The environment of the experiments is presented in Fig. 1b. As discussed previously, the brightness influences the robot's ability to detect a face and emotions appropriate to that face. It is important to make sure that the room in which the experiments are performed is bright enough. Therefore, the experiments were performed during the day additionally with lamps light. The interactions were recorded to be analyzed later. After the interaction, all subjects were asked to fill in a questionnaire.

The order of the acted emotions was previously defined. In the first session, the order was as follows: *happiness, sadness, surprise, neutral* and *anger*. Throughout the first session, we noticed that acting **sad** after acting **happy** was difficult for the most of the subjects. Therefore, in the second session, the order of emotions was changed into the following sequence: *happiness, neutral, sadness, surprise* and *anger*.

4.2   The Ability to Recognize Emotions

In this subsection, the robot's ability to recognize emotions is discussed. One of the key influences on the emotion recognition is the user's ability to act emotions. Therefore, if the subject does not act the emotion properly, the robot would not be able to recognize it.

By examining the recordings of the sessions, it can be seen that not all subjects are good at acting emotions. There were cases when the emotion was not acted good enough, leading to a confusion of emotions. An example is recognizing **neutral** for some of the subjects, although the expected emotion is **sad**. When evaluated by a human, the emotion was also classified as **neutral**. For such situations, we gave a second chance for acting the same emotion, which in the most cases led to better results.

The most difficult emotion to act, and therefore to be recognized, was **surprise** with less than 40% recognition rate in the first session and less than 25% in the second session. This emotion was mostly confused with **neutral**. The percentage of correctly identified emotions in each session is displayed in Fig. 7a. These percentages are calculated as follows. For all emotions acted by the subjects, the emotion to be acted and the corresponding recognized emotion by the robot are compared, assuming that all of the emotions were acted properly. But, as stated previously, there were subjects whose acting was not good enough. The assessment of the acting was done with human judgment by re-examining all previously filmed videos. Removing the results of subjects whose acting was not good enough caused an increase in the emotion recognition rate, as presented in Fig. 7b. The recognition rate for **surprise** increased from 37% to 42% in the first session and from 22% to 33% in the second session. For **happiness**, the recognition rate in both sessions increased up to 100%. However, despite all circumstances, there were subjects for which most of the emotions were successfully identified. For 75% of the subjects in the first session and 78% in the second session, three or more emotions were successfully recognized by the robot.

Further, to test whether the emotion acting influences the emotion recognition, the *Pearson's chi-squared test of independence* was performed and the *Pearson's correlation coefficient* was calculated. Two different scenarios were evaluated:

1. Two binary random variables are defined as follows. The first random variable represents the goodness of acted emotion - *"Yes"* if the emotion is acted properly or *"No"* otherwise. The second random variable shows the correctness of the recognized emotion - *"Yes"* if the emotion is detected properly or *"No"* if not.
2. For this scenario, the first random variable illustrates the goodness of acted emotions per subject (*"Yes"* if the emotions are correctly acted or *"No"* otherwise). The second random variable is numerical with values in the range [1,4] that represent the number of correctly identified emotions per subject.

For both scenarios, the null hypothesis is *"The occurrence of the outcomes is statistically independent"*. It is rejected according to a significance level of 0.05 with a p-value of 0.009 for the first scenario and 0.021 for the second. This indicates that the series are not independent. The correlation analysis confirms the conclusion that dependence between detecting and acting emotions exists, i.e. we cannot give a concrete justification for the detection of emotion because it is not independent of the acting. For both scenarios the correlation coefficient

(a) All subjects.

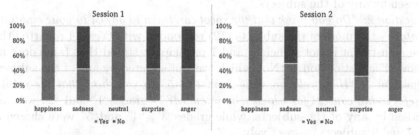

(b) Removed subjects whose acting was not good enough.

Fig. 7. Percentage of recognized emotions. "Yes" - the emotion is recognized, "No" - the emotion is not recognized.

is positive - 0.313 for the first and 0.742 for the second scenario, indicating a relation between the acting and detection of the emotions.

Besides the ability of emotion recognition, the ability to adjust to the user's emotional state can be also evaluated with the performed experiments. The robot's reward is increased if the emotion is *positive* (happiness, neutral, surprise) and decreased if the emotion is *negative* (sadness, anger). In the beginning, the robot is in an initial stage of learning to adjust to the subject's emotional state. Consequently, the next action is randomly selected. As the interaction continues, the robot starts to learn the emotional state of the subject. Hence, if the emotion is positive, the state of the robot is not changed. On the contrary, when the emotion is negative, the state changes with the decreasing of the reward. This part of the model behaves as expected - adapting to the user's emotional state. One possible negative influence may be caused by detecting emotions. In fact, if the emotion is not correctly determined, the robot will get an inappropriate reward and therefore choose an irrelevant action. The errors of the emotion recognition model, including its weaknesses like room illumination, may produce erroneous action selection.

4.3   The Experience of Interacting with the Robot

This subsection presents the results obtained by the questionnaire. The main purpose of this questionnaire is to evaluate the experience of interaction with the robot. It consists of five questions:

- *Question 1: "How well, on a scale of 1 to 5, do you think that the robot can recognize your emotion?"* - for this question, 10 of the subjects gave "4" as a grade, 3 of the subjects evaluated the ability to recognize emotions with grade "5" and "3". One subject reported a grade "2", while grade "1" was not chosen by any of the subjects.
- *Question 2: "Do you think that the robot can learn to adjust to your emotional state?"* - the most of the subjects (12) responded with "Yes", 4 reported that the adjustment is not sufficient and 1 participant stated that (s)he does not know. The last option is "No" which was not selected by any of the subjects.
- *Question 3: "How well, on scale from 1 to 5, do you think that the robot is useful in the context for which it is intended?"* - grades "1" and "2" were not chosen by any of the subjects, while grades "3", "4" and "5" were chosen by 1, 7 and 9 subjects, respectively.
- *Question 4: "Do you think that the robot Nao can have a role of social assistant/teacher?"* - to this question, 13 of the subjects responded with "Yes" and 4 with "Maybe". Options "No" and "Don't know" were also available, but not selected by any subject.
- *Question 5: "Would you, if possible, use a robot as a teacher instead of a human?"* - most of the participants (9), to this question, answered with "Maybe". The rest of the participants responded as follows: 5 with "Yes", 3 with "No" and no one with "Don't know".

From the results, it can be concluded that the overall experience is positive. Relatively large percentage of the subjects consider that the robot is able to recognize and adjust to their emotional state. Moreover, most of the subjects consider that the robot is suitable in the context for which it is intended and can have a role of social assistant. The results from this questions are summarized in Fig. 8.

5   Future Work

As mentioned previously, the emotion detection module has several weaknesses. Therefore our future work will include designing a more robust solution for solving the problem with greater accuracy. The aspired direction is building models with deep learning networks, as they show satisfactory results for facial expression emotion detection in many research studies including [4,23,31]. Another thing to consider is the implementation of Q-learning and reinforcement learning with deep neural networks. This approach is favorable if we increase the number of states and actions, making the robot more capable of intelligent interaction

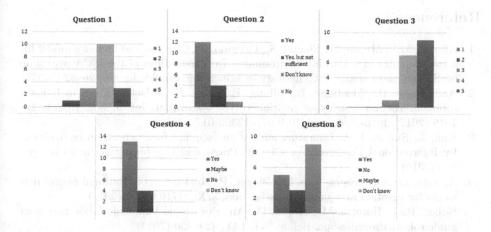

Fig. 8. Questionnaire results.

and multi-task accomplishment. Implementations that utilize this method are presented in [32, 33, 46].

Social assistive robots are already adopted and tested for the child-robot interaction [13, 16, 24]. Our future work can include research where the target group are children, that is, Nao would interact with the child as an educational tutor or as an entertainer.

6   Conclusion

This research addresses the use of the Nao robot in the role of an emotionally aware teacher. The learning of the robot is implemented as a reinforcement Q-learning. For this implementation, the robot needs to be able to determine a particular reward according to which it will update its knowledge. The corresponding reward of the robot is calculated through the emotional state of the user determined by the facial expression. Hence, the robot should adapt to the current state of the user and learn how to act at a specific time. The Q-learning method enables the robot to learn whether its behavior at some point was good, that is, whether it managed to adapt to the emotional state of the user.

An experimental study was performed to evaluate the proposed behavior of the robot. The results show that the overall experience is positive. The qualitative analysis points out that some of the subjects in the experiments were not able to act certain emotions. Despite the difficulties of recognizing emotions in some cases, in general, the robot is able to detect the correct emotion from the facial expression and therefore to adapt to the user's emotional state. Solutions of the problems in this model are considered in our future work that also includes ideas for upgrading the model.

Acknowledgement. The authors would like to thank the Faculty of Computer Science and Engineering - Skopje for partially financing this work.

References

1. Ahmad, M.I., Mubin, O., Shahid, S., Orlando, J.: Emotion and memory model for a robotic tutor in a learning environment. In: Proceedings of the ISCA workshop on Speech and Language Technology in Education, Stockholm, pp. 26–32 (2017)
2. Andreasson, R., Alenljung, B., Billing, E., Lowe, R.: Affective touch in human-robot interaction: conveying emotion to the Nao robot. Int. J. Soc. Robot. **10**, 1–19 (2017). https://doi.org/10.1007/s12369-017-0446-3
3. Anil, J., Suresh, L.P.: Literature survey on face and face expression recognition. In: International Conference on Circuit, Power and Computing Technologies, pp. 1–6 (2016)
4. Arriaga, O., Valdenegro-Toro, M., Plöger, P.: Real-time convolutional neural networks for emotion and gender classification. arXiv:1710.07557 (2017)
5. Boboc, R.G., Horaţiu, M., Talabă, D.: An educational humanoid laboratory tour guide robot. Procedia-Soc. Behav. Sci. **141**, 424–430 (2014)
6. Bozinovski, S.: A self-learning system using secondary reinforcement. In: Cybernetics and Systems Research Proceedings of the 6th European Meeting on Cybernetics and Systems Research, pp. 397–402 (1982)
7. van Breemen, A., Yan, X., Meerbeek, B.: iCat: an animated user-interface robot with personality. In: Proceedings of the 4th International Joint Conference on Autonomous Agents and Multiagent Systems, pp. 143–144. ACM (2005)
8. Broadbent, E.: Interactions with robots: the truths we reveal about ourselves. Annu. Rev. Psychol. **68**, 627–652 (2017)
9. Callejas-Cuervo, M., Martínez-Tejada, L.A., Alarcón-Aldana, A.C.: Emotion recognition techniques using physiological signals and video games-systematic review. Revista Facultad de Ingeniería **26**(46), 19–28 (2017)
10. Canales, L., Martínez-Barco, P.: Emotion detection from text: a survey. In: Proceedings of the Workshop on Natural Language Processing in the 5th Information Systems Research Working Days (JISIC), pp. 37–43 (2014)
11. Chen, A., et al.: Developing a robotic tutor to teach math skills to children (2017)
12. Dautenhahn, K.: Kaspar-a minimally expressive humanoid robot for human-robot interaction research. App. Bionics Biomech. **6**(3–4), 369–397 (2009)
13. Davison, D., Schindler, L., Reidsma, D.: Physical extracurricular activities in educational child-robot interaction. arXiv preprint arXiv:1606.02736 (2016)
14. El Ayadi, M., Kamel, M.S., Karray, F.: Survey on speech emotion recognition: features, classification schemes, and databases. Pattern Recognit. **44**(3), 572–587 (2011)
15. Fong, T., Nourbakhsh, I., Dautenhahn, K.: A survey of socially interactive robots. Robot. Auton. Syst. **42**(3–4), 143–166 (2003)
16. Fridin, M., Angel, H., Azery, S.: Acceptance, interaction, and authority of educational robots. In: IEEE Workshop on advanced robotics and its social impacts, California, USA (2011)
17. Fung, P., et al.: Towards empathetic human-robot interactions. arXiv preprint arXiv:1605.04072 (2016)
18. Gockley, R., et al.: Designing robots for long-term social interaction. In: IEEE/RSJ International Conference on Intelligent Robots and Systems, pp. 1338–1343. IEEE (2005)
19. Goodrich, M.A., Schultz, A.C.: Human-robot interaction: a survey. Found. Trends HRI **1**(3), 203–275 (2007)

20. Hegel, F., et al.: Understanding social robots. In: 2nd International Conferences on Advances in Computer-Human Interactions, pp. 169–174. IEEE (2009)
21. Huijnen, C.A., Lexis, M.A., de Witte, L.P.: Matching robot kaspar to autism spectrum disorder (ASD) therapy and educational goals. Int. J. Soc. Robot. 8(4), 445–455 (2016)
22. Ishiguro, H.: Robovie: an interactive humanoid robot. Ind. Robot: Int. J. 28(6), 498–504 (2001)
23. Jeon, J., et al.: A real-time facial expression recognizer using deep neural network. In: Proceedings of the International Conference on Ubiquitous Information Management and Communication, p. 94. ACM (2016)
24. Kanda, T., Sato, R., Saiwaki, N., Ishiguro, H.: A two-month field trial in an elementary school for long-term human-robot interaction. IEEE Trans. Robot. 23(5), 962–971 (2007)
25. Kirandziska, V., Ackovska, N.: Human-robot interaction based on human emotions extracted from speech. In: TELFOR, pp. 1381–1384. IEEE (2012)
26. Kirandziska, V., Ackovska, N.: A concept for building more humanlike social robots and their ethical consequence (best paper award). In: International Conferences ICT Society and Human Beings, pp. 19–37 (2014)
27. Kleinsmith, A., Bianchi-Berthouze, N.: Affective body expression perception and recognition: a survey. IEEE Trans. Affect. Comput. 4(1), 15–33 (2013)
28. Kory Westlund, J., et al.: Tega: a social robot. In: The Eleventh ACM/IEEE International Conference on Human Robot Interaction, pp. 561–561 (2016)
29. Kozima, H., Michalowski, M.P., Nakagawa, C.: Keepon: a playful robot for research, therapy, and entertainment. Int. J. Soc. Robot. 1(1), 3–18 (2009)
30. Leite, I., Castellano, G., Pereira, A., Martinho, C., Paiva, A.: Empathic robots for long-term interaction. Int. J. Soc. Robot. 6(3), 329–341 (2014)
31. Mavani, V., Raman, S., Miyapuram, K.P.: Facial expression recognition using visual saliency and deep learning. arXiv preprint arXiv:1708.08016 (2017)
32. Mnih, V., et al.: Playing atari with deep reinforcement learning. arXiv preprint arXiv:1312.5602 (2013)
33. Mnih, V.: Human-level control through deep reinforcement learning. Nature 518(7540), 529 (2015)
34. Paiva, A., Leite, I., Ribeiro, T.: Emotion modeling for social robots. The Oxford handbook of affective computing pp. 296–308 (2014)
35. Pekrun, R.: Emotions and Learning. International Academy of Education/International Bureau of Education (2014)
36. Pot, E., Monceaux, J., Gelin, R., Maisonnier, B.: Choregraphe: a graphical tool for humanoid robot programming. In: Robot and Human Interactive Communication, The 18th IEEE International Symposium on, pp. 46–51. IEEE (2009)
37. Preckel, K., Kanske, P., Singer, T.: On the interaction of social affect and cognition: empathy, compassion and theory of mind. Curr. Opin. Behav. Sci. 19, 1–6 (2018)
38. Sabelli, A.M., Kanda, T., Hagita, N.: A conversational robot in an elderly care center: an ethnographic study. In: Proceedings of the 6th International Conference on HRI, pp. 37–44. ACM (2011)
39. Shamsuddin, S., et al.: Humanoid robot NAO interacting with autistic children of moderately impaired intelligence to augment communication skills. Procedia Eng. 41, 1533–1538 (2012)
40. Shen, J., Rudovic, O., Cheng, S., Pantic, M.: Sentiment apprehension in human-robot interaction with NAO. In: International Conference on Affective Computing and Intelligent Interaction, pp. 867–872. IEEE (2015)

41. Sutton, R.S., Barto, A.G.: Reinforcement learning: An introduction, vol. 1. MIT press, Cambridge (1998)
42. Tanevska, A., Ackovska, N., Kirandziska, V.: Assistive robotics as therapy for autistic children. In: 13th International Conference for Electronics, Telecommunications, Automation and Informatics (2016)
43. Tanevska, A., Ackovska, N., Kirandziska, V.: Robot-assisted therapy: considering the social and ethical aspects when working with autistic children. In: Proceedings of the 9th International Workshop on Human-Friendly Robotics, Genova, Italy, pp. 57–60 (2016)
44. Tielman, M., Neerincx, M., Meyer, J.J., Looije, R.: Adaptive emotional expression in robot-child interaction. In: Proceedings of the ACM/IEEE International Conference on HRI, pp. 407–414. ACM (2014)
45. Tyng, C.M., Amin, H.U., Saad, M.N., Malik, A.S.: The influences of emotion on learning and memory. Front. Psychol. **8**, 1454 (2017)
46. Van Hasselt, H., Guez, A., Silver, D.: Deep reinforcement learning with double q-learning. In: AAAI, vol. 16, pp. 2094–2100 (2016)
47. Veloso, M.M., Rybski, P.E., Lenser, S., Chernova, S., Vail, D.: CMRoboBits: Creating an intelligent AIBO robot. AI Mag. **27**(1), 67 (2006)
48. Wada, K., Shibata, T.: Robot therapy in a care house-its sociopsychological and physiological effects on the residents. In: Robotics and Automation, Proceedings of the International Conference on, pp. 3966–3971. IEEE (2006)
49. Wada, K., Shibata, T.: Living with seal robots-its sociopsychological and physiological influences on the elderly at a care house. IEEE Trans. Robot. **23**(5), 972–980 (2007)
50. Watkins, C.J., Dayan, P.: Q-learning. Mach. Learn. **8**(3–4), 279–292 (1992)

The Application of an Air Pollution Measuring System Built for Home Living

Andrej Ilievski[✉], Dimitri Dojchinovski[✉], Nevena Ackovska,
and Vesna Kirandziska

Faculty of Computer Science and Engineering, Skopje, Republic of Macedonia
{andrej.ilievski,dimitri.dojcinovski}@students.finki.ukim.mk,
{nevena.ackovska,vesna.kirandziska}@finki.ukim.mk

Abstract. Air pollution in recent years has become alarmingly a part of our everyday life. One out of nine deaths worldwide are caused by this type of pollution. The first step towards solving this problem, is being able to measure the air particles in the air we breathe and not only outdoors, but indoors as well. Having in mind the uprising of Internet of Things (IoT) and voice interaction with devices, this paper presents a future-proof solution to this problem. It consists of a cheap and compact outdoor station and an indoor station with an ability to be connected with an Alexa enabled device and accessed using voice. These stations were made using a Raspberry Pi 3, a Wi-Fi - enabled micro controller and three different types of sensors, which can also measure the smallest standardized air pollutants - PM1, particles with less than $1\mu m$ in size. This creates an opportunity for a detailed analysis of the available data from the indoor and the cheap outdoor air pollution measuring system spread across a city.

Keywords: Air pollution · Air quality · Amazon alexa
Particulate matter

1 Introduction

Health is one of the most important aspects of our lives. But, ever since the industrial revolution, air pollution both indoor and outdoor, has emerged as the deadliest form of pollution, officialized in 2015, accounting for 7 million deaths worldwide [21]. One out of nine deaths worldwide are caused by this type of pollution. The worst part is that it affects our health from our fetal life. A recent study published in Biological Psychiatry Journal, noted that children exposed to high levels of particulate matter during their fetal life had long-term issues with impulsive behavior, self-control over temptations and even some mental health issues [20]. This can be additionally supported by two studies made in 2015 in Barcelona, Spain on 39 schools [29] and in 2011 conducted in Michigan, USA on 3660 schools [25]. Both have done tests on the cognitive performance of students from primary school up to high-school, in more and less polluted indoor

© Springer Nature Switzerland AG 2018
S. Kalajdziski and N. Ackovska (Eds.): ICT 2018, CCIS 940, pp. 75–89, 2018.
https://doi.org/10.1007/978-3-030-00825-3_7

and outdoor environments. The studies came to the same conclusion that children from highly polluted schools had smaller growth in cognitive development than children from the less polluted schools. Another recent study suggests that air pollution [24] has a positive correlation and significant impact on human behavior - more specifically on the overall crime level. The authors of this study suggest that improving air quality in urban areas may reduce crime. These are only some of the most recent research results for the effect of the air pollution aside from the well-documented premature causes of death, such as lung cancer, pulmonary disease, various heart diseases and respiratory infections, done by the World Health Organization [16]. All of this adds up to having an economic impact costing the global economy $ 225B according to the World Bank [17].

Unfortunately, Macedonia also joins this story with Skopje being declared one of the most polluted cities in Europe. The Macedonian Institute of Public Health has reported that the air pollution accounts for up to 1300 deaths per year countrywide [10]. With the authors living in Skopje, this was essentially the motivation for the creation of such system that would have two important features. First, there should be a cheap solution for precise indoor air quality measurement. Second, the system should have a personal and compact air quality station for the outdoor air quality in the area close to the user's home. Making the user's information not dependent on various online sources with low update frequency. Furthermore, multiple such outdoor devices scattered across a city present an opportunity for much more detailed view of a city's air quality in the future. Such detailed data could lead to novel insights and approaches towards battling the air pollution.

This paper presents a compact and cost-effective solution that consists of a pair of stations, one for the outdoor and one for the indoor air quality measurement. This furthermore is done by providing measurements of the smallest standardized air particles PM1, which, as of writing this paper, are not measured by any station in Macedonia. These PM1 particles, by numerous recent researches [18, 30], are shown to be the deadliest and most detrimental to people's health. With the aim of having a future-proof interface, Amazon Voice Service is integrated that enables voice control, a user-friendly interface of our system. In the following section, all the details about how the data is collected within the system is covered. Next, Sect. 3 describes how Amazon Voice Service is integrated with the project. And finally, the results from the primary experiments are shown and a conclusion is made in the final section.

2   Dataflow to IoT Database

Before the start of the project, there was a simple idea of making an indoor station for measuring just the particulate matter (PM). After defining which components would be needed and being sought out on popular e-commerce web sites, it was concluded that the station can be assembled of high quality sensors for a price that everyone can afford. Having that in mind, there was an opportunity to also develop an outdoor station. Both stations measure dust particles and upload data to the ThingSpeak database as shown on Fig. 1.

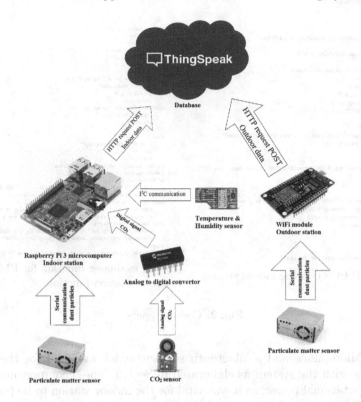

Fig. 1. Indoor and outdoor station sending information to the IoT database

ThingSpeak is a cloud platform specifically designed for IoT devices, where data is collected and then retrieved by any device capable of communicating with the database. The database in our project contains three tables. The "Indoor_PM" table collects the PM information of the indoor station and it has "PM1", "PM2.5" and "PM10" as columns. The "indoor_gas" collects the temperature, humidity and CO_2 information from the indoor station. The "outdoor_pm" collects the outdoor PM information from the outdoor station. It includes "PM1", "PM2.5" and "PM10" as attributes. The data in these tables is uploaded through HTTP Request method POST. In Python, this is done using the urllib2 library (as seen on Fig. 2a). You connect to the database providing the function with the unique database key, which gives a permission to write down the information to the corresponding columns. A detailed elaboration on the indoor and outdoor station and the data that is collected by these stations is given in the following subsections.

2.1 Indoor Setup

The indoor station is composed of the following sensors: a Particulate matter (PM) sensor, a CO_2 sensor and a temperature and humidity sensor. But also

```
def publish():

    CO2, temp, humidity, dewpoint = getData()

    now = datetime.datetime.now()

    params = urllib.parse.urlencode({
    'field1':int(CO2),
    'field2': temp,
    'field3': humidity,
    'field4': dewpoint,
    'key': DATABASE_KEY })

    headers = {
    "Content-type": "application/x-www-form-urlencoded",
    "Accept": "text/plain"
    }

    conn = http.client.HTTPCOnnection("api.thingspeak.com:80")

    try:

        conn.request("POST", "/update", params, headers)
        response = conn.getresponse()
        data = response.read()
        conn.close()

    except:

        print("connection failed")
```

(a) HTTP POST in the indoor station

```
def PM1_evaluator( value):
    name = 'PM1'
    value = value
    if value >= 0 and value < 14:
        text = 'Good'
        effect = 0.2
        index = 0
    elif value >= 14 and value < 34:
        text = 'Moderate'
        effect = 1.2
        index = 1
    elif value >= 34 and value < 95:
        text = 'Unhealthy'
        effect = 2.2
        index = 2
    else:
        text = 'Very Unhealthy'
        effect = 3.2
        index = 3
    return [name,value,text,effect,index]
def recommendation(text):
    if text=='Good':
        return 'I recommend opening the windows, so that the fresh air'
        +' can come in, or maybe some outdoor activity'
    elif text=='Moderate':
        return 'I don''t recommend opening the window or taking any'
        +' outdoor activity if you are sensitive to air pollution'
    elif text=='Unhealthy':
        return 'I highly don''t recommend opening the window or taking'
        +'outdoor activities, sensitive group may suffer'
    else:
        return 'I recommend staying indoors, outside is very hazardous'
```

(b) PM evaluator function for PM1 and a recommendation function

Fig. 2. Code snippets

it has a Microphone and a Bluetooth speaker which are used for the human interaction with the system as elaborated in Sect. 3. These components require high computational power, so it was vital for the indoor station to be positioned on a microcomputer. Today one of the most popular microcomputers used in IoT is the Raspberry Pi 3 Model B [26]. Its miniature form, affordable price and the possibility to install many operating systems (OS), most of them being open source, made it ideal for the authors' project and thus it was used in the indoor station. Since it is mostly used in the IoT world [22], the Raspbian OS [13], was chosen for this project. It is a Unix-like OS and comes with pre-installed Python, Java, Mathematica etc.

PM Sensor. When searching for the right PM sensor, the authors were looking for one with good characteristics, but for an affordable price. The sensor Plantower PMS5003, mentioned in a highly respected article written by the World Air Quality organization [3], was chosen for this project. The sensor uses laser scattering to radiate suspending particles in the air, and then it collects scattering light to obtain the curve of scattering light change with time. The microprocessor within calculates equivalent particle diameter and the number of particles with different diameter per unit volume.

PMS5003 is hooked up to the Raspberry Pi via the serial line (pins TXD and RXD). When the sensor is triggered to measure, it takes at least 30 seconds to heat up, and in the next 15 seconds it collects data. In the Python code, the serial library is used to read the information from the data pin. Each data from each measurement is formatted into a frame. Every frame, besides the list of measurements, includes a time stamp that stores information for the time of the measurement. After an array of 15 separate measurements is filled, an average of all measurements in the array is calculated and that data is send to the cloud. A library is used to help the sensor obtain the PM values [14]. The data that the sensor managed to collect in a period of a few days were analyzed in order to find the best time period between two consecutive measurements so that we have less power consumption, yet the errors in the measurements to be as low as possible. In order to find that, the mean square error estimator was ran to calculate the risk of expanding the time to measure between two consecutive measurements. On Fig. 3, the x-axis represents the time in minutes when the sensor is activated to measure and upload the data, while the y-axis represents the mean square error of the measured value after a certain time compared to the value measured every minute. Observing the result, it was concluded that the sensor should measure every 8 min which results in more efficient power consumption, while simultaneously keeping a low level of error.

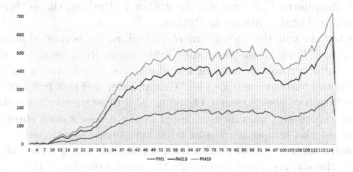

Fig. 3. Mean square error analysis

Carbon Dioxide Concentration Level Sensor.

The carbon dioxide concentration level sensor used in this project is a Mikroelektronika clickBoard (CO_2 sensor). This is the cheapest solution for a CO_2 sensor on the market that the authors could find, because it is as much as 6 times cheaper than a specialized CO_2 sensor. Even more, the sensor was chosen because the authors have a positive experience with Mikroelektronika's hardware [27]. Another advantage of this sensor is that it is sensitive to various other natural gases, whereas the more expensive ones are focused only on CO_2. However, this doesn't make it significantly incorrect, since out of all the gases that it is sensitive to, CO_2 is the most prevalent one. On the other side, being sensitive to other gases makes it a decent sensor for sensing abnormalities in the indoor air composition that arise in extreme cases such as fire by detecting the smoke. The sensor works based on the conductivity of the sensitive material made out of Tin dioxide (SnO_2), which has a different resistance depending on the quality of the air. For more precise operation of the sensor, there is an additional electrode that preheats the air around the sensitive material. It was recommended in the MQ-135 data sheet that the electrode and the sensor altogether should be running on for at least 24 h before making correct measurements. The conductive material is sensitive to NH_3, NO_x, alcohol, gasoline, smoke and CO_2.

This board is designed to integrate easily with any microcontroller that supports the mikroBUS socket (used for interfacing microcontrollers created by Mikroelektronika). But, that would require an additional Raspberry Pi add-on board to be used and it would take up more space, so it was decided to make a custom workaround. Since Raspberry Pi is lacking any analog pins, the CO_2 sensor board sends an analog signal to the 12-bit resolution Analog to Digital Converter (MCP 3204). There are 4 analog input pins and one output pin, which sends information to the Raspberry Pi about the corresponding voltage for each input pin. An already available library [15] was used to read the values of the ADC. The voltage values, are converted to analog and then used in another very detailed library [23], written in C/C++ for Arduino. Since all of this should be used on the Raspberry Pi 3 with a code written in Python, the authors needed to translate this Arduino library in Python.

In order to take valuable measurements, a calibration is needed because each sensor is unique. That is, the resistance value varies from sensor to sensor, so it has to be adjusted to the outdoor air which has a well known CO_2 level of around 407 ppm (parts per million) [4]. We can later use this point of reference for the calculation of the current CO_2 ppm. The resistance value is dependent on the temperature and humidity, as noted in the sensor's data sheet. For this purpose we use the temperature and humidity sensor mentioned in the next section. Finally, after the sensor calibration is done, we can get the values of the CO_2 ppm in the air. As noted previously, the main function of this sensor is to read the amount of CO_2 in the air, but it can also read the general air quality. Note that it is instantly sensitive to an increased amount of alcohol, gasoline, gas, and other harmful gases.

Temperature and Humidity Sensor. The MikroElektronika's SHT1X board was used as a temperature and humidity sensor and it was far simpler to integrate than the CO_2 sensor. It uses an I^2C serial interface, where one pin is used to transfer data and the other one to control the clock. For the Raspberry Pi 3 the data is transmitted via pin P3, while the clock pin is on P2. The board itself has a 14-bit resolution ADC, so the Raspberry can directly connect with this sensor, thus no further work is needed. A library [7] was used to obtain the temperature and humidity values in the indoor setup. As mentioned previously these are needed to re-adjust the resistance of the CO_2 sensor.

2.2 Outdoor Setup

The outdoor setup consists of a microcontroller NodeMCU and a PM sensor, which are connected serially. The PM sensor was described previously, so only the microcontroller will be reviewed here. NodeMCU is the name for the platform for IoT that covers the firmware running on NodeMCU devices (ESP8226 WI-FI SOC) and the devices itself. The authors' task was first to install that firmware on the device so that it can be programmable with the Arduino IDE. The plug of the PMS5003, that was originally incompatible with NodeMCU, was connected with a workaround by simply removing the plug altogether and connecting each separate wire as required with the NodeMCU.

A library written to work on Arduino that is also supported by the NodeMCU, was used for the programming. The Arduino code is a C/C++ code structured in such a way that two global functions exist. One function, called setup(), is run only at the start and the other function, called loop(), runs repeatedly until the device is powered off or reset. That is the logic of an Arduino code, but since in this case we are using a deep sleep capability of the NodeMCU [19], we only use the setup function. This is because in the setup function we are setting up the Wi-Fi, getting the measurements from the PMS5003, uploading the data to Thingspeak via HTTP request and putting the device in this deep sleep mode. This mode shuts down the Wi-Fi capabilities, CPU and only keeps a timer running, with a specified time of 8 min, after which it resets the whole device - and begins once again with the setup function. Deep sleep enables us to dramatically lower the energy consumption of the device.

3   Amazon's Alexa Voice Assistant

The data for indoor and outdoor pollution, which is stored in the Thingspeak database, is communicated with the user using the Amazon's Alexa voice assistant, which was implemented on the Raspberry Pi - the indoor station. Having USB ports and integrated Bluetooth, the Raspberry is ideal for using a microphone and a Bluetooth speaker for the voice assistant. Amazon has made it easy for developers to install Amazon Voice Services (AVS) (including Alexa), and to use it to develop all kinds of skills for Alexa [8]. Using Amazon's guide, the authors successfully installed the AVS on the Raspberry Pi, making it an Alexa enabled device. Alexa skills, are the applications that are developed so that users can have a custom and personalized interaction with the voice assistant. To develop a skill, it was inevitable to get to know the whole Alexa Skill concept [9]. The next paragraph discusses the data exchange between the user, the station, Amazon Web Service and the cloud. This data flow can be viewed on Fig. 4.

Alexa uses an invocation command, a word or a sentence, spoken by the user when he/she has some request. Each invocation command corresponds to an Alexa skill that is triggered to be executed. To go further in the complexity of the Alexa ecosystem developed for this project, lets follow an example of a sentence that would be spoken by the user: "Alexa ask air quality xXxXx". In the example, the word "Alexa" is a standard invocation word that lets Alexa listen to you, "air quality" is the custom invocation command that corresponds to the custom Alexa skill developed for this project. All possible strings that can stand for "xXxXx" in the example are defined in the Interaction model that is consisted of a table with intents (sentences) that can be asked by the user for the air pollution in the surrounding. With the right intent, the right skill function is triggered to process the data from the AWS cloud and fulfill the request by sending back the information to the station so that Alexa can inform the user. The intents include the following user requests: "air pollution"; "indoor PM information"; "outdoor PM information"; "indoor carbon dioxide level"; "indoor temperature and humidity". Each intent can be triggered with the statement "how is the xXxXx". For an example: "Alexa ask air quality how is the air pollution".

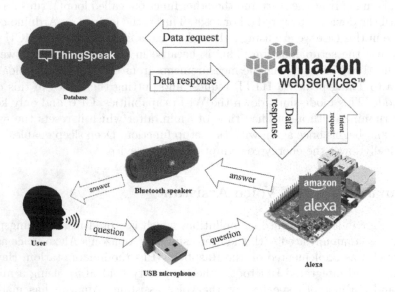

Fig. 4. Data flow: User asking Alexa about the air pollution

Lambda is a part of Amazon web services (AWS), which lets you run code on the cloud without worrying about the maintenance of the server [5]. The program can be written in a variety of languages and our project was developed in Python. Alexa uses a custom syntax, thus several tutorials with test programs [12] for learning exist. Lambda is programmed to receive a request by the AVS depending on the intent, processes the request and responds back with some information. This processing part is where the whole logic of the project is developed so that the user can be informed about the air pollution. For the purpose of this project, several functions were coded to respond to different intents sent by the AVS. The intents "indoor PM information", "outdoor PM information", "indoor carbon dioxide level" and "indoor temperature and humidity" when recognized in a sentence, trigger their corresponding Lambda function, where the parameters from the cloud are received, processed and sent back to the station where Alexa reads them to the user.

In the lambda code, the "urllib2" library is used to retrieve the data from ThingSpeak in a JSON format as shown on Fig. 6. This data is then converted to a plain string, which the custom functions process and output PM values used for evaluation. The most important functions that use this data are the PM evaluation and recommendation functions. Examples of these functions are given on Fig. 2b. In the PM evaluation, the particulate matters are ranked by their effects on human health, and when determining the pollution status they are compared by their effect index. The PM description words, depending on the PM values, are: Good, Moderate, Unhealthy and Very Unhealthy. The pollution evaluator functions for PM2.5 and PM10 works according to the European AQI scale [6], and the PM1 recommendation is based on the Dutch AQI scale, which is the only index scale for this size of particulate matter [1] (Fig. 5).

	Good	Moderate	Unhealthy	Very unhealthy
PM1	0-14	14-34	34-95	95+
PM2.5	0-20	20-25	25-50	50+
PM10	0-35	35-50	50-100	100+

Fig. 5. Classification of air quality based on PM concentration in parts per million [1,6] (Color figure online)

In this table, the first colored row represents the AQI descriptors used to help the users understand the PM pollution. Whereas, in the first column we have the three sizes of Particulate matter measured by the PMS5003 sensor. For each of these the particles, we have the scope of the quantity in ppm (particles per million) defined for each AQI descriptor. This information is used as input in the recommendation functions to generate a response for what activities the user can take at that moment, i.e. should he/she open a window to exchange air or should he/she take an activity outside. In fact, the most advanced function is created for the intent: "how is the air pollution". It requests a comparison between the indoor and outdoor particulate matter pollution and gives a recommendation for a preferred action the user should take considering the air pollution inside and outside of the user's home. This concludes the whole cycle from the user's question, to the reply he/she gets from an Alexa enabled device. The next section reviews the results from the measurements received from the devices.

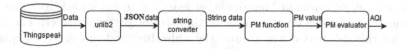

Fig. 6. Data interaction scheme

4   Results

In this section the successful implementation and results from the tests on the entire system are presented. The system, composed of the indoor and outdoor station, is shown on Fig. 7. It was built from scratch by combining all the components described in the previous sections. The communication between each of the components as well as uploading all data on the database is successful, thus the system works as expected.

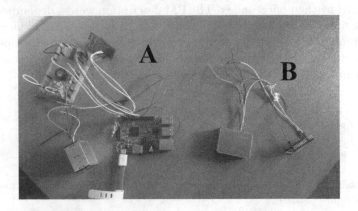

Fig. 7. The indoor station (A); The outdoor station (B)

The first step in determining the validity of our system was to compare some air pollution values between our outdoor station and the station owned by the Ministry of environment and physical planning of Macedonia located in the Municipality of "Centar" in Skopje. On Fig. 8 a graph for the PM values over a period of time is showcased. The results show that the "Centar" station shows 20–30% lower values for both PM2.5 and PM10. Note that, there are no measurements for PM1 values to be compared. This leads to an important difference in the classification of the air quality according to the index presented previously. For example in cases when our station shows an unhealthy air, the "Centar" station shows a moderately polluted air. Since the measurements are done only in a 2 h window (17:30–19:30), the authors believe that a comparison on a longer period is needed to obtain a better analysis. Another interesting observation from this test is that PM1 particulate matter values that are measured in our system are not measured in Macedonia, at least not to our knowledge. The official ministry's stations do not measure this size of particle matter. The PM1 values for this 2-h period are around the unhealthy and moderate air quality range. As noted previously the PM1 values are the most detrimental to humans health and our system enables an improved information for the pollution.

Figure 9 shows the per minute values of the indoor station PM sensor, combined with the hourly measurements from the air quality station owned by the Ministry of environment and physical planning of Macedonia located in "Centar" in Skopje [2]. As shown on Fig. 9 the indoor station PM 2.5 values are higher than the values in the "Centar" station. One can note that staying in a closed environment can sometimes be harmful for our health even if we want to avoid

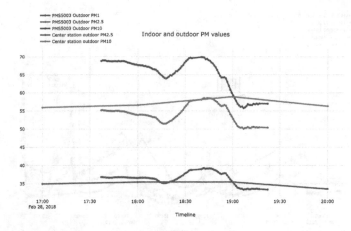

Fig. 8. Outdoor station and "Centar" station PM values over 2 h

the outdoor pollution. This is because the indoor PM2.5 levels are more influenced by human activities than outdoor PM 2.5 levels. This was also discussed in another paper following the indoor PM 2.5 correlation [28]. On the other side, the outdoor PM 10 levels are always higher than the indoor values, but they seem to be correlated. This is due to the way the experiment is performed. During this experiment the window was opened 3 times (at 20:00 on 10 March, 8:00 on 11 March and 00:00 on 12 March) for approximately 30 min. It resulted with all three PM particles experiencing a sharp increase of their values (Fig. 9). The peaks at night have a bigger value due to the fact that the outdoor air pollution was the highest in this period, and the "peak" in the morning has a lower value due to the fact that the pollution outside settles down, but still it is in the unhealthy range. After each event of opening the window first a "peak" in the PM values from the indoor station is noticed and than these values decreased slowly. This shows that opening a window has an effect on the air quality measurements of the indoor station. It suggests that the recommendation function implemented for our system would improve the air quality in home living since it would correctly recommend the user when to open the window. In this way the windows would not be opened in the period when it could have a negative effect on the indoor air quality. In the experiment two other events of cleaning and vacuuming the room happened at 8:00 and 12:00 on 12 March (Fig. 9). The goal was to see the human influence on the air pollution indoors. Both events resulted with sharp increases in the PM values. The "peak" in the morning has a slight lower value when comparing with the one at midday, due to the fact that through the day the pollution constantly increases because of other human activities.

Figures 10 and 11 show the measurements for the indoor CO_2 levels that resulted from two experiments that proved the validity and reactiveness of the CO_2 sensor. The first experiment was conducted by constantly exhaling air to the CO_2 sensor. This resulted with sharp peaks in the normal indoor carbon

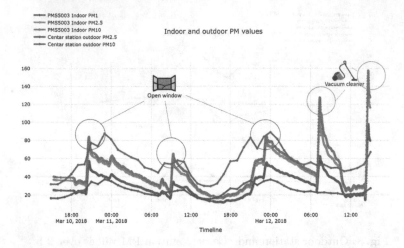

Fig. 9. Indoor station and "Centar" station PM values over 3 days

dioxide level of 500 ppm [11], jumping up to 4000 ppm and then slowly stabilizing back to its normal values. The second experiment was conducted using a Liquid Petroleum Gas lighter and directly applying it on the sensor. This had a similar effect, but with a much larger magnitude - a value of 1.6 million ppm. It has to be noted that due to this value the normal indoor carbon dioxide cannot be clearly seen on Fig. 11. The values in the millions are acceptable because the main purpose of the sensor is to monitor CO_2 levels, so when we get values out of the maximum range of CO_2 (250–5000 ppm) that should be a sign of heavy exposure to some other natural gas, because as noted in the sensor's description it is sensitive to different gases.

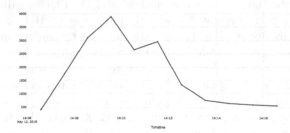

Fig. 10. Results from exposing the CO_2 values to CO_2 by exhaling

To sum up the experiments, when comparing our outdoor station with that of the "Centar" station, our results have more data due to the additional PM1 particles not covered by the "Centar" station, while giving more frequent information for the air quality right outside the user's home, and thus providing a more accurate pollution evaluation. The indoor station showed the importance of

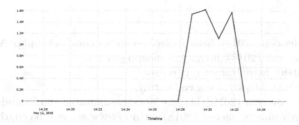

Fig. 11. Results from exposing the CO_2 values to LPG from a lighter

the human factor when it comes to the overall indoor air pollution. And finally, the two-part CO_2 experiment prove the validity of the sensor and its reactiveness to a sudden change in the air.

5 Conclusion

In this paper an affordable air pollution measuring system build for home living is presented. The system consists of both indoor and outdoor station. The outdoor station is measuring PM1, PM2.5 and PM10, while the indoor station is measuring PM1, PM2.5, PM10, CO_2, temperature and humidity. Both of the stations proved to be a success since they give information about the air we breathe not only outdoors, but also indoors where we spent most of our time. This is especially important in winter when the outdoor pollution is the highest. In addition to the measurements performed and storing the data, the voice-assistant was incorporated. It enabled easy and natural interaction between the users and the presented system. The voice assistant can inform the users about the air quality inside and outside and also it can recommend some actions for the users that would improve the air they breathe. As the technology advances, we can see a future where these sensors are integrated or connected with the personal assistants, like the Amazon Alexa devices, which will become a part of our daily life and it would be a "must" accessory of our living space.

In the future, it would be interesting to implement a node to node communication between the stations in a closed space, developing a web and mobile application which could further improve the access to information to the users' for the indoor station, but also the online community by making the outdoor stations open to everyone. By having multiple such stations across one city and with measuring the wind speed and direction, since the precise location of the station are known, we can help build a 3D pollution map of a city. It would show how the pollution is roaming around us, how outdoor pollution affects indoor pollution, and furthermore we can use machine learning to predict the pollution.

Acknowledgement. The authors would like to thank the Faculty of Computer Science and Engineering - Skopje for partially financing this work. All authors contributed equally to this paper.

References

1. Aireas.scapeler.com (2018). http://aireas.scapeler.com/index.php/AiREAS_AQI
2. Air.moepp.gov.mk (2018). http://air.moepp.gov.mk/
3. aqicn.org (2018). http://aqicn.org/sensor/
4. Co2.earth (2018). https://www.co2.earth/
5. Docs.aws.amazon.com (2018). https://docs.aws.amazon.com/lambda/latest/dg/
6. European environment agency (2018). http://www.eea.europa.eu/themes/air/
7. Github (2018). https://github.com/drohm/pi-sht1x
8. Github (2018). https://github.com/alexa/alexa-avs-sample-app/wiki/Raspberry-Pi
9. Hacker noon (2018). https://hackernoon.com/my-first-alexa-custom-skill-6a198d385c84
10. Iph.mk (2018). http://www.iph.mk/en/information-about-the-air-pollution-in-cities-in-the-republic-of-macedonia-and-possible-risks-to-health/
11. Kane international limited (2018). https://www.kane.co.uk/knowledge-centre/what-are-safe-levels-of-co-and-co2-in-rooms
12. Pythonprogramming.net (2018). https://pythonprogramming.net/headlines-function-alexa-skill-flask-ask-python-tutorial/?completed=/
13. Raspberrypi.org (2018). https://www.raspberrypi.org/documentation/installation/installing-images/README.md
14. Rigacci.org (2018). https://www.rigacci.org/wiki/doku.php/doc/appunti/hardware/raspberrypi_air
15. Webiopi.trouch.com (2018). http://webiopi.trouch.com/
16. Who.int (2018). http://www.who.int/mediacentre/news/releases/2014/air-pollution/en/
17. World bank (2018). http://www.worldbank.org/en/news/press-release/2016/09/08/air-pollution-deaths-cost-global-economy-225-billion
18. Chen, G.: Effects of ambient pm 1 air pollution on daily emergency hospital visits in china: an epidemiological study. Lancet Planet. Health 1(6), e221–e229 (2017). https://doi.org/10.1016/s2542-5196(17)30100-6
19. Foxworth, T.: Making the esp8266 low-powered with deep sleep (2018). https://www.losant.com/blog/making-the-esp8266-low-powered-with-deep-sleep
20. Guxens, M., et al.: Air pollution exposure during fetal life, brain morphology, and cognitive function in school-age children. Biol. Psychiatry 84(4), 295–303 (2018)
21. Institute for Health Metrics, E.G.B.o.D.P.: The state of global air (2018). https://www.stateofglobalair.org/sites/default/files/SOGA2017_report.pdf
22. Klosowski, T.: Lifehacker.com (2018). https://lifehacker.com/the-best-operating-systems-for-your-raspberry-pi-projec-1774669829
23. Kraujutis, V.: Viliuskraujutis/mq135 (2018). https://github.com/ViliusKraujutis/MQ135
24. Malvina, B., Roth, S., Sager, L.: Crime is in the air: the contemporaneous relationship between air pollution and crime - grantham research institute on climate change and the environment (2018). http://www.lse.ac.uk/GranthamInstitute/publication/crime-is-in-the-air-the-contemporaneous-relationship-between-air-pollution-and-crime/
25. Mohai, P., Kweon, B.S., Lee, S., Ard, K.: Air pollution around schools is linked to poorer student health and academic performance. Health Aff. 30(5), 852–862 (2011). https://doi.org/10.1377/hlthaff.2011.0077

26. MSV, J., Bhartiya, S.: 10 diy development boards for iot prototyping - the new stack (2018). https://thenewstack.io/10-diy-development-boards-iot-prototyping/
27. Ristov, S., Ackovska, N., Kirandziska, V., Martinovikj, D.: The significant progress of the microprocessors and microcontrollers course for computer science students. In: 2014 37th International Convention on Information and Communication Technology, Electronics and Microelectronics (MIPRO), pp. 818–823, May 2014. https://doi.org/10.1109/MIPRO.2014.6859679
28. Song, P., Wanga, L., Hui, Y., Li, R.: PM2.5 concentrations indoors and outdoors in heavy air pollution days in winter. Procedia Eng. **121**, 1902–1906 (2015). https://doi.org/10.1016/j.proeng.2015.09.173
29. Sunyer, J., et al.: Association between traffic-related air pollution in schools and cognitive development in primary school children: a prospective cohort study. PLOS Medicine **12**(3), e1001792 (2015). https://doi.org/10.1371/journal.pmed.1001792
30. Zwozdziak, A., Sowka, I., Willak-Janc, E., Zwozdziak, J., Kwiecinska, K., Balinska-Miskiewicz, W.: Influence of pm1 and pm2.5 on lung function parameters in healthy schoolchildren-a panel study. Environ. Sci. Pollut. Res. **23**(23), 23892–23901 (2016). https://doi.org/10.1007/s11356-016-7605-1

Framework for Human Activity Recognition on Smartphones and Smartwatches

Blagoj Mitrevski[1], Viktor Petreski[1], Martin Gjoreski[2],
and Biljana Risteska Stojkoska[1(✉)]

[1] Faculty of Computer Science and Engineering,
Ss. Cyril and Methodius University, Rugjer Boshkovik 16,
PO Box 393, Skopje, Macedonia
{mitrevski.blagoj,
petreski.viktor.1}@students.finki.ukim.mk,
biljana.stojkoska@finki.ukim.mk
[2] Department of Intelligent Systems, Jozef Stefan Institute,
Jamova 39, Ljubljana, Slovenia
martin.gjoreski@ijs.si

Abstract. As activity recognition becomes an integral part of many mobile applications, its requirement for lightweight and accurate techniques leads to development of new tools and algorithms. This paper has three main contributions: (1) to design an architecture for automatic data collection, thus reducing the time and cost and making the process of developing new activity recognition techniques convenient for software developers as well as for the end users; (2) to develop new algorithm for activity recognition based on Long Short Term Memory networks, which is able to learn features from raw accelerometer data, completely bypassing the process of generating hand-crafted features; and (3) to investigate which combinations of smartphone and smartwatch sensors gives the best results for the activity recognition problem, i.e. to analyze if the accuracy benefits of those combinations are greater than the additional costs for combining those sensors.

Keywords: Smartphone · Smartwatch · Activity recognition

1 Introduction

In the era of pervasive and ubiquity computing, various applications of activity recognition are evident in many real-life, human-centric problems such as eldercare, healthcare, sports etc. [1]. In most developed countries, demographical trends tend toward more and more elderly people, usually left to their own means in receiving healthcare and other services. The effects of these trends are dramatic on public and private healthcare, as well as on the individuals themselves. Therefore, it is economically and socially beneficial to enhance prevention, by shifting from a centralized, expert-driven model to one that permeates home environments, with focus on advanced homecare services dedicated to personal medical assistance [2].

© Springer Nature Switzerland AG 2018
S. Kalajdziski and N. Ackovska (Eds.): ICT 2018, CCIS 940, pp. 90–99, 2018.
https://doi.org/10.1007/978-3-030-00825-3_8

Regular daily exercises reduce the risk and progression of chronic diseases and improve functional abilities, cardiorespiratory fitness and metabolic health in patients with frequent diseases, such as cardiovascular, lung, and neurodegenerative diseases. With the stimulation of physical activity, the risks for these conditions can be greatly reduced. In particular, [3, 4] showed the effect of physical activity on coronary heart diseases and on the risk of hypertension known as high blood pressure. Additionally, in [5] it is confirmed the effect of the physical activity on diabetes, while [6] prescribes exercise therapy against diabetes. The benefits from regular physical activities for healthy population include reductions in body weight and fat, resting heart rate, increased high density lipoprotein cholesterol and an improved maximal oxygen uptake.

Health benefits associated with physical activity depend on the activity duration, intensity and frequency, therefore it is important to monitor and distinguish the physical activities. Activity recognition (AR) attempts to recognize the actions of an agent in environment from sequence of observations. AR has the potential to address the emerging health conditions such as obesity, heart conditions, diabetes, etc., since physical inactivity is the main factors for those conditions or at least strongly coupled with them. Additionally, human activity recognition can help to develop patient recovery trainings or even provide early detection of diseases, strokes, falls, etc.

Therefore, we can identify three main advantages of AR: *(i)* early detection of falls and other abnormalities in elderly; *(ii)* help in the process of recovery after an accident/stroke and (iii) prevention of diseases.

The in-depth data analyses can lead to a broader range of societal challenges. However, most of the studies that investigate the benefits from different physical activities are expensive and require complex process of monitoring. In most epidemiological studies, participants are equipped with special sensors, therefore, the cohort size is limited, and the conclusions cannot be simply mapped to broader population [7]. It would be more convenient more participants to be included in the studies, but without financial implications by means of hardware requirements and manual data analyses. In such scenarios, a common framework is needed that is capable not only to gather data form many participants, but also to provide automatic activity recognition.

This paper has three main goals: (1) to design a system architecture and organization for activity recognition using smartphones and smartwatches as sensor devices which gather data and recognize activities, along with a remote cloud system which is in charge of training and improving the models for recognition; (2) to develop a new lightweight algorithm for activity recognition that is easily implementable for smartphone and smartwatch applications; and (3) to evaluate the accuracy of the algorithm on smartphone and smartwatch sensors and to test which sensor combination gives the best results for the activity recognition problem, i.e. if the accuracy benefits of those combinations are greater than the additional costs of combining those sensors.

Our algorithm is based on neural network, i.e. Long Short Term Memory networks. Although this algorithm has been previously used for activity recognition on wearables [8] and smartphones [9], to the best of our knowledge, this is the first research that evaluates the algorithm on data collected from smartwatch sensors.

The remaining sections of this paper are organized as follows. The second section describes and illustrates the architecture for development of AR methods in details. In the third section we describe and test a neural network which can be used as activity recognition model, as well as the best combination of sensors that outperforms all the others. The paper is concluded in the fourth section.

2   System Architecture for Efficient AR Tools Development

The traditional approach for development new AR tools and techniques includes collecting sensors data to be further used for training and testing. The process of data collection can be made in a special laboratory conditions, or under field conditions. In both cases, all data are stored in a central server and the process of developing AR techniques is made offline.

In the first case, participants equipped with sensors are guided to perform particular activities for a relatively short period. Therefore, most of the collected datasets are with small number of participant, which is eventually leading to development of inconvenient models.

In the second case, participants are wearing the sensors during their normal daily life, reporting the activities in an activity diary. Typical sensors are accelerometers attached to an elastic belt and placed at the hip or at the dominant ankle. The main drawback of this data collection scenario is the lack of accuracy with respect to time spent performing particular activity. Additionally, wearing the sensors for a long time can even reduce participants' compliance to take part in the study [7]. Still, the approaches from the literature that investigate the AR accuracy on data collected under field condition are satisfying, performing almost equally good as on data collected under field conditions. However, this is expensive, labor and time consuming task for the application developers.

In this section we will describe a system architecture that will overcome both problems, i.e. the development of AR tools would be made online on large datasets collected under field condition. The system we purpose consists of three main parts: server (cloud), normal users and users-contributors.

- The server (cloud) is a central part in the architecture. Its function is to store data, build AR models, update AR models and distribute them to the users.
- Normal users get the AR models from the server (cloud) on their wearable (smartphone). They can contribute only by sending their experience back to the server (cloud).
- Users contributors sent labeled data to the server (cloud), so they contribute to the process of data collection. Additionally, they can act as normal users.

The contributors provide labeled data to be used for creating new AR models or improving the existing ones. Before a specific activity, the contributor can turn her device to "contribute mode" to record labeled data to be sent to the server-side for further processing. Data transferring to the server (cloud) should be done in real-time or near real-time. To save energy for data transmission, different techniques can be used for data reduction, like delta compression (for near real-time transfer) or data prediction based on dual prediction scheme [10, 11]. Additionally, some application specific heuristics can be investigated for this purpose, like sending data only when the phone is connected to the charger, or if the available battery is above a predefined threshold (Fig. 1).

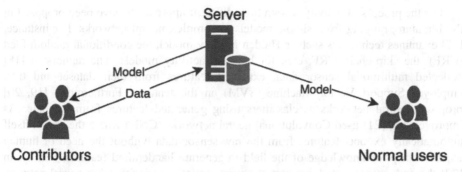

Fig. 1. General framework of the system for online data collection and processing

The architecture we propose is not only suitable for applications based exclusively for smartphone sensors, but can be extended to systems that use different wearables. In this case, the sensors from the wearables send data measurements to the smartphone, which acts as a gateway or hub to retransmit the data to the server (cloud). In this case, it is important for both devices (smartphone and wearable) to be able to communicate using common protocols with low energy requirement, like Bluetooth Low Energy (BLE), etc. Although there are many different protocols for smart devices to transfer data (ZigBee, Z-Wave, Insteon, etc.), the state-of-the-art smartphones are not supporting them [12], therefore this scenario is usually not feasible.

The need for such architecture is not new, but previously it was not feasible since most of the tools for AR require in-depth data analyses usually performed by a team that includes both data scientists and experts in this field to be jointly involved in the process of generating hand-crafted features. Recently, there have been new tools that include automated feature engineering techniques to extract features from the raw accelerometer readings and to select a subset of the most significant features.

3 Designing Accurate and Lightweight Algorithms for AR

The activity recognition at the present-day mainly is sensor-based, implemented with the help of smartphones and wearable devices acting like wearable sensors and computational (recognition) devices. In [13], artificial intelligence (AI) techniques are used to develop daily activity reminders for elders with memory impairments. Moreover, in [14] the authors built abnormal human-activity detection models which can be used to detect and notify of abnormal behavior and early detection of dementia.

Body-worn sensors can be used in sports and physical activities in order to assess and improve the overall sport performance and fitness. In [15] the authors achieved to learn the daily activity pattern of the users and assess the daily energy expenditures in order to help users improve their lifestyle. There are also commercial devices for monitoring sport activities such as Nike + [16] sensor which is placed inside a shoe to keep track (duration) of running and jogging exercises. If connected to a smartphone application, it enables the user to set training goals or to challenge friends.

For the process of activity recognition, different approaches have been proposed in the literature, ranging from simple models to complex neural networks. For instance, [17] examines techniques such as Hidden Markov model, the conditional random filed (CRF), the skip-chain CRF, etc., for building activity models. The authors in [18] collected multimodal sensor data, extracted features from their dataset and then employed Support Vector Machine (SVM) on the features. Furthermore, [19, 20] proposed neural networks as classifiers using generated features from the data. As improvement, [21] used Convolutional neural networks (CNN) where the model itself automatically extracts features from the raw sensor data without the need of human expert with prior knowledge of the field to generate handcrafted features. Finally, in [22] the authors suggested a recurrent neural model superior to other neural network models. This model is able to capture temporal correlations from the sensor data, therefore it is applicable to wide range of problems while providing acceptable accuracy. In [7], automated feature engineering technique is performed to extract features from the raw accelerometer readings from epidemiological studies, and four machine learning algorithms are used for classification, showing that only one accelerometer is sufficient for accurate activity recognition.

Apart from the accuracy, other major problems in deploying AR models and applications is the computational cost and time complexity of the algorithm, since it should operate in real-time on devices with limited energy [23–25]. Even in the modern smartphones with performances comparable to those of the computers, the power remains a challenging problem, since battery technology has not kept pace with information and communication technologies. Other issues regarding energy consumption is sampling frequency, as it is an important parameter for the accuracy of the algorithm.

In this section we describe the dataset used in our study, the model we developed for activity recognition, as well as the results of our analyses.

3.1 The Dataset

We use the AR dataset [26, 27] which consists of around 9 million entries, 4 million accelerometer data entries and 4 million gyroscope entries recorded in laboratory conditions. There are recordings of 9 users performing the following activities: 'Biking', 'Sitting', 'Standing', 'Walking', 'Stair Up' and 'Stair down' while data is recorded via two embedded sensors: accelerometer and gyroscope. Four smartwatches (two LG and two Samsung Galaxy Gear) and eight smartphones (two Samsung S3 mini, two Samsung S3, two LG Nexus 4 and two Samsung S +) were used. The data was split into four subsets: smartphone accelerometer data, smartwatch accelerometer data, smartphone gyroscope data and smartwatch gyroscope data.

3.2 Long Short Term Memory Neural Network

Generally, AR techniques first segment the time series data with sliding windows, then apply signal processing and statistical methods for feature extraction from the raw accelerometer measurements and then train machine learning algorithms for classification of different activities.

Among many techniques for activity recognition from the literature, we investigated a neural network based approach known as Long Short Term Memory (LSTM) networks. The main advantages of LSTM can be summarized as: *(i)* easily implementable on mobile applications; *(ii)* outperforms other approaches from the literature by means of accuracy; and *(iii)* robust enough to perform almost equally good on data collected under field conditions as on data collected in a controlled environment [8, 9].

LSTM network as a deep learning system appropriate for temporal modeling was initially proposed by Hochreiter [28] and later improved in 2000 by Gers [29]. It has shown improvements over Deep Neural Networks for speech recognition problem [30]. Since 2016, LSTM became integral part of many applications and services delivered by Google, Microsoft and Apple, including personalized speech recognition on the smartphone [31] and gesture typing decoding [32].

LSTM networks are a special type of neural networks that remember information from further back in the past. Given a sequence of inputs $X = \{x_1, x_2, ..., x_n\}$, LSTM associates each time step with an input gate, memory gate and output gate, denoted respectively as i_t, f_t and o_t. The information from the past is remembered using the state vector c_{t-1}. The forget gate decides how much of the previous information are going to be forgotten. The input gate decides how to update the state vector using the information from the current input. The l_t vector consists of the information from the current input added to the state. Finally, the output gate decides what information to output at the current time step. This process is formalized as in (1),

$$
\begin{aligned}
i_t &= \sigma(W_i \cdot [h_{t-1}, x_t]) \\
f_t &= \sigma(W_f \cdot [h_{t-1}, x_t]) \\
o_t &= \sigma(W_o \cdot [h_{t-1}, x_t]) \\
l_t &= \tanh(W_l \cdot [h_{t-1}, x_t]) \\
c_t &= f_t \cdot c_{t-1} + i_t \cdot l_t \\
h_t &= o_t \cdot \tanh(c_t)
\end{aligned}
\tag{1}
$$

where W_i, W_f, W_o and W_l have dimensions $D \times 2N$, D is the number of memory cells and N is the dimension of the input vector. These matrices represent the parameters of the network. LSTM is local in space and time since its computational complexity per time step and weight is $O(1)$ [28].

As an algorithm for learning models for the task of activity recognition we use LSTM Network implemented in TensorFlow [33]. For every task we tuned the parameters (learning rate, hidden layers, structure of the network) to obtain optimal results. Our basic model consists of two fully connected and two LSTM layers with 64 units each and we use L2 regularization to avoid overfitting. We train the model for 70 epochs.

3.3 Experiments

First the data was split into training and testing sets. Because our dataset contains entries of 9 users, we separated the data from two users and we used it for testing while the remaining data was used for training. With this kind of splitting we avoid overfitting on user specific data and the results are unbiased. Before feeding the data to the models, we do sliding window of size 200 and step of 50.

The experimental results are presented in Fig. 2. The bars represent the sensor combination used as input to the models. It can be seen that in general the accuracy is higher when the smartphone sensors' data is used. The highest accuracy of 94% is achieved for the combination accelerometer-gyroscope form the smartphones' sensors.

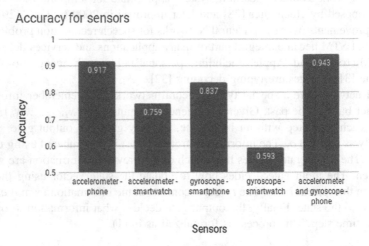

Fig. 2. Accuracy for each sensor combination

Figures 3 and 4 present the confusion matrices for the accelerometer and for the gyroscope data. The rows represent the true class and the columns represent the predicated class. From the confusion matrices it can be seen that misclassifications from the gyroscope phone data and the misclassifications from the accelerometer phone data are contradictory. For example, from Fig. 3 can be seen that the models that use only acceleration data, mostly mix the classes Sitting and Standing. On the other hand, it can be seen in Fig. 4 that the models that use only gyroscope data mix the classes

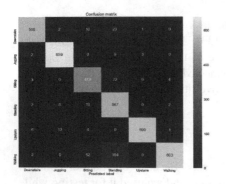

Fig. 3. Confusion matrix for phone accelerometer data

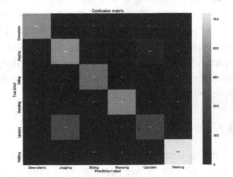

Fig. 4. Confusion matrix for phone gyroscope data

Jogging and Upstairs. To exploit the model variability, we used the accelerometer and the gyroscope data for building the final models.

4 Conclusion

Activity recognition is integral part of many wearable devices, therefore development new tools and algorithms will remain a challenging problem for the research community in the next few years. The process of data collection is usually expensive labor task that is time consuming for both developers of wearable applications and participants that assist in the process of data collection. In this paper we propose a system architecture that can decrease the time to develop new and more accurate methods and tools for activity recognition.

We develop a LSTM technique that performs accurate AR using smartphone and smartwatch sensors. From our experiments, it has been proved that it is better to combine the accelerometer and the gyroscope sensors from the smartphone in order to increase the accuracy of the model, since models that use smartphone acceleration data make different misclassification errors compared to the models that use smartphone gyroscope data.

Acknowledgment. This work was partially financed by the Faculty of Computer Science and Engineering, University "Ss. Cyril and Methodius", Skopje, Macedonia.

References

1. Avci, A., Bosch, S., Marin-Perianu, M., Marin-Perianu, R., Havinga, P.: Activity recognition using inertial sensing for healthcare, wellbeing and sports applications: a survey. In 2010 23rd International Conference on Architecture of Computing Systems (ARCS), pp. 1–10. VDE (2010)
2. Risteska Stojkoska, B., Trivodaliev, K., Davcev, D.: Internet of things framework for home care systems. Wireless Commun. Mobile Comput. **2017** (2017)
3. Weinstein, A.R., et al.: The joint effects of physical activity and body mass index on coronary heart disease risk in women. Arch. Intern. Med. **168**(8), 884–890 (2008)
4. Hu, G., Barengo, N.C., Tuomilehto, J., Lakka, T.A., Nissinen, A., Jousilahti, P.: Relationship of physical activity and body mass index to the risk of hypertension: a prospective study in Finland. Hypertension **43**(1), 25–30 (2004)
5. Haapanen, N., Miilunpalo, S., Vuori, I., Oja, P., Pasanen, M.: Association of leisure time physical activity with the risk of coronary heart disease, hypertension and diabetes in middle-aged men and women. Int. J. Epidemiol. **26**(4), 739–747 (1997)
6. Jia, Y.: Diatetic and exercise therapy against diabetes mellitus. In: ICINIS 2009. Second International Conference on Intelligent Networks and Intelligent Systems, 2009, pp. 693–696. IEEE (2009)
7. Zdravevski, E., Stojkoska, B.R., Standl, M., Schulz, H.: Automatic machine-learning based identification of jogging periods from accelerometer measurements of adolescents under field conditions. PLoS ONE **12**(9), e0184216 (2017)
8. Ordóñez, F.J., Roggen, D.: Deep convolutional and LSTM recurrent neural networks for multimodal wearable activity recognition. Sensors **16**(1), 115 (2016)

9. Milenkoski, M., Trivodaliev, K., Kalajdziski, S., Jovanov, M., Stojkoska, B.R.: Real time human activity recognition on smartphones using LSTM Networks. In: MIPRO (2018)
10. Stojkoska, B.R., Nikolovski, Z.: Data compression for energy efficient IoT solutions. In: 2017 25th Telecommunication Forum (TELFOR), pp. 1–4. IEEE (2017)
11. Stojkoska, B.R., Trivodaliev, K.: Enabling internet of things for smart homes through fog computing. In: 2017 25th Telecommunication Forum (TELFOR), pp. 1–4. IEEE (2017)
12. Stojkoska, B.L.R., Trivodaliev, K.V.: A review of internet of things for smart home: challenges and solutions. J. Cleaner Prod. **140**, 1454–1464 (2017)
13. Pollack, M.E., et al.: Autominder: an intelligent cognitive orthotic system for people with memory impairment. Robot. Auton. Syst. **44**(3), 273–282 (2003)
14. Yin, J., Yang, Q., Pan, J.J.: Sensor-based abnormal human-activity detection. IEEE Trans. Knowl. Data Eng. **20**(8), 1082–1090 (2008)
15. Long, X., Yin, B., Aarts, R.M.: Single-accelerometer-based daily physical activity classification. In: Annual International Conference of the IEEE Engineering in Medicine and Biology Society, 2009, EMBC 2009, pp. 6107–6110. IEEE (2009)
16. Official website for Nike + . http://www.nikeplus.com/
17. Kim, E., Helal, S., Cook, D.: Human activity recognition and pattern discovery. IEEE Pervasive Comput. **9**(1), 48 (2010)
18. Liu, X., Liu, L., Simske, S. J., Liu, J.: Human daily activity recognition for healthcare using wearable and visual sensing data. In: IEEE International Conference on Healthcare Informatics (ICHI), 2016, pp. 24–31. IEEE (2016)
19. Yang, J.Y., Wang, J.S., Chen, Y.P.: Using acceleration measurements for activity recognition: an effective learning algorithm for constructing neural classifiers. Pattern Recognit. Lett. **29**(16), 2213–2220 (2008)
20. Kwapisz, J.R., Weiss, G.M., Moore, S.A.: Activity recognition using cell phone accelerometers. ACM SigKDD Explor. Newslett. **12**(2), 74–82 (2011)
21. Zeng, M., et al.: Convolutional neural networks for human activity recognition using mobile sensors. In: 2014 6th International Conference on Mobile Computing, Applications and Services (MobiCASE), pp. 197–205. IEEE (2014)
22. Murad, A., Pyun, J.Y.: Deep recurrent neural networks for human activity recognition. Sensors **17**(11), 2556 (2017)
23. Anguita, D., Ghio, A., Oneto, L., Llanas Parra, F.X., Reyes Ortiz, J.L.: Energy efficient smartphone-based activity recognition using fixed-point arithmetic. J. Univ. Comput. Sci. **19**(9), 1295–1314 (2013)
24. Gordon, D., Czerny, J., Miyaki, T., Beigl, M.: Energy-efficient activity recognition using prediction. In: 2012 16th International Symposium on Wearable Computers (ISWC), pp. 29–36. IEEE, June 2012
25. Oneto, L., Ortiz, J.L., Anguita, D.: Constraint-aware data analysis on mobile devices: an application to human activity recognition on smartphones. In: Adaptive Mobile Computing, pp. 127–149 (2017)
26. Stisen, A., et al.: Smart devices are different: assessing and mitigating mobile sensing heterogeneities for activity recognition. In: Proceedings of the 13th ACM Conference on Embedded Networked Sensor Systems (SenSys 2015), Seoul, Korea (2015)
27. UCL link to the dataset. https://archive.ics.uci.edu/ml/datasets/Heterogeneity+Activity+Recognition
28. Hochreiter, S., Schmidhuber, J.: Long short-term memory. Neural Comput. **9**(8), 1735–1780 (1997)
29. Gers, F.A., Schmidhuber, J., Cummins, F.: Learning to forget: continual prediction with LSTM. Neural Comput. **12**(10), 2451–2471 (2000)

30. Sainath, T.N., Vinyals, O., Senior, A., Sak, H.: Convolutional, long short-term memory, fully connected deep neural networks. In: 2015 IEEE International Conference on Acoustics, Speech and Signal Processing (Icassp), pp. 4580–4584. IEEE (2015)
31. McGraw, I., et al.: Personalized speech recognition on mobile devices. In: 2016 IEEE International Conference on Acoustics, Speech And Signal Processing (ICASSP), pp. 5955–5959. IEEE (2016)
32. Alsharif, O., Ouyang, T., Beaufays, F., Zhai, S., Breuel, T., Schalkwyk, J.: Long short term memory neural network for keyboard gesture decoding. In: 2015 IEEE International Conference on Acoustics, Speech and Signal Processing (ICASSP), pp. 2076–2080. IEEE (2015)
33. TensorFlow Homepage. https://www.tensorflow.org/

Parallel Decoding of Turbo Codes

Dejan Spasov$^{(\boxtimes)}$

Faculty of Computer Science and Engineering, Skopje, Macedonia
dejan.spasov@finki.ukim.mk

Abstract. Given a turbo code generated by parallelly concatenated recursive systematic convolutional encoders, the turbo decoder comprises MAP decoders coupled in a serial connection, where each MAP decoder decodes a recursive systematic convolutional code. We propose a turbo decoding algorithm that, on a logical level of abstraction, is made of several turbo decoders working in parallel. Each turbo decoder is initialized with different recursive convolutional code. Practical implementation of the proposed algorithm may be achieved with a single turbo decoder, where MAP decoders are working concurrently.

Keywords: Turbo codes · MAP decoding · MAX-Log-MAP decoding
Turbo decoding · Convolutional codes

1 Introduction

Turbo Codes are a class of forward error correction codes invented by Berrou and first published in [1]. Turbo codes were the first practical system that achieved signal-to-noise ratio of 0.7 dB above the Shannon's limit while providing bit error probability of 10^{-5} [1]. Turbo codes are special sub-type of convolutional codes. In general, a turbo encoder is any combination of two or more identical convolutional encoders connected via interleavers. Traditionally, a turbo code comprises two recursive systematic convolutional (RSC) encoders coupled in a parallel concatenation scheme (Fig. 1). Turbo codes are systematic codes, which means that the input sequence appears unmodified at the output as sequence x. Figure 1 shows two recursive systematic convolutional encoders that output two coded sequences (x, y_1) and (x, y_2), where sequences y_1 and y_2 represent the parity bits. The turbo code on Fig. 1 initially has a code rate of 1/3; however, higher code rates may be achieved by applying various puncturing patterns. An example of puncturing pattern may be alternating between the parity bits y_1 and y_2. The interleaver is a device that outputs a random permutation of the received sequence. By providing random permutation of the input sequence, the interleaver allows identical recursive encoders to be used in the hardware design. Thus, two identical recursive systematic convolutional codes coupled with a random interleaver behave as two different recursive systematic convolutional codes.

Decoding of a recursive systematic convolutional code may be done with the Viterbi algorithm [2] or with Maximum A Posteriori Probability (MAP) algorithm [3]. Decoding of the turbo codes, which are made of two parallel systematic recursive convolutional codes, involves separate decoding of each of the systematic recursive convolutional codes (Fig. 2).

© Springer Nature Switzerland AG 2018
S. Kalajdziski and N. Ackovska (Eds.): ICT 2018, CCIS 940, pp. 100–106, 2018.
https://doi.org/10.1007/978-3-030-00825-3_9

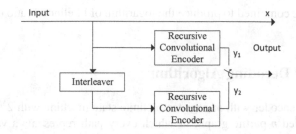

Fig. 1. Turbo encoder

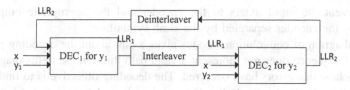

Fig. 2. Turbo decoder

The turbo decoder (Fig. 2) is made of two decoders DEC_1 and DEC_2, which are modified MAP decoders, also known as BCJR decoders [1]. The two decoders DEC_1 and DEC_2 may share information several times before the turbo decoder outputs estimates for each bit. This information sharing is known as iteration. Thus, the turbo decoder performs several iterations before outputting decision for each bit. The turbo decoder may be configured to perform two iterations simultaneously [4]. For example, when DEC_1 is working on the i-th iteration, DEC_2 may be working on the (i−1)-th iteration. The two decoders DEC_1 and DEC_2 may be Max-Log-Map decoders [6, 7], which is a simplified variant of the MAP decoder that involves the Viterbi algorithm [2]. More on implementation issues may be found in [5, 8]. Decoders DEC_1 and DEC_2 are configured to output the logarithm of likelihood ratio (LLR) for each bit x_k

$$LLR(x_k) = \log \frac{\Pr\{x_k = 1|observation\}}{\Pr\{x_k = 0|observation\}} \qquad (1)$$

where $\Pr\{x_k = 1 \, or \, 0|observation\}$ is a posteriori probability of the bit x_k. The decoder DEC_1 is activated first and it decodes the encoded sequence (x, y_1) and outputs LLR_1 quantities for each bit x_k. The LLR_1 quantities are then fed to DEC_2 that decodes the encoded sequence (x, y_2) to produce its own estimates LLR_2. The turbo decoding process continues in iterative fashion, and in the next iteration LLR_2 quantities are fed into DEC_1.

From Fig. 2, it may be observed that the turbo decoder is initialized with the first recursive convolutional code (x, y_1). In this paper we propose a turbo decoding scheme made of two serial turbo decoders that operate in parallel. The first turbo decoder is initialized with the first recursive convolutional code (x, y_1) and the second turbo decoder is initialized with the second recursive convolutional code (x, y_2). Results from both decoders are combined to produce the logarithm of likelihood ratio (LLR) for each bit x_k.

2 The MAP Decoding Algorithm

A convolutional encoder with M registers is finite state machine with 2^M states. Trellis diagram is labelled n-partite graph, in which every path represents a valid codeword (Fig. 3). Vertices of the $n + 1$ disjoint sets in the trellis represent all possible 2^M states of the encoder. Vertices are labelled as decimal numbers, such that the content of the leftmost register corresponds to the most significant bit in the decimal number. Edge labels represent the input letters to the encoder and the appropriate output letters produced by the encoder separated by the slash symbol.

Trellis diagram of convolutional codes gives a hint about the decoding process; if the received sequence does not represent a valid path through the trellis diagram, then we can conclude that errors have occurred. The decoding objective is to find the most probable valid path though the trellis. Several decoding algorithms exist for decoding convolutional codes. The most famous are the Viterbi algorithm [2] and the BCJR algorithm [3]. The Viterbi algorithm is universally used and is highly parallelizable.

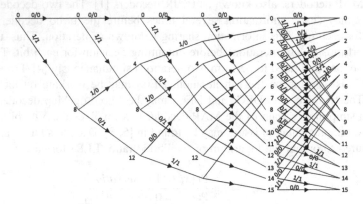

Fig. 3. Trellis diagram of a convolutional code

The BCJR algorithm can be envisioned as two stage process. In the first stage, known as *the forward α recursion*, the decoder moves through the trellis in left to right fashion and with each state s it associates a probability function $\alpha_i(s)$ that is recurrently computed. In the second stage, known as the *backward β recursion*, the BCJR decoder recurrently computes additional probability function $\beta_i(s)$ and then using the stored $\alpha_i(s)$ outputs the logarithm of likelihood ratio for each bit x_k.

Let $\alpha_i(s), s = 0, 1, \ldots, M - 1$, be a set of state metrics on the n-partite trellis at the time i. The computation of the forward α probabilities starts from the initial conditions

$$\begin{cases} \alpha_0(s) = 1 & s = 0 \\ \alpha_0(s) = 0 & s \neq 0 \end{cases} \tag{2}$$

and following the edges of the trellis with non-zero branch probabilities $\gamma_i(s, s')$, the decoder, at each iteration stores all $\alpha_i(s)$ and computes $\alpha_{i+1}(s)$, according to

$$\alpha_{i+1}(s') = \sum_{s=0}^{M-1} \alpha_i(s)\gamma_i(s, s'), \tag{3}$$

where $s' = 0, 1, \ldots, M - 1$. Stored $\alpha_i(s)$ are used during the backward computation in order to compute *the log-likelihood-ratio* (1) for the i-th information bit x_i.

Let $\beta_i(s), s = 0, 1, \ldots, M - 1$, be another set of state metrics on the n-partite trellis at the time i. The computation of the backward β probabilities starts from the initial conditions $\beta_{N-1}(s) = \frac{1}{M}$ and following the edges of the trellis with non-zero branch probabilities $\gamma_i(s, s')$, the decoder, at each iteration computes $\beta_i(s)$, according to

$$\beta_i(s') = \sum_{s=0}^{M-1} \beta_{i+1}(s)\gamma_i(s, s'), \tag{4}$$

where $s' = 0, 1, \ldots, M - 1$. Then the logarithm of likelihood ratio (LLR) for each bit x_k is computed as

$$LLR(x_k) = Log\left(\frac{\sum_s \alpha_i(s) \sum_{s'} \gamma_i(s, s')\beta_{i+1}(s')}{\sum_s \alpha_i(s) \sum_{s'} \gamma_i^{-1}(s, s')\beta_{i+1}(s')}\right) \tag{5}$$

3 Design of Parallel Turbo Decoder

Figure 4 shows a block diagram of a turbo decoder. The turbo decoder is coupled to receive channel information of a turbo code. The received turbo code is made of two parallel recursive systematic convolutional codes RSC1 and RSC2. Principle of operation of the turbo decoder is described on Fig. 2. The turbo decoder is configured first to decode the first code RSC1, then the code RSC2. The decoder repeats this sequence for predefined number of iterations.

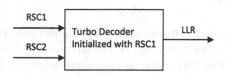

Fig. 4. Block diagram of a turbo decoder initialized with RSC1

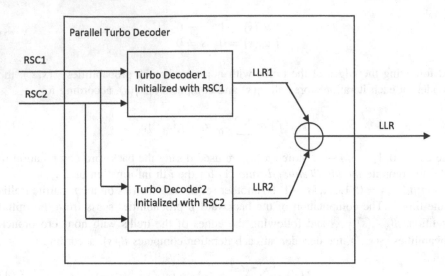

Fig. 5. Parallel turbo decoder

Parallel decoding of the turbo code made of two codes RSC1 and RSC2 may be achieved with two serial turbo decoders (as in Fig. 4) working in parallel, such that each turbo decoder is initialized with different RSC code (Fig. 5).

The parallel turbo decoder (Fig. 5) is coupled to receive channel information for two parallel recursive systematic convolutional codes RSC1 and RSC2. The parallel turbo decoder is made of two serial turbo decoders Turbo Decoder1 and Turbo Decoder2. Each of the serial turbo decoders start operation with different RSC code. The logarithm of likelihood ratio (LLR) for each bit x_k on the output of the parallel decoder is sum of logarithm of likelihood ratio (LLR) for each bit x_k on the output of each serial decoder. In general, the parallel turbo decoder may be generalized for any turbo code made of arbitrary number of recursive systematic convolutional codes in parallel connection. In practice, the parallel turbo decoder (Fig. 5) may be implemented with one serial turbo decoder (Fig. 2), where, at any moment, one MAP decoder computes the LLR1 coefficients from the Turbo Decoder1 and the other MAP decoder computes LLR2 coefficients from the Turbo Decoder2, simultaneously.

4   Practical Results

In our simulation of parallel turbo decoder, we use turbo code made of two recursive systematic convolution codes (as in Fig. 1). The turbo code is with code rate 1/2, which means that one of the two RSC codes is punctured out at any bit interval. The convolutional encoders are recursive with 16 states. Encoded sequence is 1025 bits long. The Turbo code is sent over Gaussian channel. To store state and branch metric we use IEEE 754 double precision format. Results are shown on Fig. 6.

Figure 6 compares performance of a regular turbo decoder (Fig. 2) and parallel turbo decoder (Fig. 5). Both turbo decoders are set to perform 8 iterations before outputting the

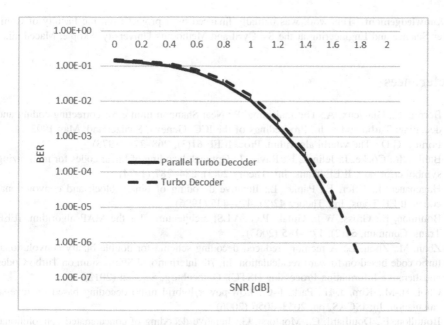

Fig. 6. Performance comparison between (regular) turbo decoder and parallel turbo decoder.

logarithm of likelihood ratio (LLR) for each bit x_k. Both the parallel turbo decoder (Fig. 5) and the regular turbo decoder (Fig. 2) are tested for bit error rate for various signal to noise ratios. Figure 6 confirms that the parallel turbo decoder (Fig. 5) performs better than the regular turbo decoder (Fig. 2), i.e. it shows improved bit error rate.

5   Conclusion

We have demonstrated a turbo decoding algorithm that improves bit error rate in decoding turbo codes by using parallel turbo decoders (Fig. 5). From the perspective of improved performance, it is obvious from Fig. 6 that the parallel turbo decoder improves the classical decoder. From the perspective of running time, using the same hardware, the parallel turbo decoding algorithm may be slower than the classical turbo decoding algorithm by a factor of two. However, if we can double the hardware resources, the parallel turbo decoding algorithm may be as fast as the classical turbo decoding algorithm.

Two lines of research may be identified in Turbo codes, with goals to develop a suboptimal turbo decoder that improves a certain bottleneck. One line of research is trying to minimize memory complexity; thus, requiring less die area and less power [6, 7]. Another line of research is to improve decoding speed of turbo decoders and to achieve the decoding speed of the LDPC codes. In this line of research, the trellis diagram of the turbo code (Fig. 3) is divided in subtrellises, which are decoded in parallel [9]. Advantage of the parallel turbo decoder (Fig. 5) that is introduced in this paper is that it can be applied on any suboptimal turbo decoder.

Acknowledgement. This work was partially financed by a project from the Faculty of Computer Science and Engineering at the Ss. Cyril and Methodius University, Skopje, Macedonia.

References

1. Berrou, C., Glavieux, A., Thitimajshima, P.: Near Shannon limit error-correcting coding and decoding: Turbo-codes. In: Proceedings of the ICC, Geneva, Switzerland, May 1993
2. Forney, G.D.: The viterbi algorithm. Proc. IEEE **61**(3), 268–278 (1973)
3. Bahl, L.R., Cocke, J., Jelinek, F., Raviv, J.: Optimal decoding of linear codes for minimizing symbol error rate. IEEE Trans. Inf. Theory **20**(3), 284–287 (1974)
4. Hagenauer, J., Offer, E., Papke, L.: Iterative decoding of binary block and convolutional codes. IEEE Trans. Inf. Theory **42**(2), 429–445 (1996)
5. Boutillon, E., Gross, W.J., Gulak, P.G.: VLSI architectures for the MAP algorithm. IEEE Trans. Commun. **51**(2), 175–185 (2003)
6. Zhan, M., Zhou, L.: A memory reduced decoding scheme for double binary convolutional turbo code based on forward recalculation. In: 7th International Symposium on Turbo Codes and Iterative Information Processing (ISTC), Gothenburg, Sweden (2012)
7. Choi, H.-M., Kim, J.-H., Park, I.-C.: Low-power hybrid turbo decoding based on reverse calculation. In: ISCAS, pp. 2053–2056 (2006)
8. Boutillon, E., Douillard, C., Montorsi, G.: Iterative decoding of concatenated convolutional codes: implementation issues. Proc. IEEE **95**(6), 1201–1227 (2007)
9. Maunder, R.G.: A fully-parallel turbo decoding algorithm. IEEE Trans. Commun. **63**(8), 2762–2775 (2015)

Optimizing the Impact of Resampling on QRS Detection

Marjan Gusev[✉] and Ervin Domazet

Ss. Cyril and Methodius University,
Faculty of Computer Science and Engineering, 1000 Skopje, Macedonia
marjan.gushev@finki.ukim.mk, ervin_domazet@hotmail.com

Abstract. QRS detection is an essential activity performed on the electrocardiogram signal for finding heartbeat features. Even though there is already a lot of literature on QRS detection, we set a research question to find the dependence of QRS detection performance on the sampling frequency, and, if possible, to find a QRS detector that will be highly efficient at different sampling rates. Our synthesis technique aims to find the optimal value of the threshold parameters that define if the detected peak is artifact, noise or real QRS peak. In addition, we conducted experimental research to find the dependence and estimate the optimal threshold values for the best QRS detection performance. Our approach results with increased QRS detection performance on the original sampling frequency by improving the original Hamilton algorithm. We tested with the MIT-BIH Arrhythmia database. Lastly, QRS detection sensitivity and positive predictive rate are used to evaluate the performance of the algorithm.

Keywords: ECG · QRS detection · Hamilton · Sampling frequency
AD conversion bit resolution

1 Introduction

An electrocardiogram (ECG) is the measured representation of the signal generated by our heart. This signal can be sensed via ambulatory twelve channel ECG sensors, six channel Holter devices or wearable one channel ECG sensors. Wearable ECG sensors, such as the one described in [6] generate more problems than standard ECG monitoring lab equipment, since the patient can move freely and perform daily activities.

Vital information about heart's health is extracted by applying various digital signal processing algorithms, which consist of at least 4 stages including Data Preprocessing, QRS Detection, QRS Classification and Rhythm identification. This has been the subject of intensive research and development for the last 50 years.

Apart from the signal generated by the contraction and relaxation of the, heart muscle, the noise produced by muscles in the human body creates many

© Springer Nature Switzerland AG 2018
S. Kalajdziski and N. Ackovska (Eds.): ICT 2018, CCIS 940, pp. 107–119, 2018.
https://doi.org/10.1007/978-3-030-00825-3_10

challenges for the algorithm that detects heartbeats and their shape. An ECG of a normal heartbeat is presented as a periodical signal with a repeating pattern of P, QRS and T waves, as illustrated in Fig. 1. The QRS wave consists of the signal starting from the local minimum Q, up to the local maximum R, and finishing at the local minimum S.

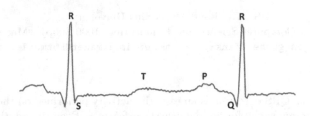

Fig. 1. Periodical nature of an ECG signal.

Hamilton [7] has created a QRS detector that works on a sampling rate of 200 Hz and tuned threshold parameters to obtain a relatively good performance, measured as QRS detection sensitivity and positive predictive rate.

Three different peaks may be classified in the QRS detection process, a real QRS beat: a pattern with identified Q, R and S points with predefined slope and amplitude; a noise peak: generated by muscles or skin, which does not follow the pattern either by the number of detected points, amplitude or length; or artifact: a peak that looks pretty much as a QRS beat, but lacks the amplitude or occurs due to muscle movements or loose contact with the electrode.

Experimental research is used to measure the performance of different sampling rates and find the optimal threshold values.

The paper is organized in the following structure. Details of an improved Hamilton's QRS detection algorithm are given in Sect. 2. The evaluation methodology is described in Sect. 3. The results from the experiments at different sampling rates are given in Sect. 4. Section 5 discusses the results including a comparative analysis with related work on the dependence of sampling frequency and bit resolution on QRS detection performance. Finally, Sect. 6 summarizes the obtained results and future directions.

2   A QRS Detection Algorithm

QRS detection algorithms generally follow a standard procedure [12]. They start with digital signal processing (DSP) filters with the aim to reduce noise and weaken the effect of other waves on QRS detection. In the next phase the signal is matched against a threshold. Last, there is a decision layer to classify peaks as real or noise.

One of the first published and most cited paper for QRS detection is the Pan and Tomkins algorithm [13]. Apart from being a real-time algorithm, it offers the ability to adapt to noises. One of the disadvantages of this algorithm is the dependence on the sampling rate, which is configured to run at 200 Hz, and its bad performance on long records with small amplitudes. Unfortunately, there is no information about bit resolution in their work.

Hamilton's algorithm is another similar derivative-based approach [7]. It uses a preprocessor similar to the Pan and Tomkins algorithm and offers a different set of complex rules for decision making. An open source implementation code has been released by EP Limited [8].

The algorithm starts processing the input signal with a 16 Hz low pass filter, followed by an 8 Hz high pass filter. The output is provided to a differentiation filter that calculates a difference between consecutive samples and determines the slope of the signal. Then, an absolute value is calculated and the average is calculated over an 80 ms window. This ends the first phase where the energy of the signal is calculated, as presented in Fig. 2(b) for the signal in Fig. 2(a).

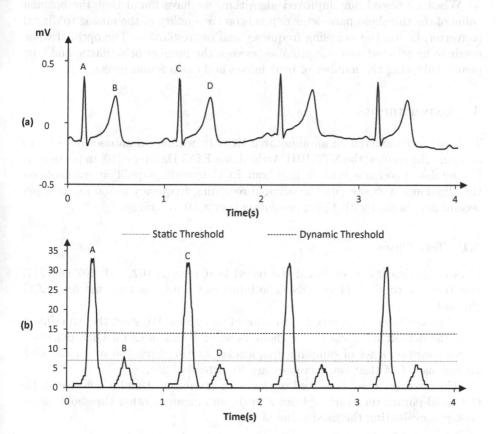

Fig. 2. Detecting artifacts, noise and real peaks based on values of static and dynamic thresholds in the original Hamilton's algorithm presented on signal extracts over record 113 (MIT-BIH Arrhythmia database).

The second phase detects if the energy output is a peak, and classifies the peak as an artifact, noise peak or real beat. Peaks are determined by calculating local maximums and comparing them to a *static* and *dynamic* threshold. The static peak determines if the analyzed local maximum is an artifact or candidate peak, while the dynamic threshold classifies real peaks by extracting noise. The effect of static and dynamic threshold filters is illustrated in Fig. 2, where the peaks A, and C are real beats (reaching a value over both thresholds), whereas B as noise peak (reaching over static threshold, but not dynamic) and D as an artifact (below both thresholds).

The dynamic threshold is calculated as a mean of the last eight candidate peaks. The default implementation uses a *fixed* static threshold equal to $MIN\_AMP\_PEAK = 7$. In our improved algorithm, we treat this parameter as modifiable. The experiments [3] show that decreasing this value will produce more artifacts and higher number of true beats, and vice versa.

When we tested our improved algorithm, we have found that the optimal value of the threshold parameter depends on the quality of the analog to digital convertor, i.e. on the sampling frequency and bit resolution. The optimal value needs to be selected as a compromise between the number of artifacts, and true peaks, balancing the number of beat misses and extra found peaks.

3   Experiments

Testing was conducted on an annotated ECG benchmark databases in order to compare the results, the MIT-BIH Arrhythmia ECG Database [10] in particular. It contains two-channels of 48 half-hour ECG recordings publicly available on the Physionet web site [5]. The original recording frequency is 360 samples per second per channel with 11-bit resolution over a 10 mV range.

3.1   Test Cases

In our experiments we excluded the paced beat records 102, 104, 107 and 217, and tested a total of 44 records. The input test data was only the first ECG channel.

The experiment is planned with a lot of test cases. We start the experiment with the default fixed static threshold value of 7 and measure QRS detection performance on a set of sampling frequencies starting from 100 up to 360 with an increase of 25 (last two increases are 30 instead of 25).

The next test cases consisted of the same environment and a different static threshold parameter, starting from 2 to 10, and included other threshold values (not just evaluating the fixed value of 7).

3.2   Test Data

We have measured the number of *true positive (TP)* detections, which correspond to correctly detected beats, the number of *false positive (FP)* errors, which is equivalent to false detection of extra peaks that are not real beats, and *false negatives (FN)* errors, which is equivalent to misses in detection of real beats. Performance measures are evaluated through QRS sensitivity QRS positive predictive rate, correspondingly calculated as (1) and (2).

$$Q_{SE} = \frac{TP}{TP + FN} \tag{1}$$

$$Q_{+P} = \frac{TP}{TP + FP} \tag{2}$$

QRS sensitivity indicates how many of the real beats are detected compared to the total number of beats; and the QRS positive predictive rate specifies how many of the detected peaks are real beats. It means that the QRS sensitivity specifies the successfulness of detecting all real beats, whereas the positive predictive rate specifies the successfulness of detecting real beats and avoiding false detections.

Given these performance measures, we can not posit what is better: to have higher QRS sensitivity and lower positive predictive rate, or vice versa. For example, it is ambiguous to compare an algorithm with a little higher value of QRS sensitivity, but much lower value of positive predictive rate than another algorithm.

In our analysis, we give a performance advantage to those that have approximately equal values of QRS sensitivity and positive predictive rate, or a higher value of sensitivity than the positive predictive rate.

4   QRS Detection Performance at Different Sampling Rates

This section presents the results from the experiments. It aims to evaluate the dependence of QRS detection performance on different sampling frequencies and to find an optimal threshold parameter for achieving the best QRS detection performance.

4.1   Fixed Threshold - Hamilton's Approach

Figure 3 presents the dependence of QRS sensitivity and positive predictive rate on different sampling frequencies. In the figure, x-axis represents different sampling frequencies (from 80 to 360) and y-axis the QRS sensitivity Q_{SE} and positive predictive rate rate Q_{+P}.

QRS sensitivity fluctuates between 99.53% for a sampling frequency of 100 Hz to 99.81% for 250 Hz. The average QRS sensitivity is equal to 99.74% with a standard deviation of 0.081.

QRS positive predictive rate fluctuates within a smaller range of values (99.71% and 99.79%) and achieves an average value of 99.75% with a standard deviation of 0.032.

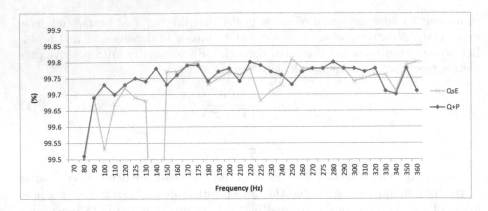

Fig. 3. QRS performance vs sampling rates for fixed static threshold equal to 7.

According to our measurements, the best performance is reached for a sampling frequency of 175 Hz. One can conclude that there is a discrepancy in QRS sensitivity and positive predictive rate at different sampling frequencies.

The best performance is, when the sensitivity and positive predictive rate reach almost the same values, and at the same time, their value is high. Note that the original Hamilton's algorithm was tuned for the static threshold parameter of 7 at a sampling frequency of 200 Hz.

Another interesting discrepancy is the dependence trend. The positive predictive rate reaches lower values at sampling frequencies that are modulo 50 than in the neighboring frequencies.

4.2 Performance Testing Different Threshold Values

To determine if there is a better performance than the one achieved for the fixed static threshold value we conducted test cases for pairs of different threshold values and sampling frequencies.

The dependence of QRS detection performance on different sampling frequencies and different threshold parameters are presented in the 3D graphs in Figs. 4 and 5. The x-axis presents the sampling frequencies, the y-axis represents different static threshold parameters and the z-axis (depth) presents the corresponding QRS sensitivity and positive predictive rate in Figs. 4 and 5.

A fixed static threshold parameter with a value of 7 reveals a relatively satisfactory performance for the QRS positive predictive rate only, even though higher static threshold values reveal higher QRS positive predictive rates. However, as we have discussed earlier, one needs to make a compromise and achieve a higher value of QRS sensitivity at the same time.

Normally, QRS sensitivity is higher for smaller static threshold parameter values.

QRS sensitivity reaches the smallest value of 98.49% for a sampling frequency of 100 Hz and a threshold parameter of 10 and the highest value of 99.87% for

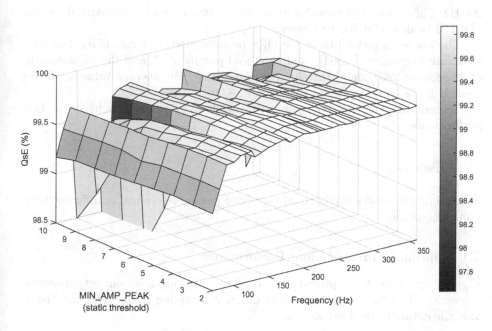

Fig. 4. QRS sensitivity values for different thresholds at different sampling rates.

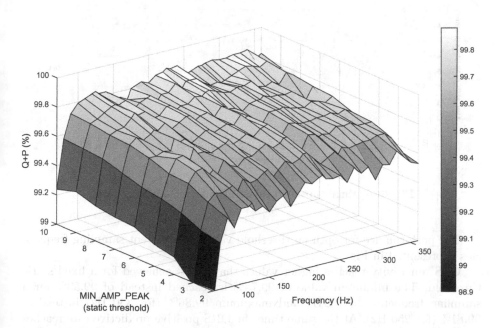

Fig. 5. QRS positive predictive rate for different thresholds at different sampling rates.

200 Hz. The average QRS sensitivity is equal to 99.71% with a standard deviation of 0.173 in all conducted test cases.

The lowest value 99.40% of the QRS positive predictive rate is reached for a sampling frequency of 100 Hz and threshold parameter 2, and the highest value of 99.88% for 225 Hz and threshold parameter 10. The average value in all test cases is 99.72% and standard deviation of 0.09.

To conclude, the threshold parameter needs to be tuned to achieve the best performance.

5   Discussion

This section discusses the optimal threshold parameter values at different sampling frequencies.

5.1   Optimal Threshold and Performance

Figure 6 presents the optimal threshold values found in our experimental research, where the x-axis represents different sampling frequencies and the y-axis the optimal threshold values.

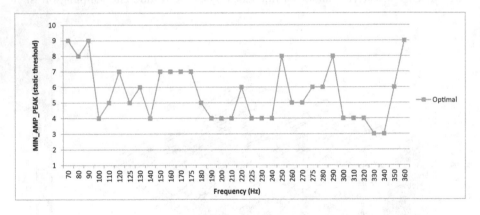

Fig. 6. Optimal threshold values at different sampling rates.

The performance for optimal threshold values at different sampling frequencies is presented in Fig. 7.

QRS sensitivity reaches higher values than those achieved for a fixed static threshold. The minimum value of 99.73% is reached instead of 99.53% for a sampling frequency of 100 Hz and maximum 99.86% (for 330 Hz) instead of 99.81% (for 250 Hz). At the same time the QRS positive predictive rate reaches a minimum of 99.67% (for 100 Hz) and maximum of 99.80% (for 250 Hz) instead of 99.79 Hz (for 225 Hz).

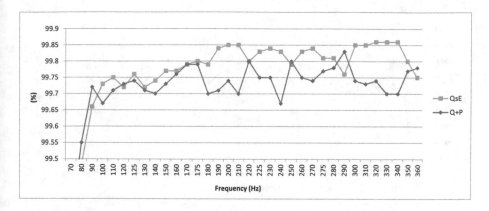

Fig. 7. Optimal performance values at different sampling rates.

The average performance values obtained for optimal threshold values are improved to 99.80% QRS sensitivity instead of 99.74% with fixed threshold keeping the same average value of the QRS positive predictive rate. At the same time the standard deviation is reduced to 0.044 instead of 0.081 for QRS sensitivity, which is almost the same as the standard deviation of the QRS positive predictive rate.

5.2   Optimal vs Fixed Static Threshold Performance

Figures 8 and 9 correspondingly compare QRS sensitivity and positive predictive rate for found optimal and fixed static threshold values at different sampling rates.

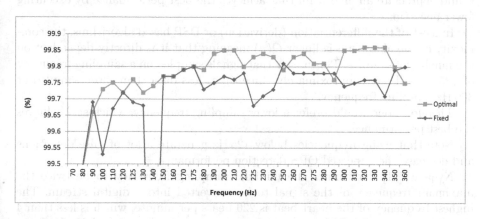

Fig. 8. QRS sensitivity at different sampling rates for optimal static threshold values.

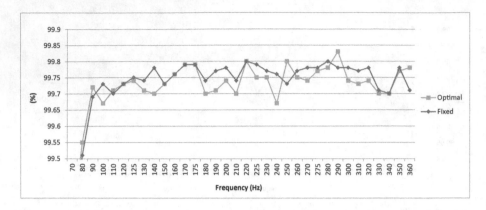

Fig. 9. QRS positive predictive rate at different sampling rates for optimal static threshold values.

In all cases, the optimal threshold parameter achieves a better performance with an exception at the sampling frequency of 250 Hz.

Although, this conclusion is also valid for the QRS positive predictive rate, it can be noted that there are cases where the fixed threshold parameter, reaches higher values, such as for 200 and 225 Hz, or 300 and 330 Hz, but these discrepancies are very small and the selected optimal threshold parameter compensates for this to achieve higher QRS sensitivity at these frequencies.

5.3   Response Time and Performance Analysis of Sampling Rates

Fast response is an important issue, especially if the algorithm is to be built in wearable ECG sensing devices. Due to limited resources and energy supply, one would appreciate an algorithm that achieves the best performance by executing fewer operations.

In most of the differentiation (derivation) or DSP based algorithms, the complexity of the algorithm is linear $O(n)$ meaning that it is directly dependent on the number of samples. This means that signals sampled on a sampling frequency of 125 Hz will have approximately three times less data than those sampled at 360 Hz sampling frequency.

Therefore, one would prefer a lower sampling frequency, which still achieves the best performance.

Note that using frequencies below 125 Hz generate a lot of false detections and decrease the expected QRS detection performance.

Nyquist [11] determines the minimal sampling frequency to be twice the maximum frequency of the signal to be converted into a digital stream. The highest frequency of the heart beat is 220 beats per minute, which is less than 4 Hz, the analysis of the QRS spectra is in the range between 5 and 20 Hz. Some features, such as short QRS peaks shorter than 40 ms may go even beyond this limit, which corresponds to 25 Hz. Therefore, the sampling frequency should be more than 50 Hz.

Table 1. QRS detector performance at different sample rates

Sample rate	MIT BIH database		AHA database	
	Q_{SE}	Q_{+P}	Q_{SE}	Q_{+P}
100	0.996856	0.997905	0.995839	0.996423
125	0.997426	0.998257	0.996660	0.996788
150	0.997458	0.998016	0.997429	0.997652
175	0.997601	0.998093	0.997119	0.997268
200	0.997426	0.998071	0.997397	0.997588
225	0.997228	0.997994	0.997268	0.997769
250	0.997502	0.998060	0.997087	0.997300
300	0.997360	0.998016	0.997450	0.997897
325	0.997448	0.997874	0.997578	0.997684
360	0.997535	0.998038	0.997503	0.997865

5.4   Comparative Analysis

Hamilton has measured the performance of the QRS detector at different sampling rates [8], and obtained the results presented in Table 1.

The performances fluctuate in the range up to 0.08% between 99.68 and 99.76% for QRS sensitivity and a relatively smaller fluctuation difference of 0.03% between 99.79 and 99.82%, for the MIT BIH Arrhythmia ECG database. In the case of the AHA database, the obtained performances are lower, and the fluctuation difference is 0.18% for QRS sensitivity and 0.13% fluctuation difference for QRS positive predictive rate.

Ajdaraga and Gusev [1] have analyzed the accuracy of QRS detection. Their study is very precise, since a correct detection (true positive) is considered to be the one that is located at most five samples from the identified one. Therefore their reported accuracy is lower than the usual published values. In this study we use a window of 150 ms where the identified peak may differ in location from the real one, which is treated as a correct QRS detection. In addition, the authors use a fixed static threshold.

Malik et al. report on a conclusion from the American Heart Association Task force: [9] low sampling rates may produce a jitter and wrong QRS detection and so the sampling frequency range needs to be between 250–500 Hz. To achieve a better performance they suggest additional resampling interpolation that refines the signal for satisfactory QSR detection.

Berntson et al. [2] also report that sampling frequencies below 100 Hz will result with decreased performance and suggest 128 samples per second as the lowest frequency to be used.

Ziemssen et al. [14] reported irrelevance in QRS detection performance at different sampling frequencies in the range between 100 and 500 Hz.

Ellis et al. [4] have conducted a series of experiments at sampling frequencies from 71.43 Hz up to 1000 Hz, and concluded a satisfactory performance even at the lowest analyzed sampling frequency, discrepancies above 125 Hz are especially negligible.

According to the above-mentioned papers, one might conclude that QRS performance may be satisfactory on a wide range of sampling frequencies without any problems. However, practical experiments confirm that threshold parameters need to be chosen carefully in order to obtain the optimal performance.

6   Conclusion

The goal of this paper was to find out what happens with QRS detection when resampling the signals and using different sampling rates. For this purpose, we conducted experimental research and found that QRS detection performance at different sampling rates can be optimized if a proper static threshold value is used as opposed to a fixed value.

Our optimization of the Hamilton's algorithm results with a higher average performance of 99.86% QRS sensitivity and 99.80% QRS positive predictive rate at different sampling frequencies as opposed to 99.81% QRS sensitivity and 99.79% QRS positive predictive rate in the original Hamilton approach with a fixed static threshold.

To conclude, this optimization reveals that a good performance can be achieved at both higher and lower sampling frequencies. Our experiments show that a good quality industry QRS detector can work even with sampling frequencies of 100 Hz to achieve a performance higher than 99.80% of QRS sensitivity and QRS positive predictive rate.

In addition, lower sampling frequencies generate less data, therefore the number of operations is lower, which reduces response time and energy requirements.

This research, along with the analysis of amplitude influence on QRS detection performance, motivates us to establish a model of QRS detection behavior for resampled and rescaled ECG signals. In addition, we wish to produce a quality QRS detector independent of the sampling frequency and rescaled amplitude.

References

1. Ajdaraga, E., Gusev, M.: Analysis of sampling frequency and resolution in ECG signals. In: 25th 2017 Telecommunication Forum (TELFOR) 2017, pp. 1–4. IEEE, November 2017
2. Berntson, G.G., Quigley, K.S., Jang, J.F., Boysen, S.T.: An approach to artifact identification: application to heart period data. Psychophysiology **27**(5), 586–598 (1990)
3. Domazet, E., Gusev, M.: Amplitude rescaling influence on QRS detection. Technical report 17/2018, University Sts Cyril and Methodius, Faculty of Computer Sciences and Engineering (2018)

4. Ellis, R.J., Zhu, B., Koenig, J., Thayer, J.F., Wang, Y.: A careful look at ECG sampling frequency and R-peak interpolation on short-term measures of heart rate variability. Physiol. Meas. **36**(9), 1827 (2015)

5. Goldberger, A.L., et al.: Physiobank, physiotoolkit, and physionet. Circulation **101**(23), e215–e220 (2000)

6. Gusev, M., Stojmenski, A., Guseva, A.: ECGalert: a heart attack alerting system. In: Trajanov, D., Bakeva, V. (eds.) ICT Innovations 2017. CCIS, vol. 778, pp. 27–36. Springer, Cham (2017). https://doi.org/10.1007/978-3-319-67597-8_3

7. Hamilton, P.S., Tompkins, W.J.: Quantitative investigation of QRS detection rules using the MIT/BIH arrhythmia database. IEEE Trans. Biomed. Eng. **12**, 1157–1165 (1986)

8. Hamilton, P.: Open source ECG analysis software documentation (2002)

9. Malik, M.: Task force of the european society of cardiology and the north american society of pacing and electrophysiology. Heart rate variability. Standards of measurement, physiological interpretation, and clinical use. Eur. Heart J. **17**, 354–381 (1996)

10. Moody, G.B., Mark, R.G.: The impact of the MIT-BIH arrhythmia database. IEEE Eng. Med. Biol. Mag. **20**(3), 45–50 (2001)

11. Nyquist, H.: Certain topics in telegraph transmission theory. Trans. Am. Inst. Electr. Eng. **47**(2), 617–644 (1928)

12. Pahlm, O., Sörnmo, L.: Software QRS detection in ambulatory monitoring: a review. Med. Biol. Eng. Comput. **22**(4), 289–297 (1984)

13. Pan, J., Tompkins, W.J.: A real-time QRS detection algorithm. IEEE Trans. Biomed. Eng. **3**, 230–236 (1985)

14. Ziemssen, T., Gasch, J., Ruediger, H.: Influence of ECG sampling frequency on spectral analysis of RR intervals and baroreflex sensitivity using the EUROBAVAR data set. J. Clin. Monit. Comput. **22**(2), 159–168 (2008)

Sarcasm and Irony Detection in English Tweets

Jona Dimovska, Marina Angelovska, Dejan Gjorgjevikj[(✉)],
and Gjorgji Madjarov

Faculty of Computer Science and Engineering,
Ss. Cyril and Methodius University in Skopje, Skopje, Macedonia
{dimovska.jona,angelovska.marina}@students.finki.ukim.mk,
{dejan.gjorgjevikj,gjorgji.madjarov}@finki.ukim.mk

Abstract. This paper describes an approach to sarcasm and irony detection in English tweets. Accurate sarcasm and irony detection in text is crucial for numerous NLP applications like sentiment analysis, opinion mining and text summarization. The detection of irony and sarcasm in microblogging posts can be even more challenging because of the restricted length of the message at hand, the informal language, emoticons and hash tags used. In our approach we combined a variety of standard lexical and syntactic features with specific features for capturing figurative content. All experiments were performed using supervised learning using different approaches for text preprocessing and feature extraction and four different classifiers. The corpus used was taken from SemEval2018 challenge containing a dataset with 3834 different tweets. The performance of the different approaches are reported and commented. The results have shown that the text preprocessing has very little impact on the results, while the word and sub-word frequencies are the most usable characteristics for determining irony in tweets. A separate experiment including a survey was also conducted in which human participants were challenged to label 20 given tweets from the dataset as ironic or not. The obtained results suggest that accurate irony detection in tweets can be a hard task even for humans.

Keywords: Irony detection · Sarcasm detection
Tweets classification · NLP

1 Introduction

When talking about irony the natural question that emerges is: How can irony be defined and what is it used for? In [8] Veale states that: "Irony is an effective but challenging mode of communication that allows a speaker to express sentiment-rich viewpoints with concision, sharpness and humour". Irony can also be defined

This research was partially funded by the Faculty of computer science and engineering, Ss. Cyril and Methodius University in Skopje.

S. Kalajdziski and N. Ackovska (Eds.): ICT 2018, CCIS 940, pp. 120–131, 2018.
https://doi.org/10.1007/978-3-030-00825-3_11

as a curious form of double-speak in which a speaker appears to say the opposite of what it actually means or implies [5], or expresses a sentiment in direct opposition to what is actually believed [7]. Though superficially like lying, irony is a form of creative mis-representation that is more sophisticated than simply saying X while meaning not-X. Contrary to lying, where the speaker is trying to deceive the audience into believing an untruth, ironic speakers' goal is actually their misrepresentations to be detected, so that the audience (or a targeted part of it) can grasp the underlying meaning.

Irony as a fundamental rhetorical device is a uniquely human mode of communication, in which the speaker says the opposite of what he or she means. Irony can be a very compact way of saying or doing two useful things at once. Irony can also be used to divide an audience into those who "get it" and those who don't; to soften a criticism with humour, to salt a wound by cloaking it in an apparent compliment that is quickly dashed [8,16].

Recently, computationally detecting irony has attracted attention from the natural language processing (NLP) and machine learning (ML) communities. This problem is very interesting by itself because there are no schemes and tools that would help us to detect and explain this phenomenon. In our everyday communication with people, we meet with sarcasm and irony which, in most of the cases, we can recognize relaying mainly on the way the message is pronounced, the context of the message itself and the character of the person/s we communicate with. However, not even in speech, we can always easily recognize a given irony or sarcasm, since it mainly depends on the context and pronunciation of the message delivered.

When a written message is in question, detecting this phenomenon becomes much more complex since the intonation of the speaker is missing and the context is not always well elaborated. On the social networks people detect sarcasm and irony mostly through various hints, such as hashtags, short exclamations (Yey, Wow) followed by a negative context or emotion. Hashtags are words or phrases preceded by a hash sign (#) to identify messages on a specific topic. Most frequently, these hashtags can be found on the social network Twitter. Therefore, most of the corpuses on which the experiments are based originate from Twitter [9]. However, these hints, which can be very helpful, are not always available.

Accurate automated irony detection is very important for opinion mining and sentiment analysis. In the past several years, many approaches for the improvement of irony and sarcasm detection have been elaborated [1,4,6,11,13,14]. In [6] the authors used Sequential Minimal Optimization (SMO) and Logistic Regression on the dataset of comments from Twitter. In the process of generating the corpus, all of the tweets were classified according to the hashtags that belong to one of the three categories: sarcasm, positive or negative. The pre-processing was done by using lexical and pragmatic features. Filatova in [4] uses a generated corpus of product reviews from Amazon based on crowd sourcing. According to the conducted experiment, the sarcastic reviews contained many positive words, when in fact the review had negative meaning. In [1] the authors were using two classification algorithms: Random Forest and Decision Tree. The results show that word frequency and sentence structure (emoticons, punctuations) are the most common characteristics for ironic tweets.

In this paper we present our experiments in detecting irony and sarcasm in tweet posts. Our work consists of two steps: the preprocessing and feature extraction step, where we implemented and tested different feature sets, and the learning step, where we investigated and optimized the performance of different learning strategies for this task. Comparison among the studied methods and also with human performance was also discussed.

The remainder of the paper is organised as follows: in Sect. 2 we describe the feature extraction and the experimental setup. In Sect. 3 the results of the experiments are presented and discussed. Section 4 presents the results of an on-line survey we conducted in an attempt to measure the realistic human performance in irony detection in tweets. Finally, we conclude in Sect. 5 with remarks on our future work.

2   Problem Definition

Irony detection is an interesting machine learning problem because, in contrast to most text classification tasks, it requires a semantics that cannot be inferred directly from word counts over documents alone. In our research we used Twitter posts. The data is taken from the Task 3 of SemEval2018: Irony Detection in English Tweets. The task was to determine if a tweet post is sarcastic or not.

2.1   Corpus Description

The corpus was prepared by the organizers of the SemEval2018 event and was conducted by searching Twitter for the hashtags #irony, #sarcasm and #not. 3000 English tweets were collected between 01/12/2014 and 04/01/2015 that represent 2676 unique users. All tweets were than manually labeled for irony. After the annotation, 2396 instances were labeled as ironic and 604 as non-ironic. In order to balance class representation in the corpus, another 1792 tweets that were manually checked to be non-ironic were added from a background corpus. The final corpus was arranged by randomly selecting 3834 instances for a training set and 784 instances for a test set.

The two datasets that were released for this task were used in our experiments. The tweets were labelled to belong to one of the two possible classes: tweets that contain irony and non-ironic tweets. In this dataset, 1911 of the tweets were labelled as ironic and 1923 were labelled as non-ironic. The second dataset, denoted as gold test set was used for evaluation and ranking the solution of the competitors and was released after the end of the competition. The gold test set contains 784 tweets of which 311 were labelled as ironic and 473 as non-ironic. The tweets were realistic and some of them also contained hashtags, URLs and emoticons. All the available tweets were taken into account regardless of the additional not pure textual content in some of them in the conducted experiments. Table 1 shows some descriptive statistics over the training and testing dataset. Except for the presence of emoji, the dataset looks pretty balanced considering the sarcasm and irony detection task.

None of the tweets contained any of the hashtags #sarcasm, #irony or #not, ones that are generally used by the users to express their non-literal intention [15]. This can sometimes be crucial to avoid misunderstanding when the number of words to pass the message is limited and the message itself usually lacks the full context, like in tweets. However some hashtags representing negation of some kind like #NO (only one), and few beginning with #no (like #notreally and #nothanks) were present. Some of the tweets also contained symbols - the so called emoticons. All the experiments conducted in order to select the model, determine its parameters and also to select pre-processing approaches were carried out using the training set only. The final results are obtained by applying the model on the test set.

Table 1. Descriptive statistics of the training and testing dataset.

	Ironic tweets		Non-ironic tweets	
	Training	Testing	Training	Testing
Total tweets	1911	311	1923	473
Total Words in tweets	26874	4575	26877	6689
Average words per tweet	14	14	13	14
Tweets containing Emoji's	175	33	236	56
Tweets containing hashtags	737	139	784	224

2.2 Data Preprocessing

We present the details of all the transformation mechanisms on the text of the tweets that were used. Inspired by some work of other researchers that suggest that sentiment and irony and sarcasm are interrelated [10], we tried to use some sentiment lexicons [15] to transform the text of the tweets prior to the feature extraction and classification.

We conducted several experiments performing different transformations on the text of the tweets, like replacing the symbols (emoji) with their corresponding description, for example ☺ was replaced with the text :smiling_face:. We also tried replacing some words like shouts and abbreviations with their proper form. Finally we tried replacing some of the hashtags with their sentiment equivalent word. We also experimented with removing the stop words from the tweets.

The possible steps that were taken for data pre-processing were:

- Removal of the stop words. The English stop words list from the NLTK library [12] was used for this step
- Removal of terms which appear in less than n documents

– Normalization, that includes (1) Replacement of shouts (like: Wow!, Yeey!) and emoji with labels for the polarity of the sentiment equivalent (positive/negative); (2) Replacement of hashtags with labels for the polarity of the sentiment equivalent (positive/negative); (3) Replacement of the URLs with label representing generic URL presence in the tweet.

We present the details of the feature extraction and transformation mechanisms we used. Our approach is based on the traditional n-gram feature extraction and the Bag-of-Words representation. For the data pre-processing, cleaning and tokenization, as well as for the learning steps, we used Python's NLTK [12] and scikit-learn [2] libraries.

2.3 Feature and Model Selection

We extracted features based on the lexical content of each tweet. In our approach we are using a bigger scope of features in order to improve the detection of sarcasm and irony in tweets. The features extracted for each tweet include: word n-grams and character n-grams. All the experiments were conducted for four different groups of features:

1. features generated from the character unigrams,
2. features generated from the character n-grams, where n was set between 1 and 4,
3. features generated from the word unigrams,
4. features generated from the word n-grams, where n was set between 1 and 3.

The features were extracted following the Bag of Words approach with some modifications clarifying what a "word" is and defining what to encode about each word in a vector. In our approach when using character features the "word" is considered to be a sequence of n consecutive characters regardless of that they can be only a part of real word. The features were then vectorized using two different vectorizers from scikit-learn Python library [2]: TF-IDF Vectorizer and Hashing Vectorizer. By vectorizing we define the general process of turning a collection of text documents into numerical feature vectors. Hash functions are an efficient way of mapping terms to features. Hashing Vectorizer applies a hashing function to term frequency counts in each document. This method converts a collection of text documents to a matrix of token occurrences. TF-IDF Vectorizer on the other hand uses the product of the term-frequency and the inverse document-frequency to represent the occurrence of a term in a document.

As for the classification task, several algorithms also from the Scikit-learn Python library were tried, but the extensive testing was performed using the following four as most promising ones:

1. k-Nearest neighbour classifier
2. Logistic regression
3. Linear Support Vector Machine (SVM)

4. Non-linear Support Vector Machine using Gaussian (also known as radial basis function or RBF) kernel.

The k-nearest neighbours (or k-NN for short) is a simple machine learning algorithm that categorizes an input by using its k nearest neighbours. The principle behind nearest neighbour methods is to find a predefined number of training samples closest in distance to the new point, and predict the label from these. In our experiments we have used the scikit-learn implementation of nearest neighbour classifier with Euclidian distance as metric, while the number of neighbours was determined by a model selection procedure.

Logistic regression is a technique borrowed by machine learning from the field of statistics and is well suited for binary classification problems. Logistic regression uses an equation as the representation, very much like linear regression except that this equation represents the logistic function which coefficient values must be estimated from the training data. This is usually done using maximum-likelihood estimation. In our experiments we have used the scikit-learn implementation of logistic regression, while the cost parameter for the regularization was determined by a model selection procedure.

Support vector machines (SVMs) are a set of supervised learning methods used for classification, regression and outlier detection. A support vector machine takes data points and outputs the hyperplane that best separates the categories maximizing the margin and thereby creating the largest possible distance between the separating hyperplane and the nearest instances on either side of it. This classifier has proven to be particularly good for problems with limited number of samples, being in the same time pretty independent of the dimensionality of the feature space. It also supports the kernel trick of mapping the data into a very high dimensional space where the training data would be linearly separable. This makes the algorithm very suitable for text classification problems, where it is common to have access to a limited dataset (couple of thousands) of tagged samples. In our experiments we have used the scikit-learn support vector classifier that uses the LibSVM [3] implementation, using two kernels: the linear one and the non-linear that uses the RBF kernel function. The regularization cost parameters C of the SVM classifier trades off misclassification of training examples against simplicity of the decision surface. The gamma parameter of the RBF kernel regulates how far the influence of a single training example reaches.

2.4   Experimental Setup

We conducted number of experiments that included four different sets of features extracted after four different ways of pre-processing the tweets giving 16 different approaches of feature extraction. In the next step each of these features were transformed in vector representation using two different approaches for vectorization and were finally fed to four different classifiers that resulted in 128 different models. In order to determine the accuracy of each of the models, we used the F1 score metrics (F1 = 2*(precision * recall)/(precision + recall)). The optimal parameters of each of the classifiers were determined in a model selection procedure using a grid search and 10-fold cross-validation procedure on the training set only. For the kNN classifier the number k of the nearest neighbours to be considered when determining an unclassified instance class was searched for all odd numbers in the range 3–73. The actual optimal values for k that were found were between 9 and 51 (with the average being around 40). For the logistic regression and the SVM classifiers, the parameter C that controls the importance of the regularization term was searched among the values 0.1, 1, 10, 100, 1000. For the RBF SVM classifiers a full grid search was performed for finding the optimal values for the regularization parameter C and the parameter gamma of the Gaussian kernel among the values 0.2, 0.4, 0.6, 0.8, 1.0, 1.2. Each classifier was then retrained using the optimal parameters, as found in the previous step, on the whole training set. The performance of the obtained classifiers was determined by a 10-fold cross-validation on the training set, and the mean of the obtained F1 score and the standard deviation are given in Tables 2 and 3. The results of trained classifiers on the test set are given in Tables 4 and 5.

Table 2. F1 score on the training set obtained by 10-fold cross validation using TF-IDF fVectorizer. The standard deviation is given in parenthesis.

Vectorizer	TdifVectorizer							
Feature extraction	Character unigram				Character n-grams (n = 1−4)			
Text preprocessing	kNN	LogR	Linear SVM	RBF SVM	kNN	LogR	Linear SVM	RBF SVM
No interventions	0.5993 (±0.0419)	0.6286 (±0.0496)	0.6292 (±0.0458)	0.6358 (±0.0506)	0.6270 (±0.0641)	0.6582 (±0.0668)	0.6590 (±0.0598)	0.6595 (±0.0619)
Stop words removed	0.5993 (±0.0419)	0.6286 (±0.0496)	0.6292 (±0.0458)	0.6358 (±0.0506)	0.6270 (±0.0641)	0.6590 (±0.0598)	0.5887 (±0.0344)	0.6595 (±0.0619)
Emoji's, hashtags, URLs replaced	0.6070 (±0.0557)	0.6260 (±0.0398)	0.6272 (±0.0375)	0.6357 (±0.0399)	0.6201 (±0.0581)	0.6573 (±0.0547)	0.6576 (±0.0572)	**0.6649** (±0.0576)
Stop words removed + Emoji's, hashtags, URLs replaced	0.6070 (±0.0557)	0.6260 (±0.0398)	0.6272 (±0.0375)	0.6357 (±0.0399)	0.6201 (±0.0581)	0.6573 (±0.0547)	0.6576 (±0.0572)	**0.6649** (±0.0576)
Feature extraction	Word unigram				Word n-grams (n = 1−3)			
Text preprocessing	kNN	LogR	Linear SVM	RBF SVM	kNN	LogR	Linear SVM	RBF SVM
No interventions	0.0.6080 (±0.0685)	0.6464 (±0.0525)	0.6466 (±0.0581)	0.6452 (±0.0529)	0.6168 (±0.0789)	0.6422 (±0.0524)	0.6447 (±0.0485)	0.6436 (±0.0494)
Stop words removed	0.6196 (±0.0491)	0.6455 (±0.0488)	0.6489 (±0.0495)	0.6490 (±0.0554)	0.6248 (±0.0618)	0.6520 (±0.0592)	0.6486 (±0.0583)	0.6595 (±0.0544)
Emoji's, hashtags, URLs replaced	0.6202 (±0.0717)	0.6439 (±0.0531)	0.6400 (±0.0517)	0.6452 (±0.0529)	0.6040 (±0.0484)	0.6406 (±0.488)	0.6426 (±0.0517)	0.6436 (±0.0494)
Stop words removed + Emoji's, hashtags, URLs replaced	0.6217 (±0.0360)	0.6453 (±0.0554)	0.6476 (±0.0564)	0.6500 (±0.0660)	0.6260 (±0.0611)	0.6477 (±0.0551)	0.6482 (±0.0622)	0.6464 (±0.0586)

3   Results

Considering the results shown in Tables 2, 3, 4 and 5 it can be noted that the preprocessing steps for changing the text of the tweets, by removing the stop words and replacing the hashtags has rather insignificant impact in the recognition.

Table 3. F1 score on the training set obtained by 10-fold cross validation using hashing vectorizer. The standard deviation is given in parenthesis.

Vectorizer	Hashing Vectorizer							
Feature extraction	Character unigram				Character n-grams (n = 1 − 4)			
Text preprocessing	kNN	LogR	Linear SVM	RBF SVM	kNN	LogR	Linear SVM	RBF SVM
No interventions	0.5763 (±0.0405)	0.6284 (±0.0452)	0.6262 (±0.0435)	0.6190 (±0.0456)	0.6047 (±0.0525)	0.6592 (±0.0434)	0.6595 (±0.0436)	0.6539 (±0.0570)
Stop words removed	0.5963 (±0.0405)	0.6284 (±0.0452)	0.6262 (±0.0435)	0.6190 (±0.0456)	0.6047 (±0.0525)	0.6592 (±0.0434)	0.6595 (±0.0436)	0.6539 (±0.0570)
Emoji's, hashtags, URLs replaced	0.5858 (±0.0500)	0.6224 (±0.0363)	0.6200 (±0.0420)	0.6274 (±0.0411)	0.5996 (±0.0610)	0.6470 (±0.0487)	0.6537 (±0.0515)	**0.6581** (±0.0533)
Stop words removed + Emoji's, hashtags, URLs replaced	0.5858 (±0.0500)	0.6224 (±0.0363)	0.6246 (±0.0400)	0.6274 (±0.0411)	0.5996 (±0.0610)	0.6470 (±0.0487)	0.6537 (±0.0515)	**0.6581** (±0.0533)
Feature extraction	Word unigram				Word n-grams (n = 1 − 3)			
Text preprocessing	kNN	LogR	Linear SVM	RBF SVM	kNN	LogR	Linear SVM	RBF SVM
No interventions	0.6006 (±0.0726)	0.6435 (±0.0529)	0.6448 (±0.0564)	0.6419 (±0.0447)	0.5986 (±0.0474)	0.6418 (±0.0581)	0.6415 (±0.0630)	0.6413 (±0.0500)
Stop words removed	0.6292 (±0.0517)	0.6524 (±0.0622)	0.6507 (±0.0606)	0.6488 (±0.0521)	0.6220 (±0.0467)	0.6465 (±0.0668)	0.6467 (±0.0644)	0.6472 (±0.0636)
Emoji's, hashtags, URLs replaced	0.5999 (±0.0661)	0.6382 (±0.0647)	0.6393 (±0.0535)	0.6387 (±0.0461)	0.5947 (±0.0629)	0.6409 (±0.0493)	0.6398 (±0.0567)	0.6423 (±0.0482)
Stop words removed + Emoji's, hashtags, URLs replaced	0.6217 (±0.0562)	0.6496 (±0.0639)	0.6454 (±0.0680)	0.6498 (±0.0597)	0.6273 (±0.0453)	0.6453 (±0.0643)	0.6464 (±0.0694)	0.6481 (±0.0650)

Table 4. F1 score on the test set for the different features and classifiers using TF-IDF vectorizer.

Vectorizer	TdifVectorizer							
Feature extraction	Character unigram				Character n-grams (n = 1 − 4)			
Text preprocessing	kNN	LogR	Linear SVM	RBF SVM	kNN	LogR	Linear SVM	RBF SVM
No interventions	0.5718	0.5807	0.5825	0.6058	0.5718	0.6701	0.6649	0.6729
Stop words removed	0.5718	0.5807	0.5825	0.6058	0.5718	0.6701	0.6649	0.6729
Emoji's, hashtags, URLs replaced	0.5708	0.5790	0.5863	0.5994	0.5708	0.6631	0.6635	0.6578
Stop words removed + Emoji's, hashtags, URLs replaced	0.5708	0.5790	0.5863	0.5994	0.5708	0.6631	0.6635	0.6578
Feature extraction	Word unigram				Word n-grams (n = 1 − 3)			
Text preprocessing	kNN	LogR	Linear SVM	RBF SVM	kNN	LogR	Linear SVM	RBF SVM
No interventions	0.6199	0.6546	0.6560	0.6652	0.6336	0.6668	0.6694	0.6714
Stop words removed	0.6152	0.6581	0.6564	0.6601	0.6426	0.6719	0.6685	**0.6732**
Emoji's, hashtags, URLs replaced	0.5957	0.6462	0.6520	0.6556	0.6270	0.6636	0.6681	0.6712
Stop words removed + Emoji's, hashtags, URLs replaced	0.6308	0.6546	0.6561	0.6601	0.6327	0.6628	0.6630	0.6654

The effect is slightly more pronounced when using simpler features like character unigrams and word unigrams than for character and word n-grams. Although the influence of these pre-processing was minor in general, of all of the mentioned

Table 5. F1 score on the test set for the different features and classifiers using hashing vectorizer.

Vectorizer	Hashing vectorizer							
Feature extraction	Character unigram				Character n-grams $(n=1-4)$			
Text preprocessing	kNN	LogR	Linear SVM	RBF SVM	kNN	LogR	Linear SVM	RBF SVM
No interventions	0.5497	0.5902	0.5930	0.5993	0.6108	0.6597	0.6561	0.6571
Stop words removed	0.5497	0.5902	0.5930	0.5993	0.6108	0.6597	0.6561	0.6571
Emoji's, hashtags, URLs replaced	0.5660	0.5723	0.5771	0.6032	0.5894	0.6534	0.6537	0.6458
Stop words removed + Emoji's, hashtags, URLs replaced	0.5660	0.5723	0.5764	0.6032	0.5894	0.6534	0.6537	0.6458
Feature extraction	Word unigram				Word n-grams $(n=1-3)$			
Text preprocessing	kNN	LogR	Linear SVM	RBF SVM	kNN	LogR	Linear SVM	RBF SVM
No interventions	0.5948	0.6592	0.6604	0.6739	0.6335	0.6703	**0.6765**	0.6711
Stop words removed	0.6300	0.6436	0.6476	0.6488	0.6395	0.6484	0.6529	0.6575
Emoji's, hashtags, URLs replaced	0.5992	0.6560	0.6543	0.6537	0.5915	0.6677	0.6720	0.6604
Stop words removed + Emoji's, hashtags, URLs replaced	0.6345	0.6363	0.6375	0.6506	0.6429	0.6558	0.6464	0.6581

Table 6. List of the tweets used in the survey.

	Tweet	Sarcastic
1	working on my birthday #yay #sucks	✓
2	It's the most wonderful time of the fiscal year 😌	✓
3	Cocktails tonight yep why not! had one in over a year!	✗
4	My grandpa just gave me a credit card for christmas...he has no idea. Lol	✗
5	IS IT FRIDAY? *heavy sigh* Nooo. It Is totally Wednesday~! What a MESS. Wait 2 more days.	✗
6	cared for 8 seconds, then I got distracted. 😩	✓
7	Aaand the hits just keep on coming. #BadDay	✗
8	I'm not really sure how I'm still alive and writing this paper right. Napping instead of sleeping on the weekends isn't cutting it #finals	✗
9	The Chicago Firehouse Restaurant caught fire this morning	✓
10	Wow I really have the best luck known to man 😌	✓
11	Isn't it great to sleep 5 hours and feel like a million bucks? #gettingold	✓
12	I'm glad H & M has employed enough staff today. #nohelp #terrible	✓
13	That's always a great way to boost self-esteem	✓
14	Riding the distraction train. ||CHOO CHOO	✗
15	#Love is far more important than #money! |The problem is, you only know it once you have money!| #life #humanity	✓
16	Home finally! What a way to start the christmas break!!! 👍 🎄	✗
17	The city that only sleeps on mass transit. #naptime #isntitironic #donchathink #imnotsure...	✓
18	I just love it when my knee clicks 😊😊	✓
19	I'm pretty sure I have a snowman problem.	✗
20	What an eventful day... 😴	✓

preprocessing steps, the emoji replacement had the biggest impact on the F1 score for irony detection.

Using character n-grams and word n-grams as features provides better performance in general. The best results on the training set were obtained with the SVM classifier utilizing RBF kernel and TF-IDF vectorization of the character n-gram features when the emoji symbols, the hashtags and the URLs were replaced, while removing the stop words in this case did not cause any change in the results.

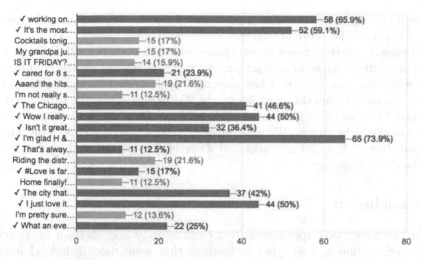

Fig. 1. Results from the conducted online survey. The green color indicates the ironic tweets, while the number next to them indicates the number of correct predictions. (Color figure online)

Considering the classifiers, it is notable that only the kNN classifier is somewhat inferior in all the cases compared to the other three classifiers that for the same set of features tend to have very similar performance. The best performance (0.6765 F1 score) on the test set was obtained by the linear SVM using the hashing vectorizer on the word n-grams with no prior interventions on the text of the tweets. Second best performance (0.6732 F1 score) was obtained using the RBF SVM classifier that utilizes TF-IDF vectorizer also on the word n-grams with no prior interventions on the text.

The obtained results are among the best on this dataset compared to the results reported by the competitors on the SemEval-2018 task 3A - Irony detection in English tweets.

4   Crowdsourcing Experiment

Finally, in order to prove our suspicion that recognizing irony and sarcasm in a given text, particularly short text messages like tweets can be difficult task even for humans, we conducted an online survey in a students' Facebook group.

The survey contained twenty tweets (taken from the training dataset on which the algorithms most frequently made mistakes). Twelve of these tweets on the survey were labeled as sarcastic. We conducted the survey among IT students between the ages of twenty and twenty-two. The text (and emoji symbols) of the tweets used in the survey are shown in Table 6.

The survey received 88 responses. Only one respondent answered all the questions on the survey correctly (i.e. correctly recognized all 12 sarcastic tweets). The percentage of correct classification of each of the tweets by the respondents is shown on Fig. 1. The calculated human performance on this small dataset of 20 tweets reveals F1 score of 0.3648.

To compare it to our system we trained our best classifier (RBF SVM, TF-IDF vectorizing, character n-grams, no text preprocessing) on the training set from which we have removed the shown 20 instances that were used in the survey. We than tested this classifier on the 20 instances from the survey and we obtained F1 score of 0.3750 that is slightly better than the performance of the respondents on the survey. We must mention that all respondents of the survey were familiar with social networks and Tweeter, but for all of them English is not their first language.

5   Conclusion

In this work, various approaches to detecting irony and sarcasm in Tweeter posts were explored. The types of features that were used include character and word unigrams and n-grams. Two approaches for vectorizing, TF-IDF and hashing were explored. Preprocessing the text of the tweets by removing the stop words and replacing the hashtags and URLs did not provide significant change in the results. Four types of classifiers were explored, each trained with optimal hyper-parameters decided by 10-fold model selection procedure. The conducted survey among the students in recognizing irony and sarcasm on the small sample of the tweets from the dataset revealed that accurate irony detection in short textual posts is a hard task even for humans. As future work, we consider using combinations of feature types instead of single feature types for classification.

References

1. Barbieri, F., Saggion, H.: Modelling irony in Twitter. In: Proceedings of the Student Research Workshop at the 14th Conference of the European Chapter of the Association for Computational Linguistics, pp. 56–64. Association for Computational Linguistics (2014)
2. Buitinck, L., et al.: API design for machine learning software: experiences from the scikit-learn project. In: ECML PKDD Workshop: Languages for Data Mining and Machine Learning, pp. 108–122 (2013)
3. Chang, C.C., Lin, C.J.: LIBSVM: a library for support vector machines. ACM Trans. Intell. Syst. Technol. **2**, 27:1–27:27 (2011). http://www.csie.ntu.edu.tw/~cjlin/libsvm
4. Filatova, E.: Irony and sarcasm: corpus generation and analysis using crowdsourcing. In: Chair, N.C.C., et al. (eds.) Proceedings of the Eight International Conference on Language Resources and Evaluation (LREC 2012). European Language Resources Association (ELRA), Istanbul, Turkey, May 2012
5. Giora, R.: On irony and negation. Discourse Process. **19**(2), 239–264 (1995)
6. González-Ibáñez, R., Muresan, S., Wacholder, N.: Identifying sarcasm in Twitter: a closer look. In: Proceedings of the 49th Annual Meeting of the Association for Computational Linguistics: Human Language Technologies: Short Papers, HLT 2011, Association for Computational Linguistics, Stroudsburg, PA, USA, vol. 2, pp. 581–586 (2011)

7. Grice, H.P.: Logic and conversation. In: Cole, P., Morgan, J.L. (eds.) Syntax and Semantics, Speech Acts, vol. 3, pp. 41–58 (1978)
8. Hao, Y., Veale, T.: An ironic fist in a velvet glove: creative mis-representation in the construction of ironic similes. Minds Mach. 20(4), 635–650 (2010)
9. Joshi, A., Bhattacharyya, P., Carman, M.J.: Automatic sarcasm detection: a survey. ACM Comput. Surv. 50(5), 73:1–73:22 (2017)
10. Kaushik, S., Barot, P.M.: Sarcasm detection in sentiment analysis. Int. J. Adv. Res. Innov. Ideas Educ. 2(6), 1749–1758 (2016)
11. Liebrecht, C., Kunneman, F., Van den Bosch, A.: The perfect solution for detecting sarcasm in tweets #not. In: Proceedings of the 4th Workshop on Computational Approaches to Subjectivity, Sentiment and Social Media Analysis, pp. 29–37. Association for Computational Linguistics (2013). http://www.aclweb.org/anthology/W13-1605
12. Loper, E., Bird, S.: NLTK: the natural language toolkit. In: Proceedings of the ACL-02 Workshop on Effective Tools and Methodologies for Teaching Natural Language Processing and Computational Linguistics, ETMTNLP 2002. vol. 1, pp. 63–70. Association for Computational Linguistics (2002)
13. Poria, S., Cambria, E., Hazarika, D., Vij, P.: A deeper look into sarcastic tweets using deep convolutional neural networks. In: COLING (2016)
14. Rajadesingan, A., Zafarani, R., Liu, H.: Sarcasm detection on Twitter: a behavioral modeling approach. In: Proceedings of the Eighth ACM International Conference on Web Search and Data Mining, pp. 97–106. WSDM 2015. ACM, New York (2015). https://doi.org/10.1145/2684822.2685316
15. Sulis, E., Farias, D.I.H., Rosso, P., Patti, V., Ruffo, G.: Figurative messages and affect in Twitter: differences between # irony, #sarcasm and #not. Knowl.-Based Syst. 108, 132–143 (2016)
16. Veale, T.: A computational exploration of creative smiles. In: Metaphor in Use: Context, Culture, and Communication, pp. 329–343. John Benjamins Publishing Company (2012)

Review of Automated Weed Control Approaches: An Environmental Impact Perspective

Petre Lameski[✉], Eftim Zdravevski, and Andrea Kulakov

Faculty of Computer Science and Engineering,
University of Sts. Cyril and Methodius in Skopje, Ruger Boskovikj 16,
1000 Skopje, Macedonia
{petre.lameski,eftim.zdravevski,andrea.kulakov}@finki.ukim.mk

Abstract. Agricultural food production is in constant struggle to meet the market demands. Weed control is used to increase the per land unit production from agricultural field. The process of weed removal is usually performed manually and is a time-consuming and labor demanding task. Since mechanical removal is a difficult process, the plantations use herbicides to remove unwanted plants. Herbicides are applied in large quantities, thus often have a degenerative effect on the land. Sometimes, they even endanger the health of the workers who apply them and the end users which consume the harvested product. We review the technologies used for automated weed control and its environmental impact, specifically on the pollution reduction. We also review the herbicides reduction reported in implemented and tested approaches for precision agriculture with emphasis on the weed control environmental impact. Based on the reviewed papers, we conclude that automated weed detection can identify unwanted plants with decent accuracy. Consequently, this can facilitate building autonomous spraying systems that can significantly reduce the quantity of applied herbicides by precisely applying the chemicals only on the plants or mechanically removing unwanted plants. We also review the challenges that need to be overcome, such as precise weed plant type detection, speed of the process and some security considerations that arise from the involvement of information and communication technologies.

Keywords: Weed control · Herbicide reduction · Clean agriculture
Pollution reduction

1 Introduction

The agricultural production has an upward trend owing to the increased area of the land used for production and due to the mechanized land and plant processing. Nevertheless, this trend is insufficient and according to [32], it will not be able to reach the demand in the future. One of the most effective way to

© Springer Nature Switzerland AG 2018
S. Kalajdziski and N. Ackovska (Eds.): ICT 2018, CCIS 940, pp. 132–147, 2018.
https://doi.org/10.1007/978-3-030-00825-3_12

increase the per land unit production is to introduce automation. The introduction of automated robots in agriculture could increase the food production [9]. The main advantage of the introduction of automation is the increased precision where the land is no longer treated as a whole, but individual small parts of the land are treated differently based on their specific needs.

According to [30] "Precision agriculture is the management of an agricultural crop at a spatial scale smaller than the individual field". Such management of the fields is a time consuming task when performed by workers, so an automated approach is needed to overcame this challenge. The introduction of precision agriculture arises some problems that need to be addressed, but on the long run, it can provide an increased production per land unit and in the same time, can reduce the cost of managing the land by reducing the amount of minerals, herbicides, pesticides and other resources that are essential for the production.

Weed control is one of the processes that can be considered as part of the precision agriculture. It is a process of removing unwanted plants from the land mechanically by manual removal, chemically by using herbicides, or by other alternative means. When used on the full fields, herbicides are needed in large amounts and have negative effect of polluting the land because they affect the wanted plants, as well. Besides the obvious land pollution from using herbicides, [24] shows that there is evident pollution in surface and ground waters too, which makes the problem of herbicide pollution even greater. The economic benefits of using precision agriculture is recognized in [7] by analyzing the specific sites for particular weed plants and applying appropriate herbicides. Other parameters, such as humidity, temperature, light etc. are also important. Several systems for precision agriculture monitoring, such as [42], have been proposed. Our main focus, however, is on the weed control, weed monitoring and herbicide reduction.

In this paper we review several state-of-the-art approaches in sensing technologies and actuators used for weed control. We also discuss the benefits of using precision agriculture for both the environment and the financial benefits for the farmers in terms of decreased expenses and increased yield from the farms. While the positive effects are well recognized in the literature, there are still many challenges that need to be addressed for applying precision agriculture for weed control in agricultural fields.

The paper is organized as follows: in Sect. 2 we describe the research methodologies of existing approaches reviewed in this paper. Next, in Sect. 3 we analyzed the related works from the impact on the environment point of view. We also examine some of the novel sensing technologies that are essential for the weed detection process. In the end, we discuss the benefits of using the proposed approaches and sensing technologies in Sect. 4 we conclude the paper.

2 Review Methodology

To provide thorough overview of the existing approaches for precision agriculture in the process of weed control, we selected papers with the search words: "precision agriculture", "herbicide reduction", and "weed control". We selected papers published after 2007. The search results showed a total of 4147 papers on which we performed quantitative analysis. From those papers, we selected 48 papers based on their relevance that we analyzed qualitatively in more detail.

The distribution of search results for papers per year from Google Scholar, using the same search words is shown in Fig. 1. Obviously, there is an increasing trend in publishing papers concerning precision agriculture for weed control.

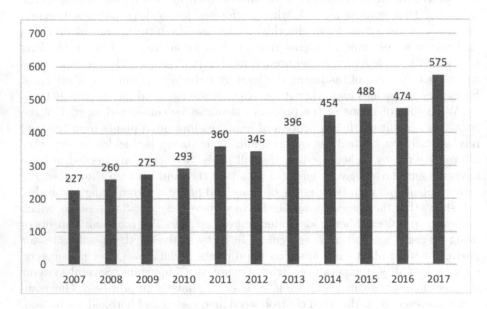

Fig. 1. Search result distribution per year of the terms: "precision agriculture", "herbicide reduction", and "weed control"

After the selection of papers, we organized them in groups that cover several aspects of weed control:

(i) Papers that explicitly report analysis of the herbicide usage reduction and of the percentage of weeds reduced by the proposed approach.

(ii) Papers that report results from using specific actuators that reduce the pollution with herbicides.

(ii) Papers that review the most important advancements of sensing technologies for application in systems for precise weed detection.

(iv) Papers that review the challenges for adoption and application of new technologies.

The organization of the reviewed papers includes the main challenges of the automated weed detection and removal. The main motivation for adopting any new technology in any production field is the cost reduction and yield increase. The quantification of these measurements is very important and in the first group we include papers that quantify the estimated weed reduction and the estimated herbicide reduction. We consider these factors as a very important motivation for further adoption of any reported approach for automated weed reduction, especially in the production phase or marketing phase of a weed control system.

The two most important things for weed control when using an automated approach are:

(i) Sensing technologies for weed detection.
(ii) Actuators for mechanical or chemical weed treatment.

The sensing technology for automated weed control is a challenge that is being intensively tackled by the scientific community. Therefore, we give a brief overview of the existing approaches. A common pattern can be observed when describing weed detection approaches where most authors use a combination of vision sensors and multi-spectral vision sensors and machine learning to solve the problem of weed detection. The actuators are commonly divided in two groups: actuators for mechanical removal and sprayers. Some of the available actuators use traditional mechanical control while others incorporate the so-called 'smart' trend, where some more complex sensing technologies are included in the control loop. As any modern technology, automated weed control needs to be adopted in the farm production process, which is a considerable obstacle. We include several papers that aim to describe and propose solutions of the existing challenges. While future challenges are hard to predict, we can, to certain degree, anticipate some of the challenges based on the way older technologies have been introduced and adopted by the agricultural industry.

3   Overview of Weed Control Approaches

Automation in the weed control process has received increased attention by the scientific community and is already being used in the agricultural industry [40]. According to [16], there is a large number of papers describing weed population mathematical models. The authors conclude that most of the prototype applications were directed towards the process of decision making in weed management. It is an important process for efficient weed removal that can reduce the costs for weed control and increase the net profit per land unit. By reducing the usage of herbicides, the production of crops becomes cleaner and healthier. There are plenty of weed control approaches described in the literature and there are also some commercially available systems that can be used for weed control. Some of them specify novel actuators that mechanically target weeds, and other consider novel sensing technologies that allow detection and measurement of weed infestation. In this paper we review several such approaches. One of the most important benefit of precise weed control is the reduction of herbicides and thus the reduction of pollution. We also review some of the currently published results that report specific percentage of herbicide reduction. Finally, we highlight some of the challenges that arise from the adaptation of the new technologies, especially in the small and mid-sized organic farms.

3.1 Herbicide Reduction Analysis

Herbicide reduction is one of the main benefits of precision agriculture. There are even approaches that aim to detect and eliminate the weed in real time. According to [33], there are both economic and ecological reasons to perform site-specific weed management.

Authors in [17] report that the precise application of the herbicides on the plants is an effective way to reduce the unwanted plants, while using only 22% of the quantity needed when applying the herbicide on the full field.

Another study presented in [11], used sensor controlled herbicide sprayers to reduce the amount of herbicides used in the fields. The described method was tested on 13 fields and the analysis of the results discovered average herbicide savings of 24.6% without yield reduction from the fields.

Authors in [38] describe an experimental automated robotic system that used micro-dosage system to spray small amounts of herbicide on detected weeds. The weeds were detected using computer vision approach and the vehicle navigated autonomously trough the field. Reduction of the recommended herbicide usage by two orders of magnitude by using the proposed system can be achieved, according to the experimental results.

In [20] authors compare site specific and uniform distribution of herbicides. They measured 69.5% herbicide savings in the site specific application compared to the uniform distribution of herbicides.

The approach presented in [6] used multi-spectral imagery of fields to discriminate weed in cruciferous field patches. Authors applied machine learning methods and color indexes to discriminate patches that contained significant amounts of weed from other patches. Using the best models on weed maps of the field, resulted in herbicide savings from 71.7% to 95.4% for the no-treatment areas and from 4.3% to 12% for the low-dose herbicide.

In [23] authors evaluated the automated boom section control on agricultural sprayers. They assessed the application inaccuracy for 21 study fields and detected a reduction of over-application from 12.4% to 6.2%. Further, they analyzed manual boom section control, which comparing to the automated approach increases the over-application.

Precision Experimental Design is proposed in [13] to model the yield loss based on the competition between weeds and crops in fields. By using precision agriculture methods for mapping the different kinds of weed species in the fields, authors were able to derive a model for yield loss based on different kinds of weed and different combinations of weeds due to the heterogeneous placement of the weed species in the field patches.

A micro sprayer system for guided sprayers using sensing technologies to discriminate weed from crops and to spray only the weeds is evaluated in [26]. The system was tested in laboratory conditions and authors report up to 94% limited growth of the weeds found in the crops fields.

The study presented in [5] analyses the potential of using images from Unmanned Aerial Vehicles (UAVs) to support patch herbicide spraying in maize crops. Authors found significant reduction of herbicide usage without significantly reducing the crop yield. Savings in herbicide were between 14% and 39.2% compared to broadcast spraying, yielding savings between 16 and 45 euro per ha.

In [8] authors studied the problem of integration between UAVs and automated ground vehicles for weed detection and removal. They found that such systems have the potential of reducing the herbicide usage significantly and in worst case, this can leave less than 40% of weed untreated.

A system for weed sensing and herbicide spraying based on real time cameras is compared to the conventional broadcast spraying in [10]. The experiment was conducted in a period of four years and herbicide savings between 30% and 43% were detected without finding any significant difference in yield.

A machine vision algorithm was applied in [2] to select patches of cereals fields that need to be sprayed in winter and spring. The approach used relative weed cover and relative mayweed cover metrics from the Weedcer algorithm to estimate the spraying decisions. Authors reported savings of 22% to 97% in different trials.

An autonomous system for precise spraying is described in [41], listing the main features of such robot. The results of the field tests demonstrated that it is accurate enough to accomplish treatment of over 99.5% of the detected weeds, thus significantly reducing herbicide usage.

In Table 1 we provide an overview of the reviewed papers related to herbicide reduction and impact on yield quantity:

Table 1. Reported herbicide reduction per study

Reference	Herbicide reduction	Yield quantity impact
[17]	88%	Insignificant
[11]	24.6%	Insignificant
[38]	Two orders of magnitude	Insignificant
[20]	69.5%	Insignificant
[6]	71.7% to 95.4%	Insignificant
[5]	14% to 39.2%	Insignificant
[2]	22% to 97%	Insignificant

All of the reviewed studies suggest that by applying precise approaches there is significant reduction in herbicide usage and there are no reported disadvantages in regards to the yield quantity from the fields. Application of precision agriculture approaches reduces the pollution of the plants and of the land, thus creating environment of cleaner agricultural food production. Lowering herbicide usage and reduction of weed are the main motivations for investments in the field of automated weed control. However, novel sensing technologies are required to achieve a precise detection of weeds in the fields.

Next, we review some of the sensing approaches that are available in the literature.

3.2 Novel Sensing Technologies

One of the main challenges in precise weed control is the process of weed detection. Plenty of approaches were adapted for weed sensing and many of them use machine learning to build models of the plants and to discriminate between crops and weeds.

Techniques based on computer vision can be used for weed mapping in the fields [37]. They could speed up the process of weed detection and this information can be used for site specific weed control measures.

The spatial pattern of weeds based on multi-species infestation maps from images was studied in [21]. Authors found out that it is not always possible to detect an aggregated spatial pattern in weed infestation in maize fields, and concluded that there is a need of techniques for assessment of weed aggregation prior to conducting site-specific weed management.

In [45] a vision-based method for detecting specific weed Avena sterilis in cereal crops is described. The method used image segmentation and decision making to select cells in the field that needed to be sprayed. The goal was to reduce the quantity of applied herbicides for effective weed removal from the fields. Vision based methods for weed detection are easily applied to well structured seedling farms, where the weed is detected between the rows of seedlings, although they can give promising results even in more traditional and low scale farming [18].

Significant improvements in precision agriculture sensing technologies that allow significant improvements in the methods used for soil and plant treatment were analyzed in [27]. New sensing technologies which use spectral imaging with high spatial accuracy and allow continuous recording of data are becoming available. They can be utilized for near real time decision making in precision agriculture.

An approach for mapping the fields using UAVs with multispectral camera in visible and near infrared range is described in [29]. The mapping process generated a weed infestation map and the average reported accuracy is 86%. Authors in [4] described a system for weed, crop and soil percentage evaluation in images under different light and weather conditions. Their system allowed choosing of the best method for evaluation based on the images. The different color indexes were evaluated in [25] for plant biomass identification from images.

In [34] authors described an approach for small-grain weed species discrimination with special regard to two types of weed: Cirsium arvense and Galium aparine. Feature ranking algorithms were applied for selection of the most informative features and three different Support Vector Machines (SVM) models were used for classification. The authors concluded that it is feasible to use image processing and classification to detect and map weeds in the field.

The accuracy of ground placed optoelectronic sensors for weed detection is evaluated in [1]. Authors compared the that data with data obtained from image processing and concluded that optoelectronic sensors could be used for inter-row weed detection and for building a cheap system for generation of maps of inter-row weeds. Machine learning approaches for weed identification from images were examined in [46]. Support Vector Machines were compared with other approaches and the results of the identification were evaluated. One of the most important tasks for specific plant targeting systems is the segmentation of the plant species. In [15] an overview of image processing techniques for plant species segmentation from images is provided.

The study presented in [48] describes the usage of UAV for mapping of weed patches using multi-spectral camera from different altitudes in Sunflower crops. Different color indexes for their ability to discriminate between plants, weed and bare soil from images recorded at different altitudes were compared. Authors conclude that it is possible to define a flight plan and a configuration to obtain optimal results in terms of weed-plant discrimination with desired spatial resolution.

Any sensing technology available should offer high enough precision to be useful in practical implementations. According to the results presented in [14], UAVs flying at altitudes of 30 m to 100 m could provide very high spatial resolution ortho-images with geo-referencing accuracy. This is required for mapping small weeds in wheat fields at very early phenological stage, which is also important for the process of early site-speciic weed management.

Another hyper-spectral sensing approach was examined in [43]. Hyperspectral sensing to detect the damages from herbicide drift on cotton growth was examined. It was concluded that hyper-spectral sensing is a good non-destructive alternative for yield prediction after simulated herbicide drift.

To overcome the differences in different fields, authors in [50] propose a self-supervised training for unsupervised learning of weed appearance model for hyper-spectral crop/weed discrimination with prior knowledge of the seeding patterns only. Authors use unmanned ground vehicles for the image acquisition and for the experiments.

Deep Convolutional Networks are a very powerful and trending method for machine learning and pattern recognition. They were employed for crop/weed discrimination in [31] based on images from multispectral camera mounted on a ground vehicle. The reported accuracy for pixel-wise classification is 97.4% and blob-wise classification is up to 98.7%. These results show that the usage of deep learning approaches could be applied for crop/weed discrimination.

Based on the analyzed studies, in Table 2 we list the main characteristics of some of the proposed sensing technologies, such as accuracy, specific weed type targeting and false positive rate, where they are applicable and available. Other reviewed papers take into consideration other important issues in sensing, such as geo-spatial detection of plants, plant/soil segmentation, etc.

All of the reviewed approaches use machine vision and machine learning for automated weed detection. The usage of UAVs is becoming a necessity for initial land mapping and information gathering. When using UAVs it is important to be able to detect the weed infestation and objectively assessing it without

Table 2. Characteristics of proposed sensing approaches

Reference	Reported max accuracy	False positives	Specific weeds	Multiple weeds
[45]	92%	Not reported, calculated	Yes	No
[29]	86%	Calculated, reported	No	Yes
[34]	97.74%	Calculated, reported	Yes	Yes
[1]	83%	Calculated, reported	No	Yes
[31]	98.7%	Not reported, calculated	No	Yes

underestimating the weed infestation in the observed regions. This factor should be reported when performing an experiment with UAVs sensing approaches.

Some approaches used automated land vehicles to directly detect and remove weed on site. All of the above mentioned approaches have certain precision reported, even though not all of the approaches report the false positive weed detection rate, where a useful plant is incorrectly recognized as weed. This rate is a must when designing a fully automated weed removal system, especially because it would mechanically or chemically target specific plants directly. This is a particularly important problem in the early growth phases of the young seedlings. Authors must address this in future weed detection approaches. Another important characteristic, especially for patch spraying based weed detection systems, is the understatement of the weed quantity.

Another challenge is the plant species detection for even more precise treatment. While distinguishing the useful plant from the weed is a challenge by itself, recognizing the plant species is even more difficult. The reason for that is because the two class classification problem becomes a multi-class classification problem. Very few approaches exist that can detect a specific weed type directly for direct plant treatment and specific plant type recognition. To perform this at different stages of growth is a very challenging task that will need to be addressed in the future. A fully automated weed control system would also need an adaptive learning capability where the system would be able to adapt to different circumstances during the detection process. Such system would need to be capable of on-line learning, as well.

3.3   Actuators that Allow Specific Weed Targeting

Several studies discuss new technologies that allow specific weed targeting using mechanical limitations of the sprayers and mechanical actuators that remove the weeds on site. One such example is presented in [12], where specialized hardware was used to limit the spraying of herbicides only in between rows, which in turn reduces the plant pollution with herbicides.

The most widely used and state of the art mechanical weeders with mechanical tools to exterminate weed in between rows of seedlings are described in [49]. According to that study, there is a variety of available commercial machines that can be used for mechanical weeding, however their speed and usefulness is limited compared to than machines that use herbicides. Additionally, they require a favorable land and cannot work on stony or thick lands due to limitations of the tools. The need for improvement of low cost vision sensors in order to make weeding systems more robust is also highlighted.

Another study presented in [47] used computer controlled hydraulic disks to eliminate weed between the plants in the plant rows. The described approach used machine vision and special kind of disks that allow the weed around the crops to be reduced by 62%–87% with minimal damage to the crops. It achieved higher speed than manual weed removal, but it still achieves lower speed than the commercial standards.

While being a potential cheaper alternative, mechanical weed removal has several drawbacks, especially when considering the irregularities of the land and the demand of strict plant seedling placement to allow the usage of such approaches.

All of the above-mentioned technological advancements are already present and some of them are in pre-production phase or even available on the market. According to [3], several robotic systems exist that are already in pre-production phase and that the market demand for agricultural robots is growing. Most of the technological challenges have already been resolved, however there are still some security and other types of challenges that need to be assessed, before the robots are placed on the market.

One such challenge is the application of actuators for mechanical treatment of plants that use active moving parts. These moving parts need to have embedded safety mechanisms that could protect unwanted damages and hazards. The automated system should be able to detect the presence of a living being and stop. Because the automated system will probably be connected to a computer network for additional data acquisition, the problem of network security should also be addressed properly. Any device connected to a network is potentially exposed to external attacks that could use the device for malicious purposes. So far, these issues have not been adequately studied by the scientific communities.

3.4 Challenges in New Technology Adoption

While in bigger corporate farms, new technologies are quickly adopted, mid-sized and small-sized organic farms usually have difficulties adopting new technologies and trends. The problems of using high-end technologies for weed detection in fields are recognized and discussed in [22]. This study identifies: that there is a need of education of farmers about the new technologies; that the technologies used are still too expensive; and recognizes the need for accurate weed maps and the use of robots for weeding, which is still a challenge. According to this study, the solution lies in education of farmers, high quality UAV maps, investments in weed robots that would commercialize the existing prototypes, lowering the technology costs by introducing cheaper approaches and standards and the need of multidisciplinary teams that would work on the problem.

According to [39], due to the rapid development in robotics, nanotechnologies, molecular biology and information technologies, it is very hard to predict the future of weed control technologies. There are very dynamic changes in all research areas and also in the field conditions which are influenced by many man-caused and natural factors such as climate change and herbicide resistance. In such dynamic environment, agricultural production and technology must adopt at a very fast pace.

The study presented in [44] reviews the diversity in weed management used in organic farms in northwest United States. The influence of several factors on the diversity of applied weed management techniques were analyzed. The authors determined that the probability of diverse techniques applied for weed controls increases with the education level of the farmers, the awareness of the farmers that weed presence decreases yield, the size of the organic fields, etc. To better assess the problem of weed control in farms, researchers must bear in mind the diverse requirements of farmers.

The challenges of applying precision agriculture on grass lands is discussed in [35]. While the economic impact of such endeavor remains uncertain, the authors recognize the need for adequate sensing technologies to be able to gather more information from grass lands, which in turn can be used in the process of decision making.

The study [51] presents the results of a simulated study on 16 fields. It also compared site specific weed management with one, two or more herbicides with broadcast application of herbicides. Authors concluded that there is a minor economic benefit of site specific management due to the increased costs for the management of the fields, but the benefit is dependent on the type of fields. They also concluded that identifying the characteristics of the weed population in the fields would be beneficial because specific herbicides can be applied to specific patches.

An analysis of the profitability of the adaptation of precision agriculture technologies is presented in [36]. According to this study, precision agriculture has an impact on the net profit, while also positively influencing the implementation of precision agriculture.

The benefits of usage of herbicides and pesticides for the increase in yield per land unit are reviewed in [19]. They conclude that most of the yield increases are due to the more efficient pest treatments. The study also identifies the implications on both the environment and the human health from increased usage of conventional herbicides and pesticides. The authors describe the new legislation of the European Union for herbicide and pesticide reduction and determine that there is an opportunity to reduce the usage significantly by means of advanced technologies and integrated management.

There are, however, certain challenges that need to be addressed especially from the evident climate changes. Further research is needed to predict the possible conditions under which the crops production would be executed. Authors in [28] discussed the main implications of climate change and proposed a method for developing Life Cycle Assessment scenarios that could deal with the uncertainty introduced by the climate changes.

4   Conclusion

The application of precision agriculture methods in the agricultural food production is evident. All of the reviewed papers show significant reduction in the quantity of herbicides used and they conclude that the application of precision agriculture approaches for weed control has several benefits. The environmental benefit is that the usage of pollutants can be significantly reduced, while the plants collected from the fields sustain reduced pollution and are healthier to consume. The weed decreasing is proven to increase the yield per land unit and is the obvious consequence of the weed control. Finally, the economic benefits are reduced costs for chemical or mechanical weed removal by applying the adequate amount of effort or quantity of herbicides on specific sites instead of broadcasting approach on the full field.

Site specific applications or even per-plant direct applications introduce additional management costs. New technologies for precise identification of specific weed types, such as image processing systems for plant identification and autonomous robot vehicles for specific plant treatment, can be utilized. They could further lead to savings in applied herbicide quantities and reducing the pollution of the useful crops, which would justify the additional costs. Several challenges still need to be overcome, such as: the price decrease of the sensing technologies and the available mechanization and increase of its effectiveness by increasing the land area they can process per hour.

Based on all of the reviewed papers, a fully automated weed control system is possible within the next decade. Most of the technologies are mature enough to be implemented in a real system. While such system is possible, several challenges still exist that need to be properly addressed. Further improvements are necessary for the weed detection systems and many security risks still exist and need to be resolved.

Overall, we can conclude that the weed control is becoming autonomous and that precision agriculture methods improve the production quality and yield cleaner and healthier products. Farmers must adapt to the changes because even the lawmakers are gaining on the trend to introduce cleaner organic farm production in every aspect of the production process. This is especially valid in the weed control process where the reduction of herbicide usage is essential, and an already set long-term target in the developed countries.

Acknowledgements. This work was partially financed by the Faculty of Computer Science and Engineering at the Sts. Cyril and Methodius University in Skopje, Macedonia.

References

1. Andújar, D., Ribeiro, Á., Fernández-Quintanilla, C., Dorado, J.: Accuracy and feasibility of optoelectronic sensors for weed mapping in wide row crops. Sensors **11**(3), 2304–2318 (2011)
2. Berge, T., Goldberg, S., Kaspersen, K., Netland, J.: Towards machine vision based site-specific weed management in cereals. Comput. Electron. Agric. **81**, 79–86 (2012). https://doi.org/10.1016/j.compag.2011.11.004
3. Bogue, R.: Robots poised to revolutionise agriculture. Ind. Robot: Int. J. **43**(5), 450–456 (2016). https://doi.org/10.1108/ir-05-2016-0142
4. Burgos-Artizzu, X.P., Ribeiro, A., Tellaeche, A., Pajares, G., Fernández-Quintanilla, C.: Improving weed pressure assessment using digital images from an experience-based reasoning approach. Comput. Electron. Agric. **65**(2), 176–185 (2009). https://doi.org/10.1016/j.compag.2008.09.001
5. Castaldi, F., Pelosi, F., Pascucci, S., Casa, R.: Assessing the potential of images from unmanned aerial vehicles (UAV) to support herbicide patch spraying in maize. Precis. Agric. **18**, 1–19 (2016). https://doi.org/10.1007/s11119-016-9468-3
6. de Castro, A.I., Jurado-Expósito, M., Peña-Barragán, J.M., López-Granados, F.: Airborne multi-spectral imagery for mapping cruciferous weeds in cereal and legume crops. Precis. Agric. **13**(3), 302–321 (2012). https://doi.org/10.1007/11119-011-9247-0
7. Christensen, S., et al.: Site-specific weed control technologies. Weed Res. **49**(3), 233–241 (2009)
8. Conesa-Muñoz, J., Valente, J., del Cerro, J., Barrientos, A., Ribeiro, Á.: Integrating autonomous aerial scouting with autonomous ground actuation to reduce chemical pollution on crop soil. Robot 2015: Second Iberian Robotics Conference. AISC, vol. 418, pp. 41–53. Springer, Cham (2016). https://doi.org/10.1007/978-3-319-27149-1_4
9. Popa, C.: Adoption of artificial intelligence in agriculture. Bull. Univ. Agric. Sci. Vet. Med. Cluj-Napoca. Agric. 68(1) (2011)
10. Dammer, K.H.: Real-time variable-rate herbicide application for weed control in carrots. Weed Res. **56**(3), 237–246 (2016). https://doi.org/10.1111/wre.12205
11. Dammer, K.H., Wartenberg, G.: Sensor-based weed detection and application of variable herbicide rates in real time. Crop Prot. **26**(3), 270–277 (2007). https://doi.org/10.1016/j.cropro.2005.08.018. Weed Science in Time of Transition
12. Davis, A.M., Pradolin, J.: Precision herbicide application technologies to decrease herbicide losses in furrow irrigation outflows in a northeastern australian cropping system. J. Agric. Food Chem. **64**(20), 4021–4028 (2016)
13. Gerhards, R., et al.: Using precision farming technology to quantify yield effects attributed to weed competition and herbicide application. Weed Res. **52**(1), 6–15 (2012)
14. Gómez-Candón, D., De Castro, A.I., López-Granados, F.: Assessing the accuracy of mosaics from unmanned aerial vehicle (UAV) imagery for precision agriculture purposes in wheat. Precis. Agric. **15**(1), 44–56 (2014). https://doi.org/10.1007/s11119-013-9335-4
15. Hamuda, E., Glavin, M., Jones, E.: A survey of image processing techniques for plant extraction and segmentation in the field. Comput. Electron. Agric. **125**, 184–199 (2016)
16. Holst, N., Rasmussen, I.A., Bastiaans, L.: Field weed population dynamics: a review of model approaches and applications. Weed Res. **47**(1), 1–14 (2007). https://doi.org/10.1111/j.1365-3180.2007.00534.x

17. Jeon, H.Y., Tian, L.F.: Direct application end effector for a precise weed control robot. Biosyst. Eng. **104**(4), 458–464 (2009)

18. Lameski, P., Zdravevski, E., Kulakov, A.: Weed segmentation from grayscale tobacco seedling images. In: Rodić, A., Borangiu, T. (eds.) RAAD 2016. AISC, vol. 540, pp. 252–258. Springer, Cham (2017). https://doi.org/10.1007/978-3-319-49058-8_28

19. Lamichhane, J.R., Dachbrodt-Saaydeh, S., Kudsk, P., Messéan, A.: Toward a reduced reliance on conventional pesticides in European agriculture. Plant Dis. **100**(1), 10–24 (2015). https://doi.org/10.1094/PDIS-05-15-0574-FE

20. Loghavi, M., Mackvandi, B.B.: Development of a target oriented weed control system. Comput. Electron. Agric. **63**(2), 112–118 (2008). https://doi.org/10.1016/j.compag.2008.01.020

21. Longchamps, L., Panneton, B., Reich, R., Simard, M.J., Leroux, G.D.: Spatial pattern of weeds based on multispecies infestation maps created by imagery. Weed Sci. **64**(3), 474–485 (2016). https://doi.org/10.1614/WS-D-15-00178.1

22. López-Granados, F.: Weed detection for site-specific weed management: mapping and real-time approaches. Weed Res. **51**(1), 1–11 (2011)

23. Luck, J., Zandonadi, R., Luck, B., Shearer, S.: Reducing pesticide over-application with map-based automatic boom section control on agricultural sprayers. Trans. ASABE **53**(3), 685–690 (2010)

24. Martinez, R.C., Gonzalo, E.R., Laespada, M.E.F., San Roman, F.J.S.: Evaluation of surface-and ground-water pollution due to herbicides in agricultural areas of zamora and salamanca (Spain). J. Chromatogr. A **869**(1), 471–480 (2000)

25. Meyer, G.E., Neto, J.C.: Verification of color vegetation indices for automated crop imaging applications. Comput. Electron. Agric. **63**(2), 282–293 (2008). https://doi.org/10.1016/j.compag.2008.03.009

26. Midtiby, H.S., Mathiassen, S.K., Andersson, K.J., Jørgensen, R.N.: Performance evaluation of a crop/weed discriminating microsprayer. Comput. Electron. Agric. **77**(1), 35–40 (2011). https://doi.org/10.1016/j.compag.2011.03.006

27. Mulla, D.J.: Twenty five years of remote sensing in precision agriculture: Key advances and remaining knowledge gaps. Biosyst. Eng. **114**(4), 358–371 (2013). Special Issue: Sensing Technologies for Sustainable Agriculture

28. Niero, M., Ingvordsen, C.H., Jørgensen, R.B., Hauschild, M.Z.: How to manage uncertainty in future life cycle assessment (LCA) scenarios addressing the effect of climate change in crop production. J. Cleaner Prod. **107**, 693–706 (2015). https://doi.org/10.1016/j.jclepro.2015.05.061

29. Peña, J.M., Torres-Sánchez, J., de Castro, A.I., Kelly, M., López-Granados, F.: Weed mapping in early-season maize fields using object-based analysis of unmanned aerial vehicle (UAV) images. PLoS One **8**(10), e77151 (2013)

30. Plant, R.E., Pettygrove, G.S., Reinert, W.R.: Precision agriculture can increase profits and limit environmental impacts. California Agric. **54**(4), 66–71 (2000)

31. Potena, C., Nardi, D., Pretto, A.: Fast and accurate crop and weed identification with summarized train sets for precision agriculture. In: Chen, W., Hosoda, K., Menegatti, E., Shimizu, M., Wang, H. (eds.) IAS 2016. AISC, vol. 531, pp. 105–121. Springer, Cham (2017). https://doi.org/10.1007/978-3-319-48036-7_9

32. Ray, D.K., Mueller, N.D., West, P.C., Foley, J.A.: Yield trends are insufficient to double global crop production by 2050. PloS one **8**(6), e66428 (2013)

33. Ritter, C., et al.: An on-farm approach to quantify yield variation and to derive decision rules for site-specific weed management. Precis. Agric. **9**(3), 133–146 (2008). https://doi.org/10.1007/s11119-008-9061-5

34. Rumpf, T., Römer, C., Weis, M., Sökefeld, M., Gerhards, R., Plümer, L.: Sequential support vector machine classification for small-grain weed species discrimination with special regard to Cirsium arvense and Galium aparine. Comput. Electron. Agric. **80**, 89–96 (2012). https://doi.org/10.1016/j.compag.2011.10.018

35. Schellberg, J., Hill, M.J., Gerhards, R., Rothmund, M., Braun, M.: Precision agriculture on grassland: applications, perspectives and constraints. Eur. J. Agron. **29**(2–3), 59–71 (2008). https://doi.org/10.1016/j.eja.2008.05.005

36. Schimmelpfennig, D., et al.: Farm profits and adoption of precision agriculture. Technical report, United States Department of Agriculture, Economic Research Service (2016)

37. Schuster, I., Nordmeyer, H., Rath, T.: Comparison of vision-based and manual weed mapping in sugar beet. Biosyst. Eng. **98**(1), 17–25 (2007). https://doi.org/10.1016/j.biosystemseng.2007.06.009

38. Søgaard, H., Lund, I.: Application accuracy of a machine vision-controlled robotic micro-dosing system. Biosyst. Eng. **96**(3), 315–322 (2007). https://doi.org/10.1016/j.biosystemseng.2006.11.009

39. Shaner, D.L., Beckie, H.J.: The future for weed control and technology. Pest Manage. Sci. **70**(9), 1329–1339 (2014). https://doi.org/10.1002/ps.3706

40. Slaughter, D., Giles, D., Downey, D.: Autonomous robotic weed control systems: a review. Comput. Electron. Agric. **61**(1), 63–78 (2008)

41. de Soto, M.G., Emmi, L., Perez-Ruiz, M., Aguera, J., de Santos, P.G.: Autonomous systems for precise spraying - evaluation of a robotised patch sprayer. Biosyst. Eng. **146**, 165–182 (2016). https://doi.org/10.1016/j.biosystemseng.2015.12.018. Special Issue: Advances in Robotic Agriculture for Crops

42. Srbinovska, M., Gavrovski, C., Dimcev, V., Krkoleva, A., Borozan, V.: Environmental parameters monitoring in precision agriculture using wireless sensor networks. J. Clean. Prod. **88**, 297–307 (2015). https://doi.org/10.1016/j.jclepro.2014.04.036. Sustainable Development of Energy, Water and Environment Systems

43. Suarez, L., Apan, A., Werth, J.: Hyperspectral sensing to detect the impact of herbicide drift on cotton growth and yield. ISPRS J. Photogram. Remote Sens. **120**, 65–76 (2016). https://doi.org/10.1016/j.isprsjprs.2016.08.004

44. Tautges, N.E., Goldberger, J.R., Burke, I.C.: A survey of weed management in organic small grains and forage systems in the northwest united states. Weed Sci. **64**(3), 513–522 (2016). https://doi.org/10.1614/WS-D-15-00186.1

45. Tellaeche, A., BurgosArtizzu, X.P., Pajares, G., Ribeiro, A., Fernández-Quintanilla, C.: A new vision-based approach to differential spraying in precision agriculture. Comput. Electron. Agric. **60**(2), 144–155 (2008). https://doi.org/10.1016/j.compag.2007.07.008

46. Tellaeche, A., Pajares, G., Burgos-Artizzu, X.P., Ribeiro, A.: A computer vision approach for weeds identification through support vector machines. Appl. Soft Comput. **11**(1), 908–915 (2011). https://doi.org/10.1016/j.asoc.2010.01.011

47. Tillett, N., Hague, T., Grundy, A., Dedousis, A.P.: Mechanical within-row weed control for transplanted crops using computer vision. Biosyst. Eng. **99**(2), 171–178 (2008)

48. Torres-Sánchez, J., López-Granados, F., De Castro, A.I., Peña-Barragán, J.M.: Configuration and specifications of an unmanned aerial vehicle (UAV) for early site specific weed management. PLOS One **8**(3), 1–15 (2013). https://doi.org/10.1371/journal.pone.0058210

49. Van der Weide, R., Bleeker, P., Achten, V., Lotz, L., Fogelberg, F., Melander, B.: Innovation in mechanical weed control in crop rows. Weed Res. **48**(3), 215–224 (2008)

50. Wendel, A., Underwood, J.: Self-supervised weed detection in vegetable crops using ground based hyperspectral imaging. In: 2016 IEEE International Conference on Robotics and Automation (ICRA), pp. 5128–5135, May 2016. https://doi.org/10.1109/ICRA.2016.7487717

51. Wiles, L.: Beyond patch spraying: site-specific weed management with several herbicides. Precis. Agric. **10**(3), 277–290 (2009)

Stories for Images-in-Sequence by Using Visual and Narrative Components

Marko Smilevski[1,2]([⊠]), Ilija Lalkovski[2], and Gjorgji Madjarov[1,3]

[1] Ss. Cyril and Methodius University, Skopje, Macedonia
marko.smilevski@webfactory.mk, gjorgji.madjarov@finki.ukim.mk
[2] Pendulibrium, Skopje, Macedonia
ilija@webfactory.mk
[3] Elevate Global, Skopje, Macedonia

Abstract. Recent research in AI is focusing towards generating narrative stories about visual scenes. It has the potential to achieve more human-like understanding than just basic description generation of images-in-sequence. In this work, we propose a solution for generating stories for images-in-sequence that is based on the Sequence to Sequence model. As a novelty, our encoder model is composed of two separate encoders, one that models the behaviour of the image sequence and other that models the sentence-story generated for the previous image in the sequence of images. By using the image sequence encoder we capture the temporal dependencies between the image sequence and the sentence-story and by using the previous sentence-story encoder we achieve a better story flow. Our solution generates long human-like stories that not only describe the visual context of the image sequence but also contains narrative and evaluative language. The obtained results were confirmed by manual human evaluation.

Keywords: Visual storytelling · Deep learning · Vision-to-language

1 Introduction

Storytelling is one of the oldest and most important activities known to mankind. It predates writing and for a long time, it was the only way to pass knowledge to the next generations. Storytelling is often used as a technique to unfold the narrative of a story, that describes scenes or activities. Storytelling is closely related to sight because visual context and narrative are the core inspiration for stories. Over the last few years, the improvements in computer vision have enabled the machines to "see" and generate labels about given images. This has allowed researchers to gain significant progress in the field of image captioning, whose goal is to generate a description for a given image, and video sequence

---

This research was partially funded by Pendulibrium and the Faculty of computer science and engineering, Ss. Cyril and Methodius University in Skopje.

S. Kalajdziski and N. Ackovska (Eds.): ICT 2018, CCIS 940, pp. 148–159, 2018.
https://doi.org/10.1007/978-3-030-00825-3_13

description, which is generating a description for a sequence of images. The next degree of reasoning, in terms of generating text from a sequence of images, is generating a narrative story about a sequence of images, that depicts events that are happening consequently. This means that we shift our focus from generating a description, to generating a story. This problem in machine learning is also known as visual storytelling. Storytelling is more complex than plain description because it uses more abstract and evaluative language for the activities in the images. This means that our goal is to generate sentences like "They enjoyed their dinner" instead of "Four people are sitting by the table".

We introduce a novel solution that is based on the Sequence to Sequence [27] model. The model generates stories, sentence by sentence with respect to the sequence of images and the previously generated sentence. The architecture of our solution consists of an image sequence encoder that models the sequential behaviour of the images, a previous-sentence encoder and a current-sentence decoder. The previous-sentence encoder encodes the sentence that was associated with the previous image and the current-sentence decoder is responsible for generating a sentence for the current image of the sequence. We also introduce a novel way of grouping the images of the sequence during the training process, in order to encapture the effect of the previous images in the sequence. Our goal with this approach was to create a model that will generate stories that contain more narrative and evaluative language and that every generated sentence in the story will be affected not only by the sequence of images but also by what has been previously generated in the story.

2   Related Work

In the last couple of years, research in the domain of vision to language has grown exponentially. In particular, research is divided into three sub-categories: Description of images-in-isolation, Description of images-in-sequence and Story for images-in-sequence.

2.1   Description of Images-in-isolation

Description of images-in-isolation is the problem of generating a textual description for an image. This category is represented by image captioning. Current caption generation research focuses mainly on concrete conceptual image descriptions of elements directly depicted in a scene [12]. Image captioning is a task whose input is static and non-sequential (an image rather than, say, a video), whereas the output is sequential (a multi-word text), in contrast to non-sequential outputs such as object labels [7]. An extensive overview of the datasets available for image captioning is provided by [3]. The three biggest datasets are MS COCO [17], SBU1M Captions [20], Deja-Image Captions [4]. Work done by [14] and [29] has achieved state-of-the-art results in image captioning. The deep architecture that these two papers suggest uses a pre-trained CNN such as

AlexNet [16] or VGG [25] to extract the features from the image that is captioned. The activation layer from the pre-trained network is then used as an input feature in the caption generator. The caption generator uses a language model to model the captions in the form of a vanilla recurrent neural network or a long-short term memory recurrent neural network. These architectures model caption generation as a process of predicting the next word in a sequence. Other significant research [2,8,19,23] has been done in image-based question answering. Datasets have been introduced by [19] and [2] where images have been combined with question-answer pairs. The proposed solutions for this problem are similar to image captioning. They have only an additional RNN for modeling the question. The combination of image features from the CNN and the question features extracted with RNN are used as an input in the RNN that generates the answers.

2.2 Description of Images-in-sequence

Description of images-in-sequence covers the problems that are related to generating a description about a sequence of images or a video. The main focus of this type of multi-frames to sentence modelling is to capture the temporal dynamics of an image sequence and map them to a variable-length of words. There is no benchmark dataset for this type of caption generation, but some of the most frequently used datasets are the Youtube2Text dataset [10], Microsoft Research Video Description Corpus [1], the movie description datasets M-VAD [28], MPII-MD [24] and the UCF101 Dataset [26]. A common solution is a sequence to sequence modelling, where a pre-trained CNN is used for feature extraction from the images (that are part of the sequence) and an RNN is used to model the temporal behaviour of the sequence of image features. The approach proposed by Yao et al. [30] employs a 3D CNN to extract local action features from every image in the sequence and an attention-based LSTM to exploit the global structure of the sequence.

2.3 Stories for Images-in-Sequence

Stories for images-in-sequence explore the task of image streams to sentence sequences. Park et al. [21] tackle the problem of describing sequences of images with more narrative language. They introduce the NY (New York) and Disney datasets that they obtained from a vast user-generated resource of blog posts as text-image data. The blogs-posts were about people's experiences while visiting New York and Disneyland. Every image from the blog-post is followed by a really long story that may or may not include information about the visual context of the image. This is the problem of these two datasets and that is why Huang et al. [13] introduce the VIST (Visual Storytelling Dataset). VIST is better then the aforementioned NY and Disney datasets, because every image in the image sequence is paired with one sentence from the story. The baseline approach is based on a Sequence to Sequence model that encodes the image features (extracted using a pre-trained CNN) with a GRU recurrent neural network [6]. In their work, they encode 5 images and try to learn the 5 sentences that are associated with them all together. Other work that uses the VIST dataset is given by Yu et al. [31] and Liu et al. [18]. In [31], the authors propose a model composed of three hierarchically-attentive Recurrent Neural Networks to encode the album photos, select representative (summary) photos and compose the story. On the other hand in [18] they propose a solution where the model learns a semantic space by jointly embedding each photo with its corresponding contextual sentence/They present a novel Bidirectional Attention-based Recurrent Neural Network (BARNN) model, which can attend on the discovered semantic relation to produce a sentence sequence and maintain its consistency with the photostream.

3   Dataset

The dataset we used to train and test our model is the Visual Storytelling Dataset (VIST). VIST consists of 210,819 unique photos and 50,000 stories. The images were collected from albums on Flickr, using Flickr API.3. The albums included 10 to 50 images and all the images in an album are taken in a 48-h span. This enables the dataset to have "storyable" images. The stories were created by workers on Amazon Mechanical Turk, where the workers were instructed to choose five images from the album and write a story about them. Every story has five sentence-stories and every sentence-story is paired with its appropriate image. The dataset is split into 3 subsets, a training set (80%), a validation set (10%) and a test set (10%). All the words and interpunction signs in the stories are separated by a space character and all the location names are replaced with the word location. Also, all the names of people are replaced with the words male or female depending on the gender of the person. One of the problems is that the stories were created by people, so not all stories necessarily have a story flow and from our overview of the dataset, some stories are not even correlated with the sequence of images. Because of this, we do not expect perfect stories in our results, but we want our model to generate stories with narrative language. We also expect the stories to contain words that will describe the visual context of the image in the sequence.

4   Architecture of Proposed Solution

In order to model storytelling with narrative and visual components accurately, we should consider the human observation of creating stories for a sequence of images. When we see the first image, we start the story with a sentence that describes and evaluates the context of that particular image. For the next image in the sequence, we analyze the current image but we also consider the influence of the previous image because that's the only way we can preserve the temporal dependencies between the events in the images. Therefore for every image in the sequence, we consider the images that have happened in the past. Besides the previous images, it is important to preserve the temporal dependencies between the sentence-story generated for the previous image in the sequence (previous sentence-story) and the sentence-story generated for the current image in the sequence (current sentence-story). We achieve better story flow by considering the previous sentence-story, while we generate the sentence-story of the current image. Dissecting the way humans create stories, helped us conclude that the problem of generating stories for a sequence of images using visual and narrative components comes down to the way we model the sequence of images and the previous sentence-story.

The architecture that we propose in this paper is based on the Sequence to Sequence model [27] described in the previous sections. It incorporates encoder and decoder modules in the same fashion as the referred model. As a novelty, our encoder module is composed of two separate encoders, one that models the behaviour of the image sequence and other that models the sentence-story generated for the previous image in the sequence of images.

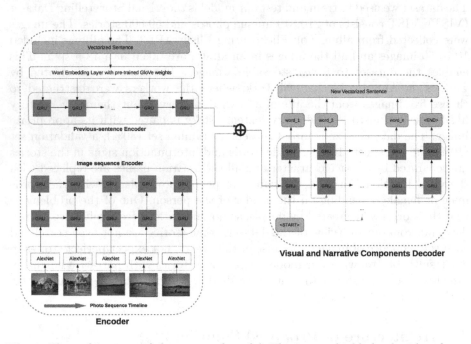

Fig. 1. The architecture of the proposed model. The images highlighted with red are the ones that are encoded and together with the previous sentence, they influence the generated sentence in the current time step.

Recurrent neural networks have been very successful in sequence modelling because they can learn temporal dependencies between the elements of sequential data and it has been proven that they are appropriate for modelling a sequence of image features vectors. The encoder that we propose for modelling the behaviour of the image sequence aligns every sentence from the story with a sequence of images. This means that a sentence-story is generated per image while considering an appropriate number of images from the sequence. Opposite to our solution, the authors of the VIST dataset [13] propose aligning of the image sequence with the entire story. This means that after the story is generated, it should be divided into sentence-stories and each sentence-story should be assigned to a particular image from the sequence, which could be a drawback.

The previous sentence-story encoder is also a recurrent neural network that learns the temporal behaviour of words in the sentence-story generated for the previous image. The two encoders of the proposed solution produce two fixed-length vector representation, one for the image-sequence and another for the previous sentence-story. In order to create a joint representation of the two encoders, we concatenate the vector representations. The concatenated vector is used as an initial hidden state of the decoder. In this way, we condition the decoder with the vector representations from the encoder module. The decoder module is a recurrent neural network that "translates" what the encoder module has produced.

During the training process, we feed the encoder model with the image sequences and the previous sentence-stories, while the story decoder model with the current sentence-stories. When it's time to generate a story, we feed the encoder model with an image sequence, a previous sentence-story and the decoder model with the <START> token. After the decoder generates a word, that word is the input in the next time step of the decoder. When we generate the whole sentence-story, we append it to the story and use it as the previous sentence-story for the next generative process. The architecture of the proposed solution can be seen in Fig. 1. The complete code and documentation of this project can be found on github[1].

5   Experimental Setup

We used the $fc7$ vectors from the AlexNet convolutional neural network [16] to describe the images. We chose AlexNet over other more precise convolutional neural networks because AlexNet is less computationally expensive than other deeper networks. First, we transformed every image from RGB to BGR and after that, we re-sized the images with respect to their ratio. Also, we cropped them centrally to fit the dimensions of the input layer of AlexNet, because we assumed that the important information in the image is placed in its centre.

The vocabulary that we created is composed of the most frequent words (words that appear at least 4 times in the stories). Also, we added <NULL>, <START>, <END> and <UNK> tokens to the vocabulary. After creating the vocabulary we decided that all the sentences would have a length of 20 words. We chose this number because most of the sentences had a length of 3 or 20 words. This meant that we would limit the longer sentences to 20 words and fill the shorter sentences with <NULL> token. We added the <START> token in front of every sentence and added the <END> token

---

[1] https://github.com/Pendulibrium/ai-visual-storytelling-seq2seq.

at their end. Every word that appears less than 4 times was substituted with the <UNK> token.

Before the sentence vectors entered the previous sentence-story encoder and the current sentence-story decoder, they passed through an embedding layer. The embedding layer used pre-trained word vectors obtained from the GloVe model [22]. This transformed the sentence, from a vector of 22 words (two words for the <START> and <END> tokens) to a vector of 22-word representations.

We experimented with an LSTM [11] and GRU [5] recurrent network. There was no difference in the results, but because GRUs are less computationally expensive our solution uses GRUs. We also used two stacked GRUs together for the image sequence encoder and decoder, because stacking recurrent neural networks helps us model more complex sequences. Recurrent neural networks are inherently deep in time since their hidden state is a function of all previous hidden states, but they benefit from increasing their depth in space just like conventional deep networks do from stacking feedforward layers [9].

After various experiments with the size of the GRU units for the components, we concluded that the best results were achieved when the GRU units for the image sequence encoder have 1024 neurons and the GRU unit for the previous-sentence encoder has 512 neurons. The encoders were set up in this way because it allowed the image sequence to have more impact on the generated sentence. Because of the concatenation of the outputs of the encoders, the GRU units for the decoder have 1536 neurons.

Categorical cross entropy was used as a loss function, because it is the preferred loss function in Neural Machine Translation. The learning rate was set to 0.0001, because with greater learning rate our network was unable to imporove during training. Adam algorithm was used as an optimization algorithm. Adam algorithm is computationally more efficient than stochastic gradient descent, has little memory requirements, and it is invariant to the diagonal rescaling of the gradients. Also, it is well suited for problems that are large in terms of data and/or parameters [15].

To reduce the overfitting of the neural network during the training process, we used dropout as regularization. After experimenting with dropout on the input layers and the layers within the recurrent neural networks, we achieved best results when we applied dropout of 0.3 on the input layer and 0.5 on the layer before the softmax layer. The last parameter we had to choose was how many images in the past we will consider given the current image. When we trained the model with all the previous images, the last sentence-story always represented a summarization of all the images in the sequence and that resulted in a very generic sentence that didn't give any sufficient information. The best results were obtained when we considered the last three images.

5.1   Evaluation Metrics

For evaluation of our generated stories, we used BLEU and METEOR score. These two metrics are usually used for evaluating Neural Machine Translation. The BLEU metric is designed to measure how close a generated translation is

Table 1. Results for the generated stories. The loss is calculated over the training set and the METEOR and BLEU score are calucated over the test set.

Models	Model1	Model2	Model3
Training loss	0.82	1.01	1.72
Number of epochs	50	30	19
BLEU score	24.5/9.0/3.2/1.3	26.0/9.7/3.6/1.5	26.4/10.1/3.8/1.6
METEOR score	23.0	23.9	23.9

to that of human reference translations. In our case, it measures how close a generated story is to the original. It is important to note that stories, generated or original, may differ significantly in word usage, word order, and phrase length. To address these complexities, BLEU attempts to match variable length phrases between the generated story and the original story. The METEOR method uses a sophisticated and incremental word alignment method that starts by considering exact word-to-word matches, word stem matches, and synonym matches. Alternative word order similarities are then evaluated based on those matches. These measures use a scale from 0 to 100 to quantify how similar the generated story is to the original based on a mechanical analysis of how many of the same words show up and how likely they are to appear in the same order. It has also been shown that a high score (as a result of a method which uses n-grams) probably indicates a good generation but a low score is not necessarily an indication of a poor generation. This was one of the major problems that we faced when evaluating our models. In reality, two people can create very different stories about the same sequence of images, and both stories can be valid because perception is subjective.

6   Results and Analysis

Results and Quantitative Analysis. In order to find the optimal solution, during the training of the models we did a quantitative analysis of the generated stories. The quantitative analysis was done by tracking the BLEU and METEOR scores of the trained models. In Table 1 we have presented the scores for three models (model1, model2, model3) that have the same network configuration (described in the previous section). The only difference between them is the number of epochs used for their training. After numerous experiments with the number of epochs used for training, we concluded that the model achieves the best scores when after training the loss (calculated over the training set) is between 0.82 and 1.72. Model1 has been trained for 50 epochs and has the smallest training loss, but it achieved the worst BLEU and METEOR scores out of the three models. This means that model1 is overfitting. Both model2 and model3 have a METEOR score of 23.9, but model3 achieved the best BLEU score. For comparison the baseline model provided in [13] achieved a METEOR score of 27.76. The difference in BLEU and METEOR scores between the model2

Fig. 2. In this figure, we can see the generated stories from the aforementioned models. The first row represents the original story, the second, third and fourth row are the generated story from the models respectively.

and model3 is very small and because of the aforementioned problems regarding the use of BLEU and METEOR scores as metrics for story generation, the results from the models had to be evaluated by human evaluators.

Qualitative Analysis. The human evaluation was done by the authors personally. From reading the generated stories and analyzing how they associate with their corresponding pictures we concluded that the best results were obtained by the model with loss of 1.01. We came to this conclusion because the generated stories by this model were better than the other models in terms of story flow and story length. Moreover, the generated stories from this model contained more words that described the visual context of the image sequence. The model with loss of 0.82 generated stories that had a lot of grammatical mistakes in them and we think that is happening because the model has over-fitted the training data. The model with a loss of 1.72 produced similar stories to the stories from the model with loss of 1.01, but it was slightly worse when it came about generating words that described the visual context of the image sequences.

Figure 2 shows the generated stories from the three models, for a given image sequence. More images with generated stories from the three models can be seen on github[2].

7 Conclusion

After a lot of experiments, we can conclude that the results from our proposed solution satisfied our expectations. The image-sequence encoder successfully learned the dependencies between the images and the proposed architecture was able to model the complex relations between the images and the stories. The improved story flow is a result of the inclusion of the previous sentence-story encoder. This encoder also contributed to the increase in the length of the generated stories. From the quantitative evaluation, it was obvious that metrics such as BLEU and METEOR are good for a distinction between really bad and supposedly good models. With the help of human evaluation, we concluded that most of the generated stories from our model made sense and looked like a story a human would tell. In order to improve our solution, in the future, we will focus on 3D convolutional neural networks for modelling the image sequences. Also, we will focus on the use of attention based models, because they will produce better alignment between the previous-sentence encoder and the decoder in our architecture.

References

1. Collecting Highly Parallel Data for Paraphrase Evaluation. Association for Computational Linguistics, January 2011. https://www.microsoft.com/en-us/research/publication/collecting-highly-parallel-data-for-paraphrase-evaluation/
2. Antol, S., et al.: VQA: visual question answering. CoRR abs/1505.00468 (2015). http://arxiv.org/abs/1505.00468

---

[2] https://github.com/Pendulibrium/ai-visual-storytelling-seq2seq/tree/master/results/images.

3. Bernardi, R., et al.: Automatic description generation from images: a survey of models, datasets, and evaluation measures. CoRR abs/1601.03896 (2016). http://arxiv.org/abs/1601.03896
4. Chen, J., Kuznetsova, P., Warren, D., Choi, Y.: Déjà image-captions: a corpus of expressive descriptions in repetition. https://doi.org/10.3115/v1/N15-1053. http://www.aclweb.org/anthology/N15-1053
5. Cho, K., van Merrienboer, B., Bahdanau, D., Bengio, Y.: On the properties of neural machine translation: encoder-decoder approaches. CoRR abs/1409.1259 (2014). http://arxiv.org/abs/1409.1259
6. Chung, J., Gülçehre, Ç., Cho, K., Bengio, Y.: Empirical evaluation of gated recurrent neural networks on sequence modeling. CoRR abs/1412.3555 (2014). http://arxiv.org/abs/1412.3555
7. Donahue, J., et al.: Long-term recurrent convolutional networks for visual recognition and description. CoRR abs/1411.4389 (2014). http://arxiv.org/abs/1411.4389
8. Gao, H., Mao, J., Zhou, J., Huang, Z., Wang, L., Xu, W.: Are you talking to a machine? Dataset and methods for multilingual image question answering. CoRR abs/1505.05612 (20150). http://arxiv.org/abs/1505.05612
9. Graves, A., Mohamed, A., Hinton, G.E.: Speech recognition with deep recurrent neural networks. CoRR abs/1303.5778 (2013). http://arxiv.org/abs/1303.5778
10. Guadarrama, S., et al.: YouTube2Text: recognizing and describing arbitrary activities using semantic hierarchies and zero-shot recognition (2013). http://www.cs.utexas.edu/users/ai-lab/pub-view.php?PubID=127409
11. Hochreiter, S., Schmidhuber, J.: Long short-term memory. Neural Comput. 9(8), 1735–1780 (1997). https://doi.org/10.1162/neco.1997.9.8.1735
12. Hodosh, M., Young, P., Hockenmaier, J.: Framing image description as a ranking task: data, models and evaluation metrics. J. Artif. Intell. Res. 47, 853–899 (2013)
13. Huang, T.K., et al.: Visual storytelling. CoRR abs/1604.03968 (2016). http://arxiv.org/abs/1604.03968
14. Karpathy, A., Li, F.: Deep visual-semantic alignments for generating image descriptions. CoRR abs/1412.2306 (2014). http://arxiv.org/abs/1412.2306
15. Kingma, D.P., Ba, J.: Adam: a method for stochastic optimization. CoRR abs/1412.6980 (2014). http://arxiv.org/abs/1412.6980
16. Krizhevsky, A., Sutskever, I., Hinton, G.E.: Imagenet classification with deep convolutional neural networks. In: Pereira, F., Burges, C.J.C., Bottou, L., Weinberger, K.Q. (eds.) Advances in Neural Information Processing Systems, vol. 25, pp. 1097–1105. Curran Associates, Inc., New York (2012). http://papers.nips.cc/paper/4824-imagenet-classification-with-deep-convolutional-neural-networks.pdf
17. Lin, T., et al.: Microsoft COCO: common objects in context. CoRR abs/1405.0312 (2014). http://arxiv.org/abs/1405.0312
18. Liu, Y., Fu, J., Mei, T., Chen, C.W.: Let your photos talk: generating narrative paragraph for photo stream via bidirectional attention recurrent neural networks. In: AAAI Conference on Artificial Intelligence, February 2017
19. Malinowski, M., Fritz, M.: A multi-world approach to question answering about real-world scenes based on uncertain input. CoRR abs/1410.0210 (2014). http://arxiv.org/abs/1410.0210
20. Ordonez, V., Kulkarni, G., Berg, T.L.: Im2Text: describing images using 1 million captioned photographs. In: Shawe-Taylor, J., Zemel, R.S., Bartlett, P.L., Pereira, F., Weinberger, K.Q. (eds.) Advances in Neural Information Processing Systems, vol. 24, pp. 1143–1151. Curran Associates, Inc., New york (2011). http://papers.nips.cc/paper/4470-im2text-describing-images-using-1-million-captioned-photographs.pdf

21. Park, C.C., Kim, G.: Expressing an image stream with a sequence of natural sentences. In: Cortes, C., Lawrence, N.D., Lee, D.D., Sugiyama, M., Garnett, R. (eds.) Advances in Neural Information Processing Systems 28, pp. 73–81. Curran Associates, Inc., Nwe york (2015). http://papers.nips.cc/paper/5776-expressing-an-image-stream-with-a-sequence-of-natural-sentences.pdf

22. Pennington, J., Socher, R., Manning, C.D.: GloVe: global vectors for word representation. In: Empirical Methods in Natural Language Processing (EMNLP), pp. 1532–1543 (2014). http://www.aclweb.org/anthology/D14-1162

23. Ren, S., He, K., Girshick, R.B., Sun, J.: Faster R-CNN: towards real-time object detection with region proposal networks. CoRR abs/1506.01497 (2015). http://arxiv.org/abs/1506.01497

24. Rohrbach, A., Rohrbach, M., Tandon, N., Schiele, B.: A dataset for movie description. CoRR abs/1501.02530 (2015). http://arxiv.org/abs/1501.02530

25. Simonyan, K., Zisserman, A.: Very deep convolutional networks for large-scale image recognition. CoRR abs/1409.1556 (2014). http://arxiv.org/abs/1409.1556

26. Soomro, K., Zamir, A.R., Shah, M.: UCF101: a dataset of 101 human actions classes from videos in the wild. CoRR abs/1212.0402 (2012). http://arxiv.org/abs/1212.0402

27. Sutskever, I., Vinyals, O., Le, Q.V.: Sequence to sequence learning with neural networks. CoRR abs/1409.3215 (2014). http://arxiv.org/abs/1409.3215

28. Torabi, A., Pal, C.J., Larochelle, H., Courville, A.C.: Using descriptive video services to create a large data source for video annotation research. CoRR abs/1503.01070 (2015). http://arxiv.org/abs/1503.01070

29. Vinyals, O., Toshev, A., Bengio, S., Erhan, D.: Show and tell: a neural image caption generator. CoRR abs/1411.4555 (2014). http://arxiv.org/abs/1411.4555

30. Yao, L., et al.: Describing videos by exploiting temporal structure (2015)

31. Yu, L., Bansal, M., Berg, T.L.: Hierarchically-attentive RNN for album summarization and storytelling. CoRR abs/1708.02977 (2017). http://arxiv.org/abs/1708.02977

Bioelectrical Impedance Technology in Sports Anthropometry: Segmental Analysis in Karate Athletes

Jasmina Pluncevic Gligoroska[1]([⊠]), Sanja Mancevska[1],
Beti Dejanova[1], and Dusana Cierna[2]

[1] Department of Physiology, Medical Faculty,
University Ss Cyril and Methodius, Skopje, Republic of Macedonia
jasnapg965@yahoo.com
[2] Phaculty of Physical Education and Sports,
Comenius University, Bratislava, Slovakia

Abstract. The modern equipment for evaluation of body composition use computerized technology to determine or estimate body components. Having a moderate amount of each component is important for healthy life. Quantification of fat has been prime focus of attention, but many coaches, sport scientists and sport physicians working with elite athletes recognize that knowledge of the amount and distribution of lean tissue, such as bone and muscle, can be just as important in determining sports performance. Bioelectrical impedance analysis (BIA) estimates the amount of total body water (TBW), fat free mass (FFM) and fat mass (FM) measuring the resistance of the body as conductor to a very small alternating electrical current. The investigated group was composed of twenty (20) elite level male karate athletes with the following characteristics (mean ± SD): age = 22.5 ±3.6 years, age span (18 to 27 years); height = 179.95 ± 2.3 cm; body mass = 77.5 ± 9.8 kg. Body composition was diagnosed with the InBody 720, multifrequency (1–1000 kHz) bioelectrical impedance analyzer (BIA). Karate athletes are obliged to maintain their body weight within certain range if they want to stay in optimal weight category. Our results showed that Macedonian karatees have symmetrical and balanced distribution between left and right side of the body. The strongest advantage of BIA methodology and InBody devices, compared to other field methods in sports anthropometry, is the segmental lean mass analysis. Monitoring the segmental analysis could help in following the quality of nutritional and training regime or rehabilitation procedure in athletes.

Keywords: Bioelectrical impedance analysis · Segmental analysis
Lean body mass · Karate

1 Introduction

Body composition represents an unbreakable unity of the humanbody basic structure elements and involves a relative representationof the various constituent elements of the humantotal body weight. The modern devices for evaluation of body composition

© Springer Nature Switzerland AG 2018
S. Kalajdziski and N. Ackovska (Eds.): ICT 2018, CCIS 940, pp. 160–171, 2018.
https://doi.org/10.1007/978-3-030-00825-3_14

use computerized technology to determine or estimate body components. Having a moderate amount of each component is important for healthy life [18]. The human body can be quantified at several levels, such as atomic, molecular or tissue, depending on the clinical concerns. Criterion methods measure certain properties of the body, like density or conductance, and measure or estimate the body structural elements [13].

The closest researchers can get to a direct measurement of body composition as far as accuracy goes is by using the multi–compartment model technique. These indirect methodsare also commonly said to be the best "reference" techniques. The reference method are by definition, the most accurate techniques for assessing body composition and have been often employed as criterion against which other methods compared [26]. Criterion methods include medical imaging techniques, computed tomography (CT), magnetic resonance imaging (MRI), densitometry and dual X ray absorptiometry (DEXA) [10]. Although some authors classified these methods as direct others define them as indirect. The group of the indirect methods for body composition analysis includes the bioelectrical impedance method. Bioelectrical impedance analysis (BIA) estimates the amount of total body water (TBW), fat free mass (FFM) and fat mass (FM) measuring the resistance of the body as conductor to a very small alternating electrical current [6, 27, 38].

The actual parameter measured with BIA is the voltage (V) that is produced between two electrodes located most often at sites near to, but different from the sites where current is introduced. The measurement normally is expressed as a ratio, V/I, who is also called impedance (Z). The measuring instrument is therefore called a bioelectrical impedance analyzer. Impedance has two components, resistance (R) and reactance (X). In BIA the resistance is nominally about 250 Ω, and reactance is about 10 percent of that amount, so the magnitude of Z is similar to that of R. In many BIA reports, Z and R are used as they are interchangeable [30]. Impedance (Z), from electrical point of view, is the obstruction to the flow of an alternating current and, is dependent on the frequency of the applied current. Bioimpedance is a complex quantity composed of resistance (R) which is caused by total body water and reactance (X_c) that is caused by the capacitance of the cell membrane [22]:

$$Z = R + jX_c \quad [22]$$

$$Z = \sqrt{R^2 + X} \quad [20, 28].$$

BIA provides a reliable estimate of total body water under most conditions. It can be a useful technique for body composition analysis in healthy individuals and in those with a number of chronic conditions such as mild-to-moderate obesity, diabetes mellitus, and other medical conditions in which major disturbances of water distribution are not prominent [23]. BIA values are affected by numerous variables including body position, hydration status, consumption of food and beverages, ambient air and skin temperature, recent physical activity, and conductance of the examining table. Reliable BIA requires standardization and control of these variables. A specific, well-defined procedure for performing routine BIA measurements is not practiced. Therefore, the scientific experts emphasized the needs of specific equations (for different population groups) and setting instrument standards and procedural methods.

1.1   The Importance of Body Composition Analysis in Athletes

Tests of anthropometry include measurements of body size, structure, and composition. There is a wide range of ideal body shapes and compositions, depending on the sports, the playing position and the fitness level. For many sports it is an advantage to be short, tall, heavy or light. In terms of ideal body composition, an athlete may wish to have anything from high muscularity to high fat levels.

In weight-sensitive sports many athletes use extreme methods to reduce mass rapidly or maintain a low body mass in order to gain a competitive advantage. Quantification of fat has been prime focus of attention, but many coaches and scientists working with elite athletes recognize that knowledge of the amount and distribution of lean tissue, such as bone and muscle, can be just as important in determining sports performance [1, 29]. The reference methods may have limited applicability for monitoring athletes. Limitations include feasibility (e.g. cadaver dissection), time and financial costs involving (e.g. MRI), a lack of published normative data and unnecessary radiation exposure (CT scanning) [10]. Several international sport federations have considered implementation of programmed aimed to discourage athletes from extreme dieting or from rapid mass loss by means of dehydration, and i n order to improve the low mass problem [32, 41].

1.2   Bioelectrical Impedance Analysis Technology

The ability of electrolyte solution to conduct an electric current contributes the human tissues to be the conductor. The tissues which are compound of greater percentage of water, consecutively electrolyte solution, are the major conductor of an electrical current. The tissues such as blood and urine are the best conductor, muscle tissue has moderate conductance properties and a fat tissue is among the poorest conductors [8]. The BIA measurements are performed using a couple of electrodes (two, four or eight) attached at certain position, usually wrists and ankles. The current which is generated from BIA device should be so "weak" to not harm the tissues and so "strong" to override the obstacles i.e. cell membranes. BIA applied small currents throughout the body and measures the voltage to get value called resistance also known as impedance. The principal behind BIA is to flow electrical currents throughout the water in the body and to measure the amount of resistance the current encounters as it travels. Simply more water will lead to lower impedance [18].

The meaning of the word impedance is the effective resistance of an electric circuit or component to alternating current, arising from the combined effects of ohmic resistance and reactance [11]. The electrical impedance is the measure of the opposition that a circuit presents to a current when a voltage is applied. Impedance extends the concept of resistant to alternating currentcircuits, and possesses both magnitude and phase, unlike resistance, which has only magnitude. When a circuit is driven with direct current (DC), there is no distinction between impedance and resistance; the latter can be thought of as impedance with zero phase angle. Impedance is a complex number, with the same units as resistance, the ohm (Ω). Its symbol is usually Z, and it may be represented by writing its magnitude and phase in the form $|Z|\angle\theta$ [32, 33]. To translate this data to a volume approximation, two basic assumptions are used. First, the body can be modeled as an isotropic cylindrical conductor with its length proportional to the subject's height (Ht). Second, the reactance (X) term contributing to the body's impedance (Z) is small, such that the resistance component (R) can be considered equivalent to body impedance. When these two assumptions are combined, it can be shown that the conducting volume is proportional to the term Ht^2/R, called the impedance index. It should be noted, however, that the human body is not a cylindrical conductor, nor are its tissues electrically isotropic, and the reactance component of the body's impedance is nonzero [2]. At 50 kHz, the body's impedance has both resistive and reactive components. The reactive component is assumed to be related to the portion of the current that passes through cells which act like capacitors that shift the voltage and current out of phase. In electrical terms, the phase angle (φ) is defined by the relationship: tan (φ) = X/R, where $Z^2 = R^2 + X^2$. In healthy adults, the phase angle at 50 kHz is usually in the range of 8–15° [7] but varies widely at high frequencies [34, 35]. Several investigators have used the phase angle to assess body composition in various clinical conditions. In renal patients, for example, the phase angle at 50 kHz is typically < 5° and has been interpreted as an indication of an expanded ECW space concurrent with a reduced ICW volume [3, 19].

Multi-Frequency Bioelectrical Impedance Analysis (MFBIA) utilizing frequencies between 1 kHz and 1000 kHz (1 MHz). An electric current less than 100 kHz cannot penetrate cell and flows through extracellular water so is used to measure extra-cellular water (ECW). An electric current over 100 kHz penetrates cell membranes and flows through cell so is used to measure total body water (TBW). Using multiple frequencies, ECW and TBW are measured separately and this can be helpful to diagnosis of body water balance, especially edema [30]. Different part of human body has significant difference on impedance. BIA use 6 testing circuits: Left Arm-Right Leg, Right Arm-Left Leg, Left Arm- Left Leg, Right Arm-Right Leg, Left Arm-Right Arm, Left Leg-Right Leg. Only this complex model can give real comprehensive impedance level of whole body.

2 Material and Method

2.1 Participants and Procedure

The sample was composed of twenty (20) elite level male karate athletes with the following characteristics (mean ± SD): age = 22.5 ±3.6 years, age span (18 to 27 years); height = 179.95±2.3 cm; body mass = 77.5 ± 9.8 kg. They have participated regularly in national and international karate eventsduring the period from 2010 until 2015 in the following weight categories: >84 kg (n = 3); <84 k (n = 12); <75 kg (n = 9); <67 kg (n = 3) and <60 kg (n = 1). The measurements were conducted at the Sport Medicine Laboratory, Institute of Physiology and Anthropology, at the Medical faculty in Skopje. The investigation protocol was conducted according to the decla-ration of Helsinki. All measurements were taken by experienced practitioners in the morning hours between 9 am and 11 am, from all subjects according to manufacturer's guideline. Ethical approval was obtained by the Ethics committee of Medical Faculty, UKIM, Skopje.

2.2 Assessment of Body Composition

Body composition was estimated by the InBody 720, multifrequency (1–1000 kHz) bioelectrical impedance analyzer (BIA). InBody 720 employs eight contact electrodes, which enable segmental analysis of the five basic body parts (upper and lower extremities and trunk). Two electrodes are positioned on the palm and thumb and another two on the front of the foot's heel. The measurement was performed under laboratory conditions according to the user manual instructions (Biospace, 2008).

2.3 Statistics

The statistical analysis was performed in Statistika 7.1 for Windows. All results were subjected to descriptive statistical analysis in order to define the basic measures of central tendency and dispersion of data (mean, SD, CI lower and upper bound, range) in numeric series. The Pearson coefficient of linear correlation (r) between measured parameters was obtained. Multiple regression analysis (R) was performed for the measured parameter "skeletal mass" as dependent variable and the parameters - right arm, left arm, trunk, right legand left legas independent variables. The level of significance was for $p < 0.05$.

3 Results

The results from the descriptive statistics for the general anthropometric characteristics and the BIA section for obesity diagnose of the karate athletes evaluated in this study are presented in Table 1. the participants' mean age was 22.5 ±3.6 years, the mean height was 179.95 ± 2.3 cm and the mean body weight was 77.5 ± 7.8 kg. Mean values for the body mass index (BMI), waist to hip ratio (WHR) and the absolute and relative body fat mass are also fully reported in Table 1.

Table 1. Body mass components and obesity diagnose BIA parameters in elite karate athletes.

	N	Mean	SD	CI -95%	CI +95%	min	max
Age (year)	20	22.5	3.6	19.13	21.84	18.0	29.0
Height (cm)	20	179.95	2.3	176.81	181.09	175.5	191.0
Weight (kg)	20	77.5	7.8	72.86	80.58	64.0	107.4
BMI	20	23.78	2.35	22.96	24.78	19.8	29.8
Fat free mass (FFM)	20	67.74	7.69	64.21	71.08	58.0	94.6
Soft lean mass (SLM)	20	64.11	7.36	60.66	67.15	54.7	89.3
Skeletal muscle mass	24	38.16	4.66	36.16	39.78	29.4	54.6
Body fat mass (kg)	20	9.84	3.65	8.39	11.25	4.8	19.4
Body fat percent (BF%)	20	13.51	4.81	11.04	14.78	6.6	21.7
Waist-to-hip ratio	20	0.83	0.04	0.81	0.85	0.75	0.90

Table 2. Segmental analysis variables in elite karate athletes (all parameters are in kg).

	N	Mean	Std. Dev.	CI −95%	CI +95%	min	max
Right arm	20	3.92	0.42	3.63	4.19	3.12	5.98
Left arm	20	3.91	0.55	3.63	4.18	3.13	5.89
Trunk	20	28.54	6.83	25.3	31.56	21.4	40.9
Right leg	20	11.12	1.27	9.95	12.04	9.18	14.3
Left leg	20	10.85	1.17	9.93	11.01	9.17	13.7

The segmental analysis obtained by InBody 720 showed lean mass distribution into trunk and left and right part of the body. Descriptive statistics for these parameters, which are available only in individuals older than 18 years, are presented in Table 3. The mean values of lower limbs show insignificantly higher values for right leg, and the mean values of right and left arm are almost the same (Table 2).

In order to find out how the one part of the body influence on the whole lean body mass we made regression analysis for skeletal mass as dependent and four limbs and trunk as independent variables. It was found that for $R = 0.99$ and $p < 0.05$ ($p = 0,000$) in the investigated relation is determined maximal positive correlation. The strongest influence on this relation has showed the left arm (Beta = 0.49), right arm ($\beta = 0.44$), left leg ($\beta = 0.09$), right leg ($\beta = 0.07$) and the weakest influence was from trunk ($\beta = -0.0006$). The Fig. 1 displayed positive linear correlation between right arm as independent and skeletal muscle as dependent variable ($R = 0.98$, $p < 0.05$).

Table 3. Correlation's analysis between segmental BIA variables as independent and skeletal muscle as dependent variable.

Regression summary for dependent variable: skeletal mass: $R = 0.9$; $F(5.15) = 155.32$ and $p < 0.001$

	Beta	Std. Err. Beta	B	Std. Err. B	t(15)	p-level
Intercept			7.41	1.87	3.95	**0.001**
Right arm	0.44	0.26	3.26	1.91	1.70	0.11
Left arm	0.49	0.23	3.82	1.78	2.14	**0.04**
Trunk	0.0006	0.04	−0.0004	0.03	−0.01	0.99
Right leg	0.07	0.05	0.001	0.0009	1.46	0.17
Left leg	0.09	0.12	0.35	0.45	0.78	0.45

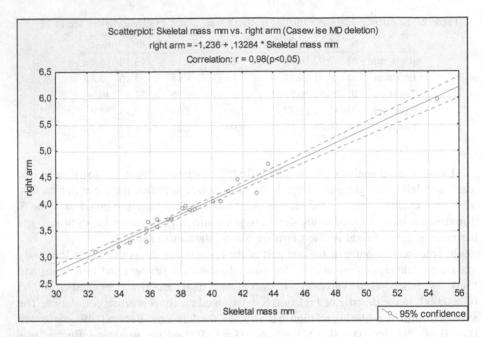

Fig. 1. Multiple regression and correlation's analysis between right arm as independent and skeletal muscle as dependent variable.

4 Discussion

Bioelectrical impedance analysis (BIA) is recognized as suitable in field studies and larger epidemiological studies because it is relatively simple, inexpensive and non-invasive technique to measure body composition. In the overview from 55 published studies of healthy population, with very broad age span 6–80 years, BIA was proved as good instrument to differentiate which type of body composition is better relates to the risk of cardiovascular diseases and all-cause mortality [19]. The older BIA devices, which have used monofrequency BIA technology and looked on subjects body as one cylinder, has been criticized as appropriate only for estimating adiposity of groups in epidemiologic and field studies but has limited accuracy for estimating body composition in individuals [31].

The unique characteristics of athletes body structure can lead to large errors when predicting fat mass (FM) and fat-free mass (FFM). Relatively new review of bioelectrical impedance body composition analysis in athletes, which overviewed the researches made until 2013, conclude that the BIA method shows potential for estimating body composition in athletes, future research should focus on the development of general athlete-specific equations using a TBW-based three- or four-compartment model [31]. Liu et al. [25] developed and cross-validate bioelectrical impedance analysis (BIA) prediction equations of total body water (TBW) and fat-free mass (FFM) for Asian pre-pubertal children from China, Lebanon, Malaysia, Philippines and. Theirs equation for the estimation of TBW was as follows: TBW = 0.231 × height(2)/resistance + 0.066 × height + 0.188 × weight + 0.128 × age + 0.500 × sex-0.316 × Thais-4 .574 (R (2) = 88.0%, root mean square error (RMSE) = 1.3 kg), and for the estimation of FFM was as follows: FFM = 0.299 × height(2)/resistance + 0.086 × height + 0.245 × weight + 0.260 × age + 0.901 × sex − 0.415 × ethnicity (Thai ethnicity = 1, others = 0) − 6.952 (R (2) = 88.3%, RMSE = 1.7 kg). No significant difference between measured and predicted values for the whole cross-validation sample was found.

4.1 The Importance of Bioimpedance Technology in Body Composition Analysis

Since 1990, BIA has become a popular method for estimation of body composition. Bioelectrical impedance analysis is simple quick and non-invasive technique which gives reliable measurements with minimal intra and inter-observer variability [39]. The results are available immediately and reproducible with <1% error on repeated measurements [4]. BIA results are influenced by factors such as the environment, ethnicity, phase of menstrual cycle, and underlying medical conditions. BIA measurements validated for specific ethnic groups and populations can accurately measure body fat in those populations. BIA may not be appropriate choice for body composition assessment for large epidemiological studies unless specific calibration equations are developed for different population groups [12]. BIA is useful in describing mean body composition for groups of individuals but large errors for an individual could limit its clinical application, especially among obese [9]. BIA is suitable for body composition monitoring in elderly people and may be good choice for detecting the prevalence of sarcopenia in this age population [42].

BIA applied to segments represents a great advance in clinical practice by being able to overcome the limitations of the traditional BIA technique [30]. It permits the analysis of body composition with edema and ascites or having muscle tissue or fat deposit depletion [43]. Another challenge for the use of BIA refers to its application in the assessment of body composition in children and adolescent, since according the different stage of growth and development there is wide variation in the various body components [40].

4.2 Segmental Analysis Discussion

The strongest advantage of BIA methodology and InBody devices compared to other field methods in sports anthropometry is the segmental lean mass analysis. InBody 720 observes the human body as a composition of five cylinders, with different lengths and widths and therefore these different segments could be partially analyzed. This kind of analysis provides us with more useful data. The lean mass distribution in upper limbs in our participants showed almost the same mean values of fat free mass in the right and in the left arm. Similar distribution of skeletal muscle mass was found in the inferiorextremities, where the right leg showed insignificantly more mass than the left leg. The trunk of the body, withits huge volume in comparison with upper and lower extremities, contributes approximately 50% of whole body mass [16].

The analysis of lean body mass distribution in athletes informs us of symmetry or asymmetry in the body composition of the athlete. The symmetry of the athlete's body depends on the motor requirements of the particular sport. The nature of during karate training imposes to karate athletes to move in various directions and attack their opponents with both upper and lower extremities in different moment of the match [5]. Unlike some other sport activity (tennis, basketball) karate mainly activates all segments of the body equally. Karateist's somatotype is characterized by higher mesomorphy and lower ectomorphy [17]. Our results showed that Macedonian karatees have symmetrical and balanced distribution between left and right side of the body.

In the investigation of the precision of BIA (InBody 720) in the body composition analysis using the DEXA as the reference method in large sample of healthy adults, Ling et al. concluded that BIA was a valid instrument for whole body composition and segmental lean mass measurements [24]. Although some investigators revealed discrepancies between BF and FFM determined with BIA and DEXA as criterion method, segmental analysis appeared to provide excellent agreement for the measurement of total body and segmental lean soft tissues [15]. An examination of the agreement of segmental multifrequency bioelectrical impedance analysis (SMF-BIA) for the assessment of whole-body and appendicular fat mass (FM) and lean soft tissue mass (LSTM) compared with dual-energy X-ray absorptiometry showed high coefficients of determination for fat mass ($R2 = 0.91$; SEE = 1.4 kg in men and $R2 = 0.94$; SEE 1.2 kg in woman [21]. Skeletal muscle mass (SMM) determined by InBody 720 showed maximal strong positive correlations with the changes in total water and its compartments TW, ECW and ICW ($r = 1$; $p < 0.05$) [37].

Karate athletes are obliged to maintain their body weight within certain range if they want to stay in optimal weight category. The body mass should be composed of optimal amount of body components to achieved a better sport performance. The need of systematic control of body composition involves a reliable and valid method for the body composition analysis and good understanding of coaches and other sport's experts for the meaning of the obtained parameters. As this study showed bioelectrical impedance analysis with InBody 720 generates a plethora of information about body components, nutritional status, obesity diagnose and fitness core of karatees. Segmental analysis of lean body mass gave us awareness into the way the LBM is distributed to main parts of athlete's body: trunk, upper and lower extremities. This analysis informed the sportsmen, his physician and coach if there is asymmetry in LBM distribution, which could be result of inappropriate training regime or injury. Monitoring the segmental analysis could help in following the quality of nutritional and training regime or rehabilitation procedure. The application of BIA technology make available abundance of information regarding the body composition of athlete which could help to sport experts to provide better health and better sport performance for athletes.

References

1. Ackland, T.R., et al.: Current status of body composition assesment in sport. Sports Med. **42** (3), 227–249 (2012)
2. Baumgartner, R.N., Chumlea, W.C., Roche, A.F.: Bioelectrical impedance phase angle and body composition. Am. J. Clin. Nutr. **49**, 16–23 (1988)

3. Bohm, A., Heitmann, B.L.: The use of bioelectrical impedance analysis for body composition in epidemiological studies. Eur. J. Clin. Nutr. **67**(1), 79–85 (2013). https://doi.org/10.1038/ejcn.2012.168
4. Bucholz, A.C., Bartok, C., Schoeller, D.A.: The validity of bioelectrical impedance models in clinical populations. Nutr. Clin. Pract. **19**, 433–446 (2004)
5. Chaabene, H., Hachana, Y., Francchini, E., Makouer, B., Chamari, K.: Physical and physiological profile of elite karate athletes. Sports Med. 1–15 (2012)
6. Chumlea, W.C., Guo, S.: Bioelectrical impedance and body composition: present status and future direction–reply. Nutr. Rev. **52**, 323–325 (1994)
7. Chumlea, W.C., Guo, S.S.: Bioelectrical impedance: a history, research issues, and recent consensus. In: Carlson-Newberry, S.J., Costello, R.B. (eds.) Emerging Technologies for Nutrition Research, pp. 169–179. The National Academies Press, Washington DC (1997)
8. Chumlea, W.C., Sun, S.S.: Biolectrical impedance analysis. In: Heymsfiled, S.B., Lohman, T.G., Wang, Z.M., Going, S.B. (eds.) Human Body Composition, 2nd edn, pp. 79–87. Human Kinetics, Champaign (2005)
9. Chumlea, W.C.: Body composition assessment of obesity. In: Bray, G.A., Ryan, D.H. (eds.) Overweight and the metabolic syndrome: from bench to bedside, pp. 23–35. Springer, New York (2006). https://doi.org/10.1007/978-0-387-32164-6
10. Clarys, J.P., Scafoglieri, A., Provin, S., et al.: A macroquality evaluation of DXA variables using whole dissection, ashing and computer tomography in pigs. Obesity **18**(8), 1477–1485 (2010)
11. Clinician Desk Reference for BIA Testing, Copyright 2003–2015 Byodinamics Corporation. www.biodyncorp.com
12. Dehghan, M., Merchant, A.T.: Is bioelectrical impedance accurate for use in large epidemiological studies? Nutr. J. **7**, 26–32 (2008)
13. Duren, D.L., et al.: Body composition methods: comparisons and interpretation. J. Diabetes Sci. Technol. **2**(6), 1139–1146 (2008). https://doi.org/10.1177/193229680800200623
14. Ellis, K.J.: Human body composition in vivo methods. Physiol. Rev. **80**(2), 647–680 (2000)
15. Esco, M.R.: Comparison of total and segmental body composition using DXA and multifrequency bioimpedance in collegiate female athletes. Strength Condit. **29**(4), 918–925 (2005)
16. Foster, K.R., Lukaski, H.C.: Whole-body impedance: what does it measure? Am. J. Clin. Nutr. **64**, 388S–396S (1996)
17. Giampietro, M., Pujia, A., Bertini, I.: Anthropometric feature and body composition of young athletes practicing karate at high and medium competitive level. Acta Diabetol. **40**, S145–S148 (2003)
18. Heyward, V.H.: ASEP methods recommendation: body composition assessment. J. Exerc. Physiol. **4**(4), 1–12 (2011)
19. Houtkooper, L.B., Lohman, T.G., Going, S.B., Howell, W.H.: Why bioelectrical impedance analysis should be used for estimating adiposity. Am. J. Clin. Nutr. **64**(3), 436s–448s (1996). PMID:8780360
20. Khalil, S.F., Mohktar, M.S., Ibrahim, F.: The theory and fundamentals of bioimpedance analysis in clinical status monitoring and diagnosis of diseases. Sensors **14**, 10895–10928 (2014)
21. Kim, M., Shinkai, S., Murayama, H., Mori, S.: Comparison of segmental multifrequency bioelectrical impedance analysis with dual-energy X-ray absorptiometry for the assessment of body composition in community-dwelling older population. Geriatr. Gerontol. Int. **15**(8), 10113–10122 (2015). https://doi.org/10.1111/ggi.12384
22. Kyle, U., Bosaeus, I., Lorenzo, A., et al.: Bioelectrical impedance analysis - part I: review of principles and methods. Clin. Nutr. **23**, 1226–1243 (2004)

23. Kyle, U., Genton, L., Pichard, C.: Low phase angle determined by bioelectrical impedance analysis is associated with malnutrition and nutritional risk at hospital admission. Clin. Nutr. 1–6 (2012)
24. Ling, C.H.Y., et al.: Accuracy of direct segmental multi-frequency bioimpedance analysis in the assessment of total body and segmental body composition in middle-aged adult population. Clin. Nutrit. **30**, 610–615 (2011)
25. Liu, A., et al.: Validation of bioelectrical impedance analysis for total body water assessment against the deuterium dilution technique in Asian children. Eur. J. Clin. Nutr. **65**(12), 1321–1327 (2011). https://doi.org/10.1038/ejcn.2011.122
26. Lohman, T.G., Harris, M., Teixeria, P.J., et al.: Assesing body composition and changes in body composition: another look at dual-energy X-ray absorptiometry. Ann. N. Y. Acad. Sci. **904**, 45–54 (2000)
27. Lukaski, H.C., Johnson, P.E., Bolonchuk, W.W., Lykken, G.I.: Assessment of fat-free mass using bioelectrical impedance measurements of the human body. Am. J. Clin. Nutr. **41**(4), 810–817 (1985)
28. Martinsen, O.G., Grimnes, S.: Bioimpedance and Biolectrical Basics. Academic Press, Waltham (2011)
29. Matias, C.N., et al.: Estimation of total body water and extracellular water with bioimpedance in athletes: a need for athlete-specific prediction models. Clin. Nutr. **35**(2), 468–474 (2016)
30. Mialich, M.S., Faccioli Sicchieri, J.M., Jordao Junior, A.A.: Analysis of body composition: a critical review of the use of bioelectrical impedance analysis. Int. J. Clin. Nutr. **2**(1), 1–10 (2014)
31. Moon, J.R.: Body composition in athletes and sports nutrition: an examination of the bioimpedance analysis technique. Eur. J. Clin. Nutr. **67**(1), 54–59 (2013). https://doi.org/10.1038/ejcn.2012.165
32. Muller, W., Groschl, W., Muller, R.: Underweight in ski jumping: the solution of the problems. Int. J. Sports. Med. **27**, 926–934 (2006)
33. National Institute of Health Bioelectrical impedance analysis in body composition measurement: National Institute of Health Technology Assessment Conference Statement. Am. J. Clin. Nutr. **64**, 524S–532S (1996)
34. Norman, K., Stobausm, N., Pirlich, M.: Bosy-Westphal.: a bioelectrical phase angle and impedance vector analyzes: clinical relevancies and applicability of impedance parameters. Clin. Nutr. **31**, 1–8 (2012)
35. Piccoli, A., et al.: Discriminating between body fat and fluid changes in the obese adults using bioimpedance vector analysis. Int. J. Obes. **22**, 76–78 (1998)
36. Piccoli, A., Rossi, B., Pillon, L., Bucciante, G.: Body fluid overload and bioelectrical impedance analysis and renal patients. Miner. Electrol. Metab. **22**, 76–78 (1996)
37. Gligoroska, J.P., Todorovska, L., Mancevska, S., Karagjozova, I., Petrovska, S.: Biolectrical impedance analysis in karate athletes: BIA parameters obtained with InBody 720 regarding the age. PESH **5**(2), 117–121 (2016)
38. Prior, B.M., Cureton, K.J., Modelsky, C.M., et al.: In vivo validation of whole body composition estimates from dual-energy X-ray absorptiometry. J. Appl. Physiol. **80**(3), 824–831 (1997)
39. Segal, K.R., Burastero, S., Chun, A., Coronel, P., Pierson Jr., R.N., Wang, J.: Estimation of extracellular and total body water by multiple frequency bioelectrical-impedance measurement. Am. J. Clin. Nutr. **54**, 26–29 (1991)
40. Silva, D.R.P., Ribeiro, A.S., Pavao, F.H., et al.: Validade dos metodos para avaliacao da gordura corporal emcriancas e adolescents pomeiode modelos multicompartimentais: una revisao systematica. Ver. Assoc. Med. Bras. **9**(5), 475–486 (2013)

41. Sundgot-Borgen, J., Tortsveit, M.K.: Aspects of disordered eating continuum in elite high-intensity sports. Scand. J. Med. Sci. Sports **20**, 112–121 (2010)

42. Wang, H., Ha, S., Cao, L., Zhou, J., Liu, P., Dong, B.R.: Estimation of prevalence of sarcopenia by using a new bioelectrical impedance analysis in Chinese community-dwelling elderly people. BMC Geriatr. **16**, 216–224 (2006)

43. Zhu, F., Leonard, E.F., Levin, N.W.: Extracellular fluid redistribution during hemolysis: bioimpedance measurement and model. Physio. Meas. **29**(6), 491–501 (2008)

Initialization of Matrix Factorization Methods for University Course Recommendations Using SimRank Similarities

Alisa Krstova⬛, Bozhidar Stevanoski(✉)⬛, Marija Mihova, and Vangel V. Ajanovski⬛

Faculty of Computer Science and Engineering, Ss. Cyril and Methodius University, Skopje, Macedonia
{krstova.alisa,bozidar.stevanoski}@students.finki.ukim.mk
{marija.mihova,vangel.ajanovski}@finki.ukim.mk

Abstract. The accurate estimation of students' grades in prospective courses is important as it can support the procedure of making an informed choice concerning the selection of next semester courses. As a consequence, the process of creating personal academic pathways is facilitated. This paper provides a comparison of several models for future course grade prediction based on three matrix factorization methods. We attempt to improve the existing techniques by combining matrix factorization with prior knowledge about the similarity between students and courses calculated using the SimRank algorithm. The evaluation of the proposed models is conducted on an internal dataset of anonymized student record data.

Keywords: Course recommendation engine
Study plan development · Collaborative filtering · Matrix factorization

1 Introduction

With the rapid development of information technology, data-driven decision making in higher education has become a global trend aiming to support universities to meet both external standards, as well as internal self-evaluation and improvement requirements. The latter point includes, among other things, improving student retention and success rates, increasing motivation and overall satisfaction during the course of their studies. An indispensable aspect of this process is the careful collection, organization and analysis of educational data, which can be generated by different sources including student data, teacher data and data gathered from the process of teaching and assessment. Higher education is getting close to a time when personalization of study plans and career paths will become a common practice. Rather than using the "one-size-fits-all" approach when constructing a study program and requiring all students to enroll

© Springer Nature Switzerland AG 2018
S. Kalajdziski and N. Ackovska (Eds.): ICT 2018, CCIS 940, pp. 172–184, 2018.
https://doi.org/10.1007/978-3-030-00825-3_15

same or similar subjects, universities begin to turn to building systems that would provide relevant, accurate course recommendations and corresponding grade predictions that are specifically tailored to each individual student. The vast amount of student and teacher-related data provides a basis for the development of intelligent systems that model the prediction of the final grades and systems that allow students to customize their degree plans to better match their career goals, personal interests and predispositions.

As more and more students choose to pursue a degree in higher education, the universities start to face the problem of having a large number of students with a wide range of abilities, skills, interests and potential, attempting to make an informed choice when choosing their elective courses. The level of freedom of choice has increased significantly over the last decade, and at the moment it amounts to 50% of the credits needed to successfully finish the undergraduate studies [9].

The lack of official guidance within the process of semester enrollment and course selection has led to a situation in which students rely predominantly on the word-of-mouth of their colleagues. Such recommendations are usually biased and do not take into account the student's personal abilities and inclinations. Motivated by these observations, our goal is to develop a prototype of a system that will assist university students in making an optimal choice when it comes to elective courses. We seek to analyze and use existing methods which have already been proven to result in accurate recommendations and predictions and explore possible modifications to the well-established algorithms. We aim to provide accurate grade predictions and recommendations over a wider set of electives best suited for the student in question.

The paper is organized as follows. Section 2 contains literature review of related work done in the past by other authors. Section 3 describes the dataset that was used in the experiments, as well as a short clarification on the data acquisition process. Section 4 defines the problem addressed in this paper. In Sect. 5 we make a short overview of the well-known algorithms: Probabilistic Matrix Factorization (PMF) [5], Bayesian Probabilistic Matrix Factorization (BPMF) [10] Alternating Least Squares (ALS) [13] and SimRank [4], and propose a novel approach for the initialization of the first three using the last one. We shortly explain the metrics used to evaluate the performance of the algorithms, and our approach. Section 6 compares the performance of the base versions of PMF, BPMF and ALS with each other, and with the enhanced versions using the SimRank initialization. Finally, Sect. 7 concludes the paper.

2   Related Work

There have been extensive academic research efforts and numerous industrial implementations of recommender systems in the past. Since predicting course grades and recommending subjects differs significantly from recommending music or movies, we will focus on reviewing the work most relevant to our context. Several authors have taken a similar approach by identifying groups of similar students when predicting a grade for a course [7,14]. These implementations also rely on neighborhood-based collaborative filtering methods. The past grades of the student's colleagues are used to make an estimate of the grade that might be obtained in a hypothetical enrollment of the course by using some similarity-weighted aggregation function.

Several course grade prediction models are proposed in [3,6,8], that use methods based on sparse linear models and low-rank matrix factorization. When using such an approach, the students' success in the past courses plays a deciding role. The factorization matrix is built so that rows represent students, columns represent the available courses and each matrix element stores the grade that the ith student obtained in the jth course. A missing value signals that the respective student has not yet enrolled/passed the respective course. As will be explained in the following sections, we will use this approach when describing the specifics of our own data.

As the algorithm described in [13] resolves scalability and handles sparseness of the data which is a major issue in recommender systems, and as the probabilistic techniques in [5,10] considerably outperform standard matrix factorization as well as other approaches, essentially motivated by their performances in the context of recommending university courses in [9], we chose to work with these methods and tried to improve them.

The rationale of using grade prediction for future course enrollments is not in the sheer possibility to accurately predict the students' future, but in the opportunities that open in the processes of guiding and advising students. Course recommender systems have been successfully integrated in educational dashboards and learning analytics systems, such as the one described in [1]. The described Virtual Student Adviser enables the students to explore different study programs and guide them on their own personalized academic path – annotating risky mandatory courses and paths on one hand, and recommended elective courses among the pool of freely selectable options on the other hand.

From the conducted literature review, it can be concluded that the problems of grade prediction and course recommendations have been tackled by various approaches. These processes are beneficial not only for increasing the satisfaction of the student after choosing a certain course, but also for increasing her chances of success during her studies. Grade prediction can also be used as a tool to perform a hypothetical (what-if) analysis on the impact of one selection from a list of courses against another, and assessing the courses which may result in risk of failure. This would provide the student with an indicator of where to direct her efforts and time.

However, most of the reviewed systems suffer from the well-known problem which often arises in constructing recommender systems – the cold start problem. There are always students with very few grades, i.e. those who are currently in the first year of undergraduate studies. This results in low amount of information on the preferences and performance of the students based upon which we would like to make predictions. Therefore, special attention must be paid in order to remedy this issue.

3   Dataset

The experiments were conducted on real, fully anonymized records of student course enrollments and grades at the authors' institution. The dataset includes data course enrollments in the period from 2011 to 2016. The process of anonymization was performed by administrative staff, prior to the process of acquisition of data for research purposes. As researchers, we did not have any access to the full official data records, but only to already anonymized replicas. For the purpose of this research, we acquired records in the form of triplets (student, course, grade) – whereas the student was represented by an anonymous identifier, the course was represented by the real name of the course, and the grade was the grade achieved by the student (5 meaning a failed course, 6 being the lowest passing grade and 10 being the highest grade, null meaning a course that was enrolled but was not yet graded by the respective teachers).

We ran the experiment over a 25-day interval in 5 cycles, having 6 batches of data records. The first batch contained data from the previous academic years, while the other subsequent batches add records that were newly input during the respective cycles. The size of the dataset for each cycle was divided into a training and test dataset, as shown in Table 1.

Table 1. Number of training and test records for each cycle of the dataset

	Cycle 1	Cycle 2	Cycle 3	Cycle 4	Cycle 5
Training records	58915	59492	60039	61544	62899
Test records	595	541	968	1261	1127

The time frame of data acquisition was chosen to be the final exams period, immediately prior to the deadline for the process of course enrollment, the reason being two-fold:

1. The finals are typically the last responsibility the students should pass in order to get their grade.
2. The end of the exam period typically concurs with the opening date of the period when the students choose the courses they would like to enroll in for the next semester. Hence, this is the period when course recommendations are most needed and sought for.

As an illustration for the outlook of the dataset, Table 2 contains all instances from the first cycle for the student 1773012. It is essential to note that the ID 1773012, as all the other IDs in our dataset, is not a real identifier. This ID does not correspond to an existing student identifier and the example we give for illustration purposes is not based on any student's real data. The complete dataset is an extension of Table 2, so that it contains similar records for all the students.

Table 2. Illustrative dataset records from the first cycle

Student	Course	Grade
1773012	Business and management	10
1773012	Introduction to Internet	9
1773012	Introduction to Informatics	7
1773012	Discrete mathematics 1	7
1773012	Marketing	8
1773012	Object-oriented programming	10
1773012	Introduction to Web design	9
1773012	Professional skills	6
1773012	Structured programming	7

4   Problem Formulation

From the student-course-grade relation, a grade matrix G can be constructed. Suppose we represent the courses and the students with integer IDs of the intervals $[1, M]$ and $[1, N]$, where M and N are the number of courses and students respectively. The grade matrix G will be such that the element in the ith row and j^{th} column represents the grade the student i obtained in the course j.

Since the number of available courses is much larger than the number of courses the students are required to pass in order to graduate, and some students are in their first years of studying, G will be sparse. In this paper, we aim to predict the missing elements of the matrix, or in other words, we predict the performance of each student in the courses she has not already been enrolled in, and based on such obtained performances, we recommend her university courses.

Having the entries from the train set represented with G, for evaluation purposes, we compare the predicted grades with the real ones from the test set. The predictions are made using commonly utilized methods, however, here we improve them by introducing a novel approach in their initialization.

5   Methodology

5.1   Matrix Factorization Methods

Matrix factorization methods have gained popularity in recent years [12] due to their good predictive accuracy. Techniques based on matrix factorization are effective because they allow the discovery of the latent features underlying the interactions between users and items. This idea can be mapped to the concept of making course grade predictions by observing students as users and their courses as items. The algorithm will be applied in order to predict the missing grades of students interested in a particular course, and afterwards to make personalized course recommendations based on the computed grade predictions.

As the name suggests, matrix factorization and its variation intends to find two low-dimensional matrices S_{DxN} and C_{DxM} that factor G, where G is the aforementioned student-course-grade matrix and D is a positive integer representing the number of latent features to be considered. In other words, the product of S^T and C should approximate G.

A convenient interpretation of the matrices S and C is to consider that each student and each course is mapped to a D dimensional latent feature space, and S and C have the corresponding feature vectors of the students and courses respectively. Having these vectors, a grade prediction for student i in the course j is just a dot product of their corresponding vectors, i.e.

$$\hat{g_{i,j}} = s_i^T c_j \tag{1}$$

where $\hat{g_{i,j}}$ is the predicted grade and s_i, c_j are the ith and j^{th} columns in S and C respectively.

To learn the factor vectors (s_i, c_j), the algorithm strives to minimize the regularized squared error on the set of known grades:

$$\min \sum_{(s,c) \in K} (\hat{g_{i,j}} - s_i^T c_j)^2 + \lambda(||s_i||^2 + ||c_j||^2) \tag{2}$$

where K is the set of student-course (s, c) pairs for which the grade is known from the training set and λ is a regularization term.

Alternating Least Squares (ALS). As described in the previous section, our goal is to minimize the loss function. Derivatives are an obvious method for minimizing functions in general, so several derivative-based methods have been developed. One of the most popular approaches is the Alternating Least Squares (ALS) algorithm [13].

ALS minimization starts with holding one set of latent vectors constant (for example, the student vectors), then taking the derivative of the loss function with respect to the other set of vectors (the course vectors), setting the derivative equal to zero and solving the resulting equation for the non-constant vectors. Afterwards, these new vectors are held constant and the derivative of the loss function with respect to the previously constant vectors is taken. The steps are repeated until convergence is achieved. This can be formulated mathematically as:

$$\frac{\partial L}{\partial x_s} = -2 \sum_i (g_{s,c} - x_s^T \hat{y_c}) y_c^T + 2\lambda_x x_s^T \tag{3}$$

$$0 = -(g_s - x_s^T Y^T)Y + \lambda_x x_s^T \tag{4}$$

$$x_s^T (Y^T Y + \lambda_x I) = g_s Y \tag{5}$$

$$x_s^T = g_s Y (Y^T Y + \lambda_x I)^{-1} \tag{6}$$

Here, the course items y_c are held constant and the derivative is taken with respect to the student vectors x_s. The symbol Y refers to a matrix consisting of all student row vectors. The row vector g_s contains all course grades for student s taken from the grades matrix and $g_{s,c}$ is the grade obtained by student s in course c. Finally, I represents the identity matrix.

Probabilistic Matrix Factorization (PMF). Minimization of the loss function described in (2) appears as a goal if we take the probabilistic way of approximating the grade matrix. One simple algorithm is the Probabilistic Matrix Factorization (PMF) [5] that outperforms many others, as we will see in the results' section. PMF assumes normal distribution for the user and item latent features as priors, as well as for the result, such that the grade predicted in (1) is taken as the mean for the distribution.

An important note to take from this algorithm is that it scales linearly with the number of records, and therefore, it outperforms the other similar methods in execution time.

Bayesian Probabilistic Matrix Factorization (BPMF). This method takes a Bayesian approach to the problem which includes integrating out the model parameters, i.e. it gives a fully Bayesian treatment to the already described Probabilistic Matrix Factorization (PMF) model. What makes Bayesian Probabilistic Matrix Factorization (BPMF) different is that it uses Markov chain Monte Carlo (MCMC) methods for approximate inference [10]. As in PMF, the prior distributions over the students and courses features are assumed to be Gaussian. Inference is, however, achieved through the Gibbs sampling algorithm that iterates through the latent variables and samples each from its distribution conditional on the current values of the rest of the variables. This algorithm is normally used when conditional distributions are easy to sample from – due to the use of conjugate priors for the parameters and hyper-parameters in the BPMF, the Gibbs algorithm is very well applicable. It has been demonstrated that the BPMF model may outperform in specific cases the classical PMF approach on user-rating datasets, such as the large Netflix dataset [10].

5.2 SimRank

The previous sections demonstrate three different variations of the matrix factorization algorithm which use different ways to initialize the matrix. The majority of the approaches used in literature use random initialization or initialization based on taking the averages of the respective rows or columns. We seek to explore the effects of adding a semantic component to matrix initialization, i.e. to augment the algorithm by providing it with some initial similarities between courses and students. However, in order to do this, we must first define course similarity. Intuitively,

– Two courses are similar if similar students are enrolled in both of them, and
– Two students are similar if they are enrolled in similar courses.

These definitions are cyclic and the intuition behind them is comparable with the one behind Google's PageRank algorithm, which ranks websites based on their importance. Hence, we turn to a somewhat related algorithm for computing similarities between items and users (in our case courses and students) – SimRank [4].

In order to use the algorithm, we form a directed bipartite graph whose nodes represent the students and the courses from the dataset described in Sect. 3. There is an edge from student s to course c, if and only if s has enrolled c.

A difference exists between the similarities of the nodes in our graph – the similarity between students s_1 and s_2, and the similarity between courses c_1 and c_2 can be calculated with the following recursive relations:

$$s\left(s_1, s_2\right) = \frac{C_1}{\left|O\left(s_1\right)\right|\left|O\left(s_2\right)\right|} \sum_{i=1}^{|O(s_1)|} \sum_{j=1}^{|O(s_2)|} s\left(O_i\left(s_1\right), O_j\left(s_2\right)\right) \qquad (7)$$

$$s\left(c_1, c_2\right) = \frac{C_2}{\left|I\left(c_1\right)\right|\left|I\left(c_2\right)\right|} \sum_{i=1}^{|I(c_1)|} \sum_{j=1}^{|I(c_2)|} s\left(I_i\left(c_1\right), I_j\left(c_2\right)\right) \qquad (8)$$

$I\left(v\right)$ and $O\left(v\right)$ are the set of in-neighbors and out-neighbors of node v, and $I_i\left(v\right)$ and $O_i\left(v\right)$ is an individual in-neighbor and out-neighbor respectively. C_1 and C_2 are decay factors and have values in the range $(0, 1)$, As originally proposed in [4], C_1 and C_2 are taken to be 0.8.

5.3   Matrix Factorization with SimRank Weights Initialization

As we discussed previously, the initial features for both users and items are chosen randomly, and then, they are tweaked with different methods, depending on the algorithm used. Under this assumption, at the beginning, we do not differentiate neither the students, nor the courses.

Having only the student-course-grade records, and no other additional information, limits the possibilities for initialization. However, they are enough to compute the SimRank results, that can be later used in the initialization of the matrices.

We propose a method of initialization of latent features using the previously computed SimRank similarities. For each of the above matrix factorization algorithms we take the resultant student and course latent features, S_0' and C_0', after one run of the algorithm with the usual random initialization. Let us denote the student-student similarity matrix with S_{sim} and the course-course one with C_{sim}. In them, the (i, j) entry shows the similarity of ith and j^{th} student or course. The new student and course latent features S_1 and C_1 are computed as weighted arithmetic mean of S_0 and C_0 with weights S_{sim} and C_{sim}, or

$$S_1 = \frac{S_0' \cdot S_{sim}}{S_{avg}} \qquad (9)$$

$$C_1 = \frac{C_0' \cdot C_{sim}}{C_{avg}} \qquad (10)$$

where S_{avg} is a DxN matrix that as i^{th} column has the mean vector of the similarities for the ith student. C_{avg} is the analogous DxM matrix for courses. The fractions in (9) and (10) denote an element-wise matrix division.

One can easily notice that new initial latent features S_2 and C_2 can be constructed in a similar way using (9) and (10), i.e. $S_2 = \frac{S_1' \cdot S_{sim}}{S_{avg}}$ and $C_2 = \frac{C_1' \cdot C_{sim}}{C_{avg}}$, using the matrices S_1' and C_1' that are resultant features after algorithm converged. In fact, the above discussion can be generalized for any non-negative integer t in the following manner

$$S_{t+1} = \frac{S_t' \cdot S_{sim}}{S_{avg}} \tag{11}$$

$$C_{t+1} = \frac{C_t' \cdot C_{sim}}{C_{avg}} \tag{12}$$

We terminate with such initializations once the latest is outperformed by a former one.

5.4   Evaluation Metrics

As a method for testing the performance of our models, we use the Root Mean Square Error (RMSE) and the Mean Absolute Error (MSE) measures. These metrics can be defined mathematically as:

$$RMSE = \sqrt{\frac{1}{F} \sum_{g_{i,j} \in T} (g_{i,j} - \hat{g_{i,j}})^2} \tag{13}$$

$$MAE = \frac{\sum_{g_{i,j} \in T} |g_{i,j} - \hat{g_{i,j}}|}{F} \tag{14}$$

$$MAE_i = \frac{1}{H} \sum_{j=1}^{H} |\hat{g_{i,j}} - g_{i,j}| \tag{15}$$

Here, F is the number of total available grades, H is the number of grades the current student has obtained until the point of computation, T is the set of the test records, $g_{i,j}$ refers to the actual grade, whereas $\hat{g_{i,j}}$ to the predicted grade.

6   Results and Discussion

In this section we present the results of the performance evaluation conducted on the dataset described in Sect. 3. First, we begin by analyzing the results of the three matrix factorization techniques in reference to the previously explained performance metrics RMSE and MAE.

Table 3 summarizes the obtained RMSE and MAE scores on the dataset from the first cycle for ALS, BPMF and PMF with 7 different values for the parameter

D, i.e. number of latent features (see Sect. 5.1). The comparative performances of the algorithms are similar to those on the data from the other cycles. The approach taken to obtain these results involves initializing the respective algorithms using their proposed random initialization - this provides us with initial insights on the performance of the different techniques and represents a reference point for our own SimRank-based initialization.

Table 3. Performance comparison of the three matrix factorization techniques

		D = 20	D = 30	D = 40	D = 50	D = 60	D = 70	D = 80
ALS	*RMSE*	1.637	1.509	1.476	1.438	1.447	1.430	1.430
	MAE	1.270	1.147	1.102	1.075	1.072	1.069	1.064
BMPF	*RMSE*	1.647	1.417	1.389	1.417	1.487	1.416	1.318
	MAE	1.236	1.081	1.040	1.088	1.126	1.086	1.007
PMF	*RMSE*	1.196	1.168	1.151	1.115	1.085	1.054	1.068
	MAE	0.837	0.808	0.796	0.768	0.736	0.721	0.737

For each algorithm, we take the value for D for which the smallest RMSE is observed, and take the student and course latent features for that D. Using them, we iteratively apply (11) and (12) as explained in Sect. 5.3. On the data from the first cycle, this initialization surpasses the simple random initialization of the algorithms. In particular, ALS notes RMSE of 1.014 in the first, and 1.011 in the second iteration, which is a significant improvement over the best value of 1.43 for the best case of $D = 80$. The RMSE for BPMF lowers to 1.17 after only 1 iteration.

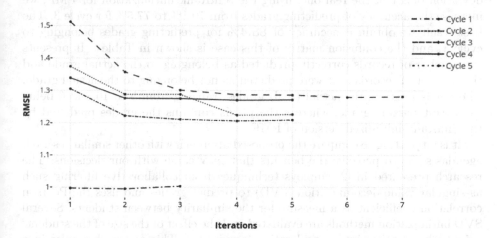

Fig. 1. Improvements of the RMSE of PMF over all dataset cycles for the corresponding best value for D. Termination is done when an iteration yields worse results than the previous one.

Table 4. Confusion matrix of PMF for the first cycle, with $D = 140$

		\multicolumn{6}{Predicted grades}
		5	6	7	8	9	10
Real grades	5	31	2	0	0	0	0
	6	9	124	113	10	2	0
	7	2	47	94	11	1	0
	8	2	14	35	13	9	0
	9	0	2	26	14	3	4
	10	0	0	5	5	13	4

PMF does not only outperform the others in the basic case, but also notices improvements for every cycle, and it continues yielding the best results after the SimRank initialization. Fig. 1 shows how RMSE drops in every iteration for PMF over all dataset cycles. Although in the first cycle, an improvement for all three algorithms is noted, BPMF and ALS algorithms have insignificant or even no benefit from this initialization in the other cycles. Therefore, we keep the focus on PMF's results.

When considering an accurately predicted grade to be one that falls within deviation of ±1 of the real one, using the SimRank initialization for PMF, we increase the accuracy of predicting grades from 70.1% to 77.5% for cycle 3. The best result we obtain is accuracy of **88.4%** for predicting grades belonging to cycle 1, and the confusion matrix of this case is shown in Table 4. It presents the number of records correctly predicted as belonging to the actual grade and the number of records that were predicted as not belonging to the actual grade. The x-axis presents the actual (real) grades, ranging from 5 to 10 and 6 being the lowest passing grade, whereas the y-axis contains the grades predicted by the SimRank-initialized version of PMF.

It is interesting to compare the proposed approach with other similar research agendas so as to perceive the benefits that may come with our decisions. The research presented in [2] employs techniques from collaborative filtering such as singular value decomposition (SVD) to predict grades and uses the Pearson correlation coefficient as a measure for the similarity between students. Several SVD initialization methods are evaluated and the effect of the size of the student-neighborhood taken into consideration is explored. This approach results in a MAE value of minimum 1.5, with some of the parameter configurations yielding an even higher value of up to 2.2. As can be seen from Table 3 and the discussion following it, the SimRank-initialized matrix factorization methods demonstrate superior performance. Another sophisticated approach is described in [9], where the authors evaluate several matrix factorization techniques for grade prediction based on a historical dataset of over 10 years of student record data. The presented results show that around 80% of the predicted records fall in a deviation of ±1 of the actual grade. The modification of the PMF algorithm presented in this research achieves an increase in accuracy of more than 8%.

It is important to note, however that the aforementioned comparisons have been done in different settings and tested on institution-specific datasets which might have less or more strict course programs and assessment procedures.

7   Conclusion and Future Work

In this paper, we presented a novel hybrid approach of initializing matrix factorization methods with student-course similarities obtained using the SimRank algorithm. We employ this approach in extending three well-known matrix factorization techniques (Alternating Least Squares, Probabilistic Matrix Factorization and Bayesian Probabilistic Matrix Factorization). The evaluation performed on a dataset of past student records showed that the Probabilistic Matrix Factorization (PMF) gives better results in all performance measures compared to ALS and BPMF. Furthermore, an overview of the algorithms' performance depending on the hyper-parameters was illustrated and discussed. We also demonstrate very promising results in accuracy achieved by the SimRank initialization method for PMF.

Our current research agenda includes re-evaluating the proposed method on an extended internal dataset that best reflects the current study programs offered at the authors' institution. This will be basis for the integration of the recommendation engine prototype in a user-friendly web application, whose purpose will be to allow the students to browse through course information, track their current study progress and display recommendations for prospective elective courses. Since the similarity matrices containing the SimRank scores are pre-computed and the matrix factorization methods are fast, the estimation of the most suitable courses for a particular student can be made almost real-time, i.e. on average for a single student the estimation lasts about 0.3 ms on a machine configuration with Intel®Core™ i7-4720HQ CPU @ 2.6 GHz, 8 GB RAM on a 64-bit Ubuntu 16.04 LTS.

Incorporating user feedback into the system, i.e. letting the user evaluate the quality of the recommended course with respect to their interests and predispositions will further improve future predictions. We would also like to explore the effect of adding additional sources of data to our models, for example records about the respective course instructors, as well as some background information of the students' past education.

Acknowledgments. This work is a result within the project SISng (Study Information Systems of the Next Generation) [11], which is currently ongoing at the Faculty of Computer Science and Engineering in Skopje. The authors would also like to thank Ljupcho Rechkoski for the provided materials.

References

1. Ajanovski, V.V.: Guided exploration of the domain space of study programs: recommenders in improving student awareness on the choices made during enrollment. In: Proceedings of the Joint Workshop on Interfaces and Human Decision Making for Recommender Systems (INTRS17), pp. 43–47. CEUR-WS, Como (2017)
2. Carballo, F.O.G.: Masters Courses Recommendation: Exploring Collaborative Filtering and Singular Value Decomposition with Student Profiling, Instituto Superior Tecnico, Lisboa (2014). https://fenix.tecnico.ulisboa.pt/downloadFile/563345090413333/Thesis.pdf
3. Hu, Q., Polyzou, A., Karypis, G., Rangwala, H.: Enriching course-specific regression models with content features for grade prediction. In: 2017 IEEE International Conference on Data Science and Advanced Analytics (DSAA), pp. 504–513 (2017)
4. Jeh, G., Widom, J.: SimRank: a measure of structural-context similarity. In: Proceedings of the Eighth ACM SIGKDD International Conference on Knowledge Discovery and Data Mining, pp. 538–543. ACM Press, Edmonton (2002)
5. Mnih, A., Salakhutdinov, R.: Probabilistic matrix factorization. In: Proceedings of the 20th International Conference on Neural Information Processing Systems, NIPS 2007, Vancouver, British Columbia, Canada, pp. 1257–1264 (2008)
6. Morsy, S., Karypis, G.: Cumulative knowledge-based regression models for next-term grade prediction. In: Proceedings of the 2017 SIAM International Conference on Data Mining, pp. 552–560. SIAM, Houston (2017)
7. O'Mahony, M.P., Smyth, B.: A recommender system for on-line course enrollment: an initial study. In: Proceedings of the 2007 ACM Conference on Recommender Systems, pp. 133–136. ACM, New York (2007)
8. Polyzou, A., Karypis, G.: Grade prediction with models specific to students and courses. Int. J. Data Sci. Anal. **2**(3–4), 159–171 (2016)
9. Rechkoski, L., Ajanovski, V.V., Mihova, M.: Evaluation of grade prediction using model-based collaborative filtering methods. In: 2018 IEEE Global Engineering Education Conference (EDUCON), pp. 1096–1103. IEEE, Tenerife (2018). https://doi.org/10.1109/EDUCON.2018.8363352
10. Salakhutdinov, R., Mnih, A.: Bayesian probabilistic matrix factorization using Markov chain Monte Carlo. In: Proceedings of the 25th International Conference on Machine Learning, pp. 880–887. ACM, New York (2008)
11. Student Information System of the Next Generation (2009/2018). https://develop.finki.ukim.mk/projects/sisng
12. Symeonidis, P., Zioupos, A.: Matrix and Tensor Factorization Techniques for Recommender Systems. Springer, Cham (2016). https://doi.org/10.1007/978-3-319-41357-0
13. Zhou, Y., Wilkinson, D., Schreiber, R., Pan, R.: Large-scale parallel collaborative filtering for the netflix prize. In: Fleischer, R., Xu, J. (eds.) AAIM 2008. LNCS, vol. 5034, pp. 337–348. Springer, Heidelberg (2008). https://doi.org/10.1007/978-3-540-68880-8_32
14. Zhuhadar, L., Nasraoui, O., Wyatt, R., Romero, E.: Multi-model ontology-based hybrid recommender system in e-learning domain. In: 2009 IEEE/WIC/ACM International Joint Conference on Web Intelligence and Intelligent Agent Technology, pp. 91–95. IEEE, Milan (2009)

Deep Learning the Protein Function in Protein Interaction Networks

Kire Trivodaliev[1(✉)], Martin Josifoski[2], and Slobodan Kalajdziski[1]

[1] Ss. Cyril and Methodius University, Skopje, Macedonia
{kire.trivodaliev,slobodan.kalajdziski}@finki.ukim.mk
[2] Ecole Polytechnique Fédérale de Lausanne EPFL, Lausanne, Switzerland
martin.josifoski@epfl.ch

Abstract. One of the essential challenges in proteomics is the computational function prediction. In Protein Interaction Networks (PINs) this problem is one of proper labeling of corresponding nodes. In this paper a novel three-step approach for supervised protein function learning in PINs is proposed. The first step derives continuous vector representation for the PIN nodes using semi-supervised learning. The vectors are constructed so that they maximize the likelihood of preservation of the graph topology locally and globally. The next step is to binarize the PIN graph nodes (proteins) i.e. for each protein function derived from Gene Ontology (GO) determine the positive and negative set of nodes. The challenge of determining the negative node sets is solved by random walking the GO acyclic graph weighted by a semantic similarity metric. A simple deep learning six-layer model is built for the protein function learning as the final step. Experiments are performed using a highly reliable human protein interaction network. Results indicate that the proposed approach can be very successful in determining protein function since the Area Under the Curve values are high (>0.79) even though the experimental setup is very simple, and its performance is comparable with state-of-the-art competing methods.

Keywords: Protein interaction network · Deep learning
Protein function prediction

1 Introduction

Proteins are the building blocks of life. Protein functions are at the core of understanding and solving crucial questions and problems in life sciences like disease mechanisms and proper drug development, design of new biochemicals, elucidation of unknown life phenomena, etc. With the upsurge in high-throughput technologies big data is taking center stage in life sciences, ranging from sequences to complex proteomic data, such as gene expression data sets and protein interaction networks (PINs). One of the main challenges is bridging the incompleteness of the data, especially in terms of building an effective and precise system for analyzing such data and uncovering their intrinsic functional meaning [5].

Protein-protein interaction (PPI) data are fundamental to biological processes [24] and in terms of single-source computational protein function prediction this data is the

© Springer Nature Switzerland AG 2018
S. Kalajdziski and N. Ackovska (Eds.): ICT 2018, CCIS 940, pp. 185–197, 2018.
https://doi.org/10.1007/978-3-030-00825-3_16

best choice. PPI data has the nature and organization of networks, with proteins and their interactions considered nodes and edges in the network. These networks are referred to as Protein Interaction Networks (PINs) and protein functions associated to a protein can be modeled as labels of the corresponding node. Taking this definition, the problem of computational function prediction of a protein is translated to a problem of proper labeling of its corresponding node in its PIN graph representation. The semantics of protein functions is usually defined using notational schemes organized as ontology, the most comprehensive one being the Gene Ontology (GO) [4]. GO defines three semantic contexts, stored as separate subontologies within the GO: Biological Process (BP), Molecular Function (MF), and Cellular Component (CC). Each subontology consists of a set of terms (GO terms), connected in a directed acyclic graph. GO is the most applied functional annotation scheme across a wide variety of biological data [14] and as such is the scheme used in this research.

Existing computational function prediction methods using PINs can be characterized based on what and how much information is used in the method: (1) neighborhood-based [10, 16], where the query protein "receives" its functions from the "dominant" terms in its immediate neighborhood, (2) global optimization-based [20, 25], where the neighboring information may be insufficient, so the functions of the query protein are inferred from the indirectly connected proteins, sometimes the entire network, (3) clustering-based [19, 24, 26, 32], where query protein's functions are chosen from "dominant" functions present in its determined network cluster, (4) association-based [11], similar to clustering approaches, but here functional modules are hypothesized from frequently occurring sets of interactions in PINs of protein complexes.

Computational function prediction has improved significantly in recent years [14], however there is lack of research taking into account the sparsity of GO annotations [27]. The problem arises due to GO providing mainly "positive" associations between a protein and a GO term (the protein has the function defined with the GO term). The specification of "negative" terms is very rare and in the context of computational function prediction the lack of a positive association can not be treated as a negative association. The binary classification problem requires for explicit positive and negative samples to learn the desired discriminative model. Recent approaches define negative examples directly or by using some heuristics. The direct approaches make assumptions based on the lack of annotations (associations) for a given term and a given protein, either by taking as negative proteins that lack a query annotation [8] or proteins that lack the annotation of sibling terms of a query term [3, 18]. The authors of [31] propose a parametrization Bayesian priors method that selects negative examples based on an approximation of the empirical conditional probability that a term will be annotated to a protein given that the protein is already annotated with another term. Two additional negative examples selection algorithms, selection of negatives through observed bias (SNOB) and negative examples from topic likelihood (NETL) are given in [30]. NegGOA [6] takes advantage of a hierarchical semantic similarity between GO terms and performs downward random walks with restart on the hierarchy and on the empirical conditional probability that two terms co-annotated to a protein, to determine the negative examples.

Recently, a lot of research has been focused on the problem of producing network embeddings. The DeepWalk [21] method uses short random walk sequences in the context of sentences from a natural language and using the Skip-Gram word representation model [17] learns a vector representation of the graph nodes. The LINE [23] method first learns node representations produced as concatenations of first- and second-order proximities. In a similar fashion, GraRep [2] uses various loss functions to capture k-order proximities and combines the learned representations. The TADW method [29] builds on the proof that DeepWalk is equivalent to matrix factorization by incorporating node level rich text information in the network representation learning. To capture the non-linear network structure, [28] proposes a deep learning model with non-linear functions and produces results that preserve first- and second-order proximities. The node2vec [7] method uses a biased random walk procedure with a flexible notion of a nodes neighborhood, which efficiently explores diverse neighborhoods. Node2vec treats the results of these walks as natural language sentences (same as DeepWalk) and using these "sentences" produces vectors that maximize the likelihood of preservation of the graph topology and semantics locally and globally.

In this paper a novel approach DeePin using deep learning is proposed for the computational protein function prediction. The problem is solved as term-centric i.e. the result of the prediction gives answer if a protein is annotated with a given term or not. In order to build the appropriate deep learning models, the PIN is first transformed into a continuous vector space (an instance per node) and for each term-model a set of positive and negative examples is chosen. The aim is to show that although very simple this approach can produce comparable results with other leading approaches.

The rest of the paper is organized as follows. Section 2 presents the steps in acquiring and building the data for the research and the technical details on the methods used. In the third section a detailed description of the performed experiments and the corresponding results is given. Discussion for the results and possible improvements of the method is also provided. Finally, the paper is concluded in the fourth section.

2   Materials and Methods

In this paper the graph representation of the PIN is used to derive a continuous vector representation. The approach of [7] is adopted and it employs a biased random walk procedure with a flexible notion of a nodes neighborhood, which efficiently explores diverse neighborhoods. The problem is formulated as a maximum likelihood optimization problem i.e. one that maximizes the log-probability of observing a neighborhood node for a target node, conditioned on its vector representation. The advantage of this approach over other algorithms arises in its scalability, as well as flexibility to easily custom-fit the representation for detecting node dependencies based on communities they belong to, structural equivalences based on the nodes role in the network, or a mixture of both. The produced vectors maximize the likelihood of preservation of the graph topology locally and globally. The next step is to binarize the PIN graph nodes (proteins) i.e. for each protein function derived from Gene Ontology (GO) determine the positive and negative set of nodes. Since positive set is predefined the only challenge is determining the negative node sets. Based on a random walk on the GO acyclic graph weighted using Lin similarity metric computed from the available annotations, probabilities of not observing a label annotated to a protein will be associated on every pair of protein function and protein [6]. Using a threshold, the negative set of proteins for a given protein function is determined and combined with the positive set are used in the process of learning the protein model for the function. A simple deep learning six-layer model is built for the protein function learning. The following subsections present the steps in acquiring and building the data for the research. Additionally, technical details on the methods used are also provided.

2.1   Protein Interaction Network Data

The Protein Interaction Network (PIN) is constructed on data from the HIPPIE (v2.0) database [22], which is a highly reliable human PPI dataset, built from multiple experimental datasets, that integrates the amount and quality of evidence for a given interaction in a normalized scoring scheme. Data is first preprocessed to remove self-interactions, zero-confidence interactions and duplicate interactions. Duplicates are removed so that only the highest confidence score interaction remains. The preprocessing results in an undirected weighted graph, having proteins as nodes, their interactions as edges, and the confidence scores of interactions as weights associated to corresponding edges. The largest connected component of this graph is the final PIN graph used in the research and is consisted of 16,769 proteins and 277,055 protein-protein interactions.

To model the problem of function prediction as a label prediction problem the PIN needs to be enriched i.e. describe every protein with all its known functional annotations. To that aim a Gene Ontology Annotation (GOA) file from the European Bioinformatics Institute is used (archived date: May 3, 2013). The GOA file provides GO annotations which associate gene products with GO terms. This file is processed so that terms labeled as 'obsolete' and annotations with evidence code 'IEA' (Inferred from Electronic Annotation) are excluded. Additionally, duplicate annotations are also excluded. The remaining annotations are associated with their corresponding nodes in the final PIN. From the initial GOA file consisting of 369,199 annotations, in the final labeled PIN there are a total of 126,367 annotations.

2.2   Vector Representation of the PIN

To apply deep learning of the protein function one first needs to construct highly informative, discriminating and mutually independent feature representations of the PIN's nodes/edges. The node2vec algorithm [7] provides a semi-supervised method for learning continuous vector representation for nodes in a network, that map every node to a d-dimensional feature space in a process that maximizes the likelihood of preserving the network neighborhood of nodes. Formally, for a network with a graph representation $G = (V, E)$ let $f : V \rightarrow R^d$ be the mapping to the feature space. For every protein node $u \in V$, a neighborhood of node u is defined with $N(u) \subset V$. Now the problem is formulated as a maximum likelihood optimization problem with the following objective function:

$$\max \sum_{u \in V} \log \Pr(N(u)|f(u)) \qquad (1)$$

Equation 1 maximizes the log-probability of observing a neighborhood node for a node u, conditioned on its vector representation, given by f. This procedure with certain parameter settings becomes equivalent to DeepWalk [21], and by transitivity incorporates the key features of other previously proposed methods for network embeddings.

Let $G = (V, E)$ be the graph representation for the PIN. The process of learning the representations starts by generating r random walks from every protein node u as a source, with fixed length l. Let c_i be the i-th protein in the walk and $c_0 = u$. The generation of a sequence of proteins is done using the distribution

$$P(c_i = x | c_{i-1} = v) = \begin{cases} \frac{\pi_{vx}}{Z} & (u, v) \in E \\ 0 & \text{otherwise} \end{cases} \qquad (2)$$

The random walks are biased to provide a representation that captures the right mix of equivalences from the graph using the parameters p and q. Suppose we are at a protein v, and we have traversed there through edge (u, v) from protein u, the next protein t in the walk is decided using the transition probability π_{vt} (Fig. 1). Let w_{vt} be the weight and π_{vt} the transition probability on the edge (v, t) directed from v, then the transition probability is set to $\pi_{vt} = \alpha_{pq}(u, t) * w_{vt}$, such that

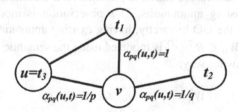

Fig. 1. Random walk transition probability illustration

$$\alpha_{pq}(u, t) = \begin{cases} 1/p & \text{if} \quad d_{ut} = 0 \\ 1 & \text{if} \quad d_{ut} = 1 \\ 1/q & \text{if} \quad d_{ut} = 2 \end{cases} \qquad (3)$$

Using p, q we can control how far a random walk can get from a source node, thus direct (bias) the walk and redefine the context of a node's neighborhood. Higher values of $p(>(\max(1, q)))$ reduce the probability of revisiting a node, and therefore forces the walk in an "exploratory mode", conversely, low values of $p(<(\min(1, q)))$ result in going back to already visited nodes, hence constraining the walk close to the source node. On the other hand, having a higher value for $q(q > 1)$ makes the walk focused on nodes closer to the source and results in sample proteins within a small locality, while having lower values for $q(q < 1)$ tend to explore interactions that lead to more distant protein nodes from the source protein.

The whole representation learning process can be summarized in three phases: preprocessing to compute transition probabilities, random walk simulations and optimization of the vector representations using stochastic gradient descent.

2.3   Negative Examples Selection

In this research the NegGOA [6] approach is adopted, and negative examples are selected based on the available protein annotations and the term hierarchy defined in GO. Initially, the pairwise semantic similarity of all terms is calculated using Lin's approach:

$$sim_H(t, s) = \frac{2 \times IC(LCA(t, s))}{IC(t) + IC(s)} \tag{4}$$

having $LCA(t, s)$ denote the lowest common ancestor of the two terms, while $IC(\cdot)$ is the information content defined with:

$$IC(t) = \left(1 - \frac{log_2(|desc(t)|)}{log_2(|T|)}\right) \tag{5}$$

$|T|$ denotes the number of terms, and $desc(t)$ is the complete set of descendants of t.

In time, more annotations are added to proteins. New annotations often correspond to descendants of existing annotations. This observation is modeled using random walks with restarts on the GO hierarchy, having existing annotations as source nodes. The transition matrix $W'_H \in R^{|T| \times |T|}$ is modeled using the semantic similarities between term pairs in the following way:

$$W'_H(t, s) = sim_H(t, s) \times G(t, s) \tag{6}$$

where $G(t, s) = 1$ if s is child of t, and 0 otherwise. The transition matrix is further normalized:

$$W_H(t, s) = W'_H(t, s) / \sum_{v \in T} W'_H(t, v) \tag{7}$$

Using this transition matrix, the probability to reach a term v starting from a term t is defined as the random walk with restart probability in the 4th iteration and is denoted with $R_H(t, v) = W_H^4(t, v)$.

To utilize the existing annotation set an empirical conditional probability is defined for terms t and s:

$$p(s|t) = \frac{|A_t \cap A_s|}{|A_t|} \tag{8}$$

where A_x is the set of proteins annotated with term x. The lower the value of the conditional probability $p(s|t)$, the higher the probability for s to be chosen as a negative example. Once again, a random walk with restart needs to be employed and the transition matrix to be defined

$$W_C(t,s) = p(s|t)/\sum_{v \in T} p(v|t) \tag{9}$$

having, $W_C^0(t,t) = 1$ and $W_C^0(t,s) = 0 (t \neq s)$.

As in the previous case the 4th iteration of the random walk with restart probability is the probability to reach term v starting from term t, i.e. $R_C(t,v) = W_C^4(t,v)$.

The probabilities derived from the GO hierarchy and the existing annotation in the dataset are used to calculate the following two metrics:

$$L_H(i,v) = 1 - \frac{1}{|T_i|}\sum_{t \in T_i} R_H(t,v) \tag{10}$$

$$L_C(i,v) = 1 - \frac{1}{|T_i|}\sum_{t \in T_i} R_C(t,v) \tag{11}$$

where T_i is the set of existing annotations of the i-th protein, including terms added with transitive closure. $L_H(i,v)$ is the predicted likelihood of term $v \notin T_i$ as a negative example of the i-th protein from $R_H(t,v)$. $L_C(i,v)$ is the predicted likelihood of negative example from $R_C(t,v)$.

The two metrics are combined in one:

$$L(i,v) = \beta L_H(i,v) + (1-\beta)L_C(i,v) \tag{12}$$

where $\beta \in [0,1]$ is a scalar used to control the influence of $L_H(i,v)$ and $L_C(i,v)$.

Finally, the i-th protein receives as negative the annotations that correspond the largest values of $L(i,\cdot) \in R^{|T|}$.

2.4   Deep Learning the Protein Function

The problem of computational protein function prediction i.e. learning the correct labels for the PIN graph nodes is modeled as a binary classification problem i.e. whether a specific functional term should be associated with a protein node or not. In this research a feed-forward deep neural network is used. A simple deep learning six layer model is built for the protein function learning: the first layer consisting of 128 fully connected ReLU (Rectified Linear Units) neurons, the second layer consisting of 128 fully connected sigmoid neurons, the third is a dropout layer (with a 0.25 rate) followed by a layer of 64 fully connected sigmoid neurons, the fifth is once again a dropout layer (0.25 rate) and the final layer is a single fully connected sigmoid neuron. The dropout layers are used since the number of instances used in the learning process is modest in terms of deep learning and overfitting needs to be avoided.

3 Results and Discussions

In the experiments performed for the proposed approach two different vector representations are used for the PIN graph. The first representation is derived by using p = 1, q = 1 in the vector representation method, and the resulting representations will be close for nodes (proteins) in the same network community. The second representation corresponds to p = 1, q = 2, and the resulting representations are close for nodes that share similar structural roles in the graph.

The next step in creating proper experimental setup is the creation of corresponding datasets for each deep learning model that needs to be built. These datasets need to contain a sufficient number of positive examples as well as a corresponding number of negative examples. To that aim the 20 most frequent GO terms for each ontology (MF, BP, CC) present in the final GOA file are chosen as the initial target terms for the deep learning models. The initial target terms are further filtered based on their specificity since the aim is to be able to predict the most specific terms possible. A specificity metric is introduced:

$$C_t = \frac{|F_t|}{|A_t|} \tag{13}$$

where $|A_t|$ is the number of nodes annotated with term t in the final GOA file, while $|F_t|$ is the number of nodes annotated with term t when the annotation sets for each node are expanded using transitive closure in GO (if a node is annotated with term t then it is also annotated with every ancestral term of t). The initial 60 target terms are filtered so that each term that has a specificity metric $C_t > 1.3$ will be discarded. Using this filtering the final target term set is composed of 22 terms.

Positive examples are chosen based on the annotations in the final GOA file. Using the negative example selection method, the examples with the highest likelihood are chosen as negative by defining a lower bound on the likelihood so that the number of negative examples that survive the cutoff is comparable to number of positive examples. This procedure of creating the complete dataset of positive and negative examples is done for each target term separately and independently.

The neural network models are implemented using the deep learning library Keras with Tensorflow as a backend. The binary cross entropy is used as a loss function in the training process, and parameters are optimized using the Adaptive gradient descend (Adagrad) algorithm.

Binary classification is performed using 10-fold cross-validation and the following are of interest: True Positives (TP) – when the functional term is predicted for a protein and is part of the annotation set of the protein, True Negatives (TN) – when the functional term is not predicted for a protein and is not part of the annotation set of the protein, False Positives (FP) – when the functional term is predicted for a protein, but it is not part of the annotation set of the protein, False Negatives (FN) – when the functional term is not predicted for a protein, but it is part of the annotation set of the protein. Using these, the following classification quality measures can be defined:

$$Sensitivity\ (True\ Positive\ Rate) = \frac{TP}{TP+FN} \qquad (14)$$

$$False\ Positive\ Rate = \frac{FP}{FP+TN} \qquad (15)$$

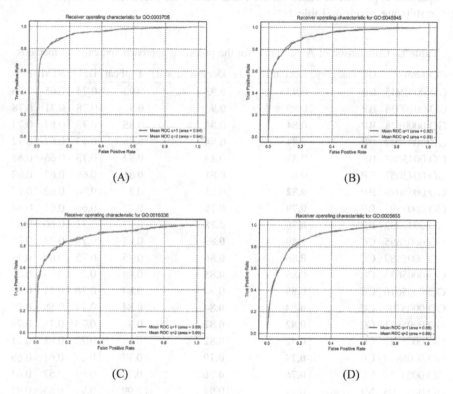

Fig. 2. Receiver operator characteristics for the four top performing deep learning models for (A) GO:0003705, (B) GO:0045945, (C) GO:0016036, (D) GO:0005655

Plotted as coordinate pairs Sensitivity and FalsePositiveRate define the Receiver Operating Curve (ROC). The ROC curve describes the model performance over the complete range of classification thresholds. The Area Under the Curve (AUC) for a classifier model is equivalent to the probability that the classifier will rank a random positive example higher than a random negative example. Figure 2 depicts the top four ROC curves for the deep learning models with best performances.

The proposed approach DeePin is compared with existing methods like topological-feature based prediction (TopFeat) [15], diffusion state distance (DSD) [1], scale-aware topological measures (STM) [13], PPI information (PPIi) [12]. TopFeat considers weighted PIN network topology features with local and global information to characterize proteins and identify protein function. DSD uses graph-diffusion based metric to capture detailed distinctions in proximity and use them in the process of transferring functional annotations in a PPI network. STM uses scale-invariant description of the topology around or between proteins with a network smoothing operation and diffusion kernels. PPIi takes into account function information in the neighborhood of a query protein and the weights of interactions with neighbors and infers specific function for the query protein using a so-called "inclined potential". Comparison is performed based on the AUC values for predicting the target GO terms and results are given in Table 1.

Table 1. Comparison of AUC values for the proposed approach and existing methods

GO term	GO category	DeePin (q = 1)	DeePin (q = 2)	TopFeat	DSD	STM	PPIi
GO:0045945	BP	0.92	**0.93**	0.92	0.74	0.65	0.76
GO:0016036	BP	0.89	**0.9**	**0.9**	0.78	0.71	0.78
GO:0000128	BP	0.84	0.84	**0.85**	0.73	0.67	0.73
GO:0007269	BP	0.84	0.84	**0.86**	0.73	0.67	0.72
GO:0045088	BP	0.83	0.83	**0.84**	0.73	0.66	0.83
GO:0043067	BP	0.8	**0.81**	**0.81**	0.69	0.62	0.69
GO:0055086	BP	**0.82**	0.81	0.8	0.68	0.62	0.67
GO:0007597	BP	**0.79**	**0.79**	0.77	0.66	0.61	0.64
GO:0044282	BP	**0.78**	**0.78**	0.77	0.71	0.66	0.69
GO:0005655	CC	**0.89**	**0.89**	0.87	0.78	0.72	0.75
GO:0005889	CC	**0.86**	**0.86**	0.85	0.75	0.69	0.74
GO:0005635	CC	**0.85**	**0.85**	0.84	0.7	0.65	0.69
GO:0005616	CC	**0.85**	0.84	0.84	0.76	0.71	0.73
GO:0005790	CC	**0.84**	**0.84**	**0.84**	0.74	0.68	0.75
GO:0005831	CC	**0.82**	**0.82**	0.8	0.74	0.7	0.74
GO:0005887	CC	**0.82**	**0.82**	0.81	0.73	0.68	0.71
GO:0005814	CC	**0.79**	**0.79**	0.77	0.67	0.61	0.69
GO:0005731	CC	**0.76**	**0.76**	0.73	0.61	0.57	0.63
GO:0003705	MF	0.94	0.94	**0.99**	0.92	0.83	0.91
GO:0003714	MF	0.88	0.88	**0.95**	0.88	0.78	0.88
GO:0005525	MF	0.75	0.76	**0.85**	0.76	0.72	0.73
GO:0042805	MF	0.72	0.71	**0.8**	0.72	0.65	0.7

As can be seen from the results in Table 1 the proposed approach DeePin gives similar performance to TopFeat, which is the top performing existing method, when targeting BP GO terms. The DeePin performance is slightly better than TopFeat when the target terms are from the CC category. TopFeat significantly outperforms the proposed approach in the MF category. This is mainly due to the fact that very few MF targets are present in the final target term set. One drawback of the proposed approach lies in its need for higher number of positive and negative examples to train the deep learning models. As seen from the filtering of the data performed in this research this is not always the case when it comes to more specific GO terms, which are much more significant in terms of computational function prediction. This is due to the incompleteness of the knowledge for all possible protein interactions and all possible functions a protein performs. However, with the rise in protein data generation in the future this problem will diminish. The current incompleteness of data can explain the "poor" performance the proposed approach has on MF terms, since the MF ontology is by far the "easiest" to predict [14]. With the data increase the proposed approach can only improve in performance since all steps are inherently sensitive to the amount of information presented.

4 Conclusion

In this paper a novel approach for computational function prediction using deep learning in protein interaction networks (PINs) i.e. DeePin is proposed. The approach is very simple since it requires a single algorithm/computation at each step of its pipeline. The PIN graph is translated in a vector space that is able to capture the topological/structural properties of the PIN and their dependencies in a single optimization problem, as opposed to extracting these features independently like in other competing methods. The quality of the prediction is enhanced by making an informed choice on the negative examples used in building the deep learning models. Finally, the deep models are very simple consisting of only six layers with total number of neurons of less than 1000. Experiments are performed using a highly reliable human protein interaction network. Results indicate that the proposed methodology can be very successful in determining protein function and its performance is comparable with state-of-the-art competing methods. The drawback of the proposed approach is its inherent sensitivity to the amount of data available. However, this may become an asset in the future, since big data is generated for proteins daily and the knowledge base for proteins, their interactions and functions, deepens.

References

1. Cao, M., et al.: Going the distance for protein function prediction: a new distance metric for protein interaction networks. PLoS ONE **8**, e76339 (2013)
2. Cao, S., Lu, W., Xu, Q.: Grarep: learning graph representations with global structural information. In: Proceedings of the 24th ACM International on Conference on Information and Knowledge Management, pp. 891–900. ACM (2015)
3. Cesa-Bianchi, N., Re, M., Valentini, G.: Synergy of multi-label hierarchical ensembles, data fusion, and cost-sensitive methods for gene functional inference. Mach. Learn. **88**, 209–241 (2012)
4. Consortium, G.O.: Expansion of the Gene Ontology knowledgebase and resources. Nucl. Acids Res. **45**, D331–D338 (2016)
5. Friedberg, I.: Automated protein function prediction—the genomic challenge. Brief. Bioinform. **7**, 225–242 (2006)
6. Fu, G., Wang, J., Yang, B., Yu, G.: NegGOA: negative GO annotations selection using ontology structure. Bioinformatics **32**, 2996–3004 (2016)

7. Grover, A., Leskovec, J.: node2vec: scalable feature learning for networks. In: Proceedings of the 22nd ACM SIGKDD International Conference on Knowledge Discovery and Data Mining, pp. 855–864. ACM (2016)

8. Guan, Y., Myers, C.L., Hess, D.C., Barutcuoglu, Z., Caudy, A.A., Troyanskaya, O.G.: Predicting gene function in a hierarchical context with an ensemble of classifiers. Genome Biol. **9**, S3 (2008)

9. Hakes, L., Lovell, S.C., Oliver, S.G., Robertson, D.L.: Specificity in protein interactions and its relationship with sequence diversity and coevolution. Proc. Natl. Acad. Sci. **104**, 7999–8004 (2007)

10. Hishigaki, H., Nakai, K., Ono, T., Tanigami, A., Takagi, T.: Assessment of prediction accuracy of protein function from protein–protein interaction data. Yeast **18**, 523–531 (2001)

11. Hu, H., Yan, X., Huang, Y., Han, J., Zhou, X.J.: Mining coherent dense subgraphs across massive biological networks for functional discovery. Bioinformatics **21**, i213–i221 (2005)

12. Hu, L., Huang, T., Shi, X., Lu, W.-C., Cai, Y.-D., Chou, K.-C.: Predicting functions of proteins in mouse based on weighted protein-protein interaction network and protein hybrid properties. PLoS ONE **6**, e14556 (2011)

13. Hulsman, M., Dimitrakopoulos, C., de Ridder, J.: Scale-space measures for graph topology link protein network architecture to function. Bioinformatics **30**, i237–i245 (2014)

14. Jiang, Y., et al.: An expanded evaluation of protein function prediction methods shows an improvement in accuracy. Genome Biol. **17**, 184 (2016)

15. Li, Z., et al.: Large-scale identification of human protein function using topological features of interaction network. Sci. Rep. **6**, 37179 (2016)

16. McDermott, J., Bumgarner, R., Samudrala, R.: Functional annotation from predicted protein interaction networks. Bioinformatics **21**, 3217–3226 (2005)

17. Mikolov, T., Chen, K., Corrado, G., Dean, J.: Efficient estimation of word representations in vector space. arXiv preprint arXiv:1301.3781 (2013)

18. Mostafavi, S., Morris, Q.: Using the gene ontology hierarchy when predicting gene function. In: Proceedings of the Twenty-Fifth Conference on Uncertainty in Artificial Intelligence, pp. 419–427. AUAI Press (2009)

19. Mukhopadhyay, A., Ray, S., De, M.: Detecting protein complexes in a PPI network: a gene ontology based multi-objective evolutionary approach. Mol. BioSystems **8**, 3036–3048 (2012)

20. Nabieva, E., Jim, K., Agarwal, A., Chazelle, B., Singh, M.: Whole-proteome prediction of protein function via graph-theoretic analysis of interaction maps. Bioinformatics **21**, i302–i310 (2005)

21. Perozzi, B., Al-Rfou, R., Skiena, S.: Deepwalk: online learning of social representations. In: Proceedings of the 20th ACM SIGKDD International Conference on Knowledge Discovery and Data Mining, pp. 701–710. ACM (2014)

22. Schaefer, M.H., Fontaine, J.-F., Vinayagam, A., Porras, P., Wanker, E.E., Andrade-Navarro, M.A.: HIPPIE: integrating protein interaction networks with experiment based quality scores. PLoS ONE **7**, e31826 (2012)

23. Tang, J., Qu, M., Wang, M., Zhang, M., Yan, J., Mei, Q.: Line: large-scale information network embedding. In: Proceedings of the 24th International Conference on World Wide Web, pp. 1067–1077. International World Wide Web Conferences Steering Committee (2015)

24. Trivodaliev, K., Bogojeska, A., Kocarev, L.: Exploring function prediction in protein interaction networks via clustering methods. PLoS ONE **9**, e99755 (2014)

25. Trivodaliev, K., Cingovska, I., Kalajdziski, S., Davcev, D.: Protein function prediction based on neighborhood profiles. In: Davcev, D., Gómez, J.M. (eds.) ICT Innovations 2009, pp. 125–134. Springer, Heidelberg (2010). https://doi.org/10.1007/978-3-642-10781-8_14

26. Trivodaliev, K., Kalajdziski, S., Ivanoska, I., Stojkoska, B.R., Kocarev, L.: SHOPIN: semantic homogeneity optimization in protein interaction networks. In: Advances in Protein Chemistry and Structural Biology, vol. 101, pp. 323–349. Elsevier (2015)

27. Valentini, G.: Hierarchical ensemble methods for protein function prediction. ISRN Bioinform. **2014**, 1–31 (2014)

28. Wang, D., Cui, P., Zhu, W.: Structural deep network embedding. In: Proceedings of the 22nd ACM SIGKDD International Conference on Knowledge Discovery and Data Mining, pp. 1225–1234. ACM (2016)

29. Yang, C., Liu, Z., Zhao, D., Sun, M., Chang, E.Y.: Network representation learning with rich text information. In: IJCAI, pp. 2111–2117 (2015)

30. Youngs, N., Penfold-Brown, D., Bonneau, R., Shasha, D.: Negative example selection for protein function prediction: the NoGO database. PLoS Comput. Biol. **10**, e1003644 (2014)

31. Youngs, N., Penfold-Brown, D., Drew, K., Shasha, D., Bonneau, R.: Parametric Bayesian priors and better choice of negative examples improve protein function prediction. Bioinformatics **29**, 1190–1198 (2013)

32. Zhang, Y., Lin, H., Yang, Z., Wang, J., Li, Y., Xu, B.: Protein complex prediction in large ontology attributed protein-protein interaction networks. IEEE/ACM Trans. Comput. Biol. Bioinform. **10**, 729–741 (2013)

Getting Engaged: Assisted Play with a Humanoid Robot Kaspar for Children with Severe Autism

Tatjana Zorcec[1](✉) ⓘ, Ben Robins[2], and Kerstin Dautenhahn[2]

[1] University Children's Hospital, Skopje, Macedonia
tzorcec@gmail.com
[2] University of Hertfordshire, Hatfield, UK

Abstract. Autism is a developmental disability defined as deficits in social communication and interaction and presence of restricted, repetitive behaviors, interests and activities. A recent study from April 2018, estimates autism's prevalence to be 1 in 59 children. The symptoms are manifested as continuum or spectrum, from mild to severe manifestations, demanding different degrees of support in the daily life. Children diagnosed with autism may benefit from early interventions when adjusted to their specific needs. The aim of the study described in this paper was to explore the possible added value of the humanoid robot Kaspar as an intervention tool, in therapeutic and educational purposes in children with autism or robot-assisted play in the context of autism therapy in a hospital setting. This paper provides case studies evaluation of some interaction aspects (i.e. learning basic emotions (happy and sad) and social skills (greetings) etc.) between two young children with severe form of autism and Kaspar. An observational analysis of the skills and the behavior of the children was undertaken across the one year trials. They interacted very fast and quite spontaneously with Kaspar which made a solid ground for achieving our goals-social communication skills as greetings, social interaction skills as eye gaze, learning emotions etc. Our preliminary conclusion is that children are learning in this robot-assisted play, enjoy interaction with a robot, gain and learn quick new knowledge, show much more communication, initiative and proactivity, improve their behavior and generalize some of the learned information in real life.

Keywords: Robot assisted play · Child-robot interaction · Autism

1 Introduction

Autism spectrum disorder (ASD) is a developmental disability defined as deficits in social communication and interaction as well as presence of restricted, repetitive behaviors, interests and activities [1].

A recent study from April, 2018, estimates autism's prevalence to be 1 in 59 children or 1 in 37 boys and 1 in 151 girls [3]. The symptoms are manifested as continuum or spectrum, from mild to severe manifestations, demanding different degree of support in the daily life. Children with autism have variety in needs,

© Springer Nature Switzerland AG 2018
S. Kalajdziski and N. Ackovska (Eds.): ICT 2018, CCIS 940, pp. 198–207, 2018.
https://doi.org/10.1007/978-3-030-00825-3_17

capacities and interventions. Therapeutic objectives differ for each child and change over time which requires tuning and personalization of the objectives. This makes autism a heterogeneous disorder.

Children diagnosed with autism benefits from early interventions when adjusted to their specific needs [20]. Although, some children achieve progress from therapies in some areas, prognosis in general is poor. Children with autism will become adults with autism with difficulties in independent living, professional and social life [10].

Technology, as a supporting tool in interventions, education and daily living of people with autism, has proven to be valuable tool in the hand of researchers and practitioners [2, 7].

1.1 Robots as Assistive Technology for Children with Autism

In recent years, many research studies have shown that the use of robots, as a supporting tool in the development of various skills and knowledge in children with special needs, is quite beneficiary [11, 16, 17]. Mobile robots like IROMEC and QueBall [12, 15] and humanoid robots like Kaspar, Robota, Milo and Nao [4, 5, 9, 13, 14, 19] are especially used to help children with autism develop various skills and mediate interactions with peers and adults. Furthermore, other robots have been used to engage children with autism in play activities like artificial pets (baby seal robot Paro and the teddy bear Huggable) [8, 18] and the small cartoon-like robot Keepon [6].

1.2 Why Using Robots for Children with Autism?

Robots can provide individualized approaches for a specific child's needs. For ASD children, they provide predictable, safe and reliable environment and this experience can be less terrifying for them than interacting with a human being. Many ASD children have difficulty in making eye contact or typical social interaction. It's not comfortable or easy for them to do it. It's a problem that affects their social life as well the lives of their families. But why are ASD children more interacting with robots than with humans? Researchers think it's because robots are not as complex as people, and can be made with human-like, but very much simplified features, i.e. simplified appearance, less complex facial expressions and simplified basic behavior. In addition, as any other children, children with ASD like toys too. Furthermore, the use of robots is based on our intrinsic desire and motivation for technology. Autistic children can learn with more ease and less effort, and social interaction will no longer be as challenging. Furthermore, robots enable social interaction that can be generalized in daily living.

1.3 The Robot Platform-Kaspar

Kaspar is a child-sized humanoid robot that is using bodily expressions (movements of the head, arms, and torso), facial expressions, gestures and prerecorded speech to interact with a human. Kaspar has been developed by the University of Hertfordshire, Adaptive Systems Research Group, a multidisciplinary group that conducts pioneering research into artificial intelligence and robotics. The Group studies how humans and machines interact with each other. Building on research that stretches back to 1998, the group has developed Kaspar into a working prototype that has met with significant success in trials with autistic children in various settings. Kaspar is a 60 cm tall robot that is fixed in a sitting position (see Fig. 1).

Fig. 1. Wireless keypad for manual control (left); The Kaspar robot (right)

Kaspar has been purposefully designed as a robot with simplified, realistic human-like features offering a predictable form of communication, making social interaction simpler, non-judgmental and more comfortable for the child. As a three-dimensional, physical object, it provides children with the opportunity for safe, physical engagement and exploration of Kaspar's features and behaviors in a non-judgmental manner.

Kaspar has 11 degrees of freedom in the head and neck, 1 DOF in the torso and 5 in each arm. Kaspar's head can tilt, move from side to side, and up and down. Kaspar's face is made from a silicon rubber mask that covers a plastic frame and includes eyes that can move from side to side, as well as and up and down, and eye lids that can open and shut. The mouth is capable of opening, smiling and portray happy and a sad expressions. In addition to the above, Kaspar's torso can move from side to side. Kaspar is mounted with several touch sensors on cheeks, torso, left and right arm, back and palm of the hands and also soles of the feet. These tactile sensing capabilities allow the robot to respond autonomously when being touched, allow it also to give feedback according to the style of the interaction, encouraging certain behaviors and discouraging inappropriate behavior. The robot could also be operated by a remote controlled keypad which can be used by the accompanying adult e.g. a therapist/teacher/parent or by the children themselves who interact with it.

Kaspar can engage children with autism in a variety of therapeutic/educational games, e.g. turn-taking, joint attention and collaborative games, cause and effect games etc. The robot can also hold objects (e.g. fork, spoon, tooth brush, comb etc.) in order to play games where children can learn about food and hygiene etc.

2 Kaspar Trials Set-Up and Procedures

The trials took place at the University Children's Hospital-Skopje, Macedonia. Prior to the trials, Kaspar was programed with movements and speech phrases in the Macedonian language by the clinician. In order to provide a safe environment where the child will feel comfortable, the trials were designed to permit free interaction with Kaspar in a familiar room, in a present of the parent and the clinician.

Some of the play scenarios included imitation games, learning emotions, turn taking games, learning animals and animal sounds, learning sounds and words, hygiene and food. Touching different robot parts causes different reactions and movements of the robot and this setting was used in all other scenarios e.g. Kaspar can react to the style of interaction with appropriate feedback and facial expressions. For example, soft touch or tickling on the left foot will cause the robot to smile and say "this is nice, it tickles me", touch on the torso will cause a loud laugh saying "ha, ha, ha", hitting or punching will cause the robot to have a 'sad' expression, turn the face and torso away to one side, cover the face with the hands and saying "ouch-this hurts".

Kaspar was placed on a chair, connected to a laptop. The clinician was seated next to the robot. The children were brought to the familiar room by their parents. If the child showed a sign of anxiety or appeared not interested for interaction, the sessions were interrupted. One video camera was recording the trials. Kaspar could respond autonomously to different tactile interactions, as well as be operated remotely via a wireless remote control (a specially programmed keypad), either by the clinician or by the child himself. Each trial lasted as long as the children were comfortable staying in the room. The average duration of each trial was approximately 20 min.

3   Observations of Interactions with Kaspar

We present a case study of two young children with autism in their engagement with Kaspar. The first child was a boy, who has been diagnosed for severe form of autism at the age of 26 months in November, 2016. He was non-verbal child, with poor eye contact, poor social communication and interaction, stereotyped behavior, restricted and repetitive behaviors, interests and activities, auto aggressive behavior, poor playing skills and no joint attention skills, and he had sensory difficulties. He started with various early interventions. The second child is a girl who has been diagnosed with a severe form of autism as well at the age of 24 in December, 2016, with similar symptoms as the boy but without sensory problems and auto aggression. They started sessions with Kaspar in January, 2017.

We will present our observations in some of the play scenarios and goals during a one year of intervention.

A. Familiarizing sessions

Kaspar was introduced to them in January 2017, with sessions once to twice per week, depending on the child's response and frequency of other therapeutic interventions. Initially we started with simple exposure to the robot and gradually moved to more complex engagement. The images in Fig. 2 below are taken in the familiarizing sessions with Kaspar.

In these first sessions children were ignoring the robot and wondering around the room. They didn't show any discomfort, just no interest to interact, so the sessions were lasting from 10–15 min where the robot was saying "Hi, my name is Kaspar, let's play together", singing child's songs, making laughing sounds and "expressing" happy

Fig. 2. Initial sessions with Kaspar

feelings. Gradually, in repeated encounters with Kaspar, those two children started interacting with the robot.

B. Imitation games

After 3–4 familiarization sessions, the children started approaching the robot with a smile, seating on a chair in front of Kaspar, ready to interact, and both showed an intensive interest in Kaspar and some initiation of physical contact with a co-present adult. We started with imitation games that were simple and fun for the children (Fig. 3).

Fig. 3. Imitation games with Kaspar

In these games Kaspar was singing a very popular Macedonian children's song with movements appropriate to the lyrics, while the children were imitating its movements. The children played with enjoyment and with happy facial expression whilst making eye contact with Kaspar. The song would not be played again until the child gave any sign of a desire for repetition (e.g. nodding for "yes" or making a sound that means "more").

C. Learning skills of greetings

One of the main goals in the trials with Kaspar was to learn social skills and make generalization in real life. Greetings as "Hello", "Hi, how are you?" and "Goodbye" were part of every session (Fig. 4).

After the children learnt to practice the greetings with Kaspar, they then practiced it with the therapist and parent that were present at the session, with a goal to practice it in real life with other people too. This behavior involves a smile and a proper eye contact (Fig. 5).

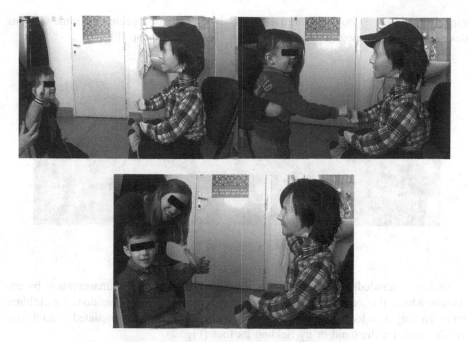

Fig. 4. Practicing "Hello", "Hi, how are you?" and "Goodbye"

After 10 sessions, the children started using greetings in their everyday life without anxiety, resistance or discomfort. At present, when entering and leaving the lab room, both children are using these gestures to salute the practitioners.

Fig. 5. Practicing handshake

D. Learning emotions

An additional and very important goal in the sessions with Kaspar was to learn to recognize basic emotions (sad, happy, hurt and afraid) and to respond appropriately.

It can be noted that in the initial tactile interactions with Kaspar those two children were behaving in some forceful way (poking Kaspar's eyes or grasping it very firmly), especially the young boy. This initial "aggressive" behavior repeated for some sessions despite Kaspar's responding with a "sad" or "hurt" expression (Fig. 6).

Fig. 6. Child's interact in forceful way

After acknowledging that this type of behavior is hurtful and unacceptable by co-present adults, this behavior drastically declined. In the following sessions the children were learning how to make Kaspar happy or laugh again. It was stimulated to do it with gentle touch on the hand or by tickling its foot (Fig. 7).

Fig. 7. Comforting Kaspar with tickling

One of the most important feelings to learn was happiness. It is interesting to notice that the children learned much easier and faster happy feelings and expressions than the negative ones (Fig. 8).

Parents of both children later reported that the children now know the difference between happy and sad, especially happy, and use this gesture in real life.

Fig. 8. Recognizing and responding to happiness

4 Discussion and Conclusion

The use of robots in therapy for ASD children is very unique approach that is receiving increasing attention from the scientific and general public. Research in this area is still very young and its pioneering work with innovative ideas on a robot use in a clinical environment.

These case studies presented that some children with a severe form of autism can interact with the humanoid robot Kaspar and learn from that interaction. They get engaged with Kaspar in several ways, developing a repertoire of new knowledge and behaviors. For example they have learned basic social communication skills like greetings (which they didn't do before), and basic emotions that resulted with more calm and happy behavior during the intervention sessions. Furthermore, during the intervention, parents stated that after 8–10 sessions of practicing, greetings as well as recognition and appropriate reactions to happy and sad emotions, are used in daily life too.

The programing and use of the Kaspar robot by the researchers was quite easy and simple. The software allows easy programming of newly developed game scenarios and therapeutic goals. The use is simple and can be adapted to the children's needs and their progress during the sessions.

As the robot assisted play for these children is ongoing, a further longitudinal study on how persistent and continuous this learning effect might be is needed and is planned.

As we mentioned before, autism is an extremely heterogeneous condition and although children with autism share the same core difficulties, each child has an individual set of manifestation. It is expected in some other groups of children to have different responses. Many challenges, due to the nature of the condition, requires a modular approach in designing the assistive technology and play scenarios for child-robot interaction in children with autism, aiming their individual needs. Rigid play scenarios and interventions would not be able to address the variety of behaviors. This paper is an example of a flexible approach in use of robots, where the robot can react autonomously but also the practitioner can guide the robot according to the child's behavior, provoking additional robot responses when needed.

There are numerous positive reasons for using robots in ASD individuals. One is to use their intrinsic interest for technology as a tool to learn, second is that robots are simple and predictable, third is that robots are rather easy for programming and using, forth is the possibility of adapting the robot's behavior making an individual protocol for each ASD child etc. Nevertheless, if we really want to actually make a difference in the lives of children with autism, we need to find a way to expand the findings from case studies to full studies.

References

1. American Psychiatric Association. Diagnostic and Statistical Manual of Mental Disorders. 5th edn. American Psychiatric Association, Arlington (2013)
2. Aresti-Bartolome, N., Garcia-Zapirain, B.: Technologies as support tools for persons with autistic spectrum disorder: a systematic review. Int. J. Environ. Res. Public Health (2014). https://doi.org/10.3390/ijerph110807767
3. Baio, J., Wiggins, L., Christensen, D.L., et al.: Prevalence of autism spectrum disorder among children aged 8 years-autism and developmental disabilities monitoring network, 11 sites, United States 2014. MMWR Surveill Summ. 67(6), 1–23 (2018). https://doi.org/10. 15585/mmwr.ss6706a1
4. Billard, M., Robins, B., Nadel, J., Dautenhahn, K.: Building robota, a mini-humanoid robot for the rehabilitation of children with autism. RESNA Assist. Technol. J. 19, 37–49 (2006)
5. Dautenhahn, K., et al.: KASPAR – a minimally expressive humanoid robot for human-robot interaction research. Appl. Bionics Biomech. 6, 369–397 (2009). Special issue on "Humanoid Robots"
6. Kozima, H., Yasuda, Y., Nakagawa, C.: Social interaction facilitated by a minimally-designed robot: findings from longitudinal therapeutic practices for autistic children. In: RO-MAN: 16th IEEE International Symposium on Robot and Human Interactive Communication, Jeju, 26–29 August, vols 1–3, pp. 598–603. IEEE, Jeju (2007)
7. Lee, H., Hyun, E.: The intelligent robot contents for children with speech-language disorder. Edu. Technol. Soc. 18(3), 100–113 (2015)
8. Marti, P., Pollini, A., Rullo, A., Shibata, T.: Engaging with artificial pets. In: Proceedings of Annual Conference of the European Association of Cognitive Ergonomics, Chania, Greece, pp. 99–106 (2005)
9. Milo. https://robots4autism.com/
10. Myers, S.M., Johnson, C.P.: Management of children with autism spectrum disorders. Pediatrics 120(5), 1162–1182 (2007). https://doi.org/10.1542/peds.2007-2362
11. Pennisi, P., et al.: Autism and social robotics: a systematic review. Autism Res. 9, 165–183 (2015)
12. Robins, B., et al.: Scenarios of robot-assisted play for children with cognitive and physical disabilities. Interact. Stud. 13(2), 189–234 (2012)
13. Robins, B., Dautenhahn, K., Dickerson, P.: From isolation to communication: a case study evaluation of robot assisted play for children with autism with a minimally expressive humanoid robot. In: Proceedings the Second International Conferences on Advances in Computer-Human Interactions, ACHI 09 (2009)
14. Robins, B., Dautenhahn, K., Dickerson, P.: Embodiment and cognitive learning – can a humanoid robot help children with autism to learn about tactile social behaviour? In: Ge, S. S., Khatib, O., Cabibihan, J.-J., Simmons, R., Williams, M.-A. (eds.) ICSR 2012. LNCS (LNAI), vol. 7621, pp. 66–75. Springer, Heidelberg (2012). https://doi.org/10.1007/978-3-642-34103-8_7
15. Salter, T., Davey, N., Michaud, F.: Designing and developing QueBall, a robotic device for autism therapy, Ro-Man (2014)

16. Sartorato, F., Przybylowski, L., Sarko, D.K.: Improving therapeutic outcomes in autism spectrum disorders: enhancing social communication and sensory processing through the use of interactive robots. J. Psychiatr. Res. **90**, 1–11 (2017)
17. Scassellati, B., Admoni, H., Mataric, M.: Robots for use in autism research. Annu. Rev. Biomed. Eng. **14**, 275–294 (2012)
18. Stiehl, W.D., Lee, J.K., Breazeal, C., Nalin, M., Morandi, A.: The huggable: a platform for research in robotic companions for pediatric care. In: Creative Interactive Play for Disabled Children Workshop at the 8th International Conference on Interaction Design and Children (IDC 2009), Como, Italy, 3–5 June, p. 8 (2009)
19. Tapus, A., et al.: Children with autism social engagement in interaction with Nao, an imitative robot - a series of single case experiments,". Interact. Stud. **13**(3), 315–347 (2012)
20. Volkmar, F.R., Paul, R., Rogers, S.J.: Handbook of Autism and Pervasive Developmental Disorders: Diagnosis, Development, and Brain Mechanisms, vol. 1. Wiley, Hoboken (2014)

Evaluation of Multiple Approaches for Visual Question Reasoning

Kristijan Jankoski and Sonja Gievska[(✉)]

FCSE, Ss. Cyril and Methodius University, Skopje, Macedonia
mail@kjanko.com, sonja.gievska@finki.ukim.mk

Abstract. Extracting meaningful patterns between objects i.e., relational reasoning is crucial element of human reasoning and still a challenging task for
artificial intelligence. Our research objective was to investigate two end-to-end
architectures augmented with a relational neural module on a challenging
Cornell NLVR visual question answering task. It was our hope that the relational reasoning capabilities on multi-modal inputs for which the relational
networks are famous for would be leveraged on the task at hand. We have
achieved state-of-the-art performance outperforming the results reported in the
related research studies conducted on the same benchmark dataset.

Keywords: Relational reasoning · Visual question answering
Deep relational neural networks · NLVR challenge

1 Introduction

The ability to reason relationally is inherent to human thinking. People use a diverse set
of skills to extrapolate meaningful patterns in information of different modalities that is
critical to a number of cognitive skills, from language understanding and visual reasoning to critical thinking and problem-solving. The ability to reason about visual input
is a challenging task for agents and some authors have argued that to learn such a
complex process, aspect of compositionality [8] or relational computation [13] must be
integrated in agent's reasoning.

Over the last decade, research on deep learning has emerged and a remarkable
number of tasks in the domain of image and natural language processing has been
accomplished. New ways on amplifying the current level of technology has advanced
deep learning models to the point of being successful on a number of complex tasks
involving both, natural language processing and visual reasoning (e.g., automatic
generation of image descriptions, visual question answering (QA), spatial reasoning,
storytelling).

The purpose of this research was to evaluate multiple end-to-end solutions to the
problem recently presented by Suhr et al. [14], describing the creation of the Cornell
Natural Language Visual Reasoning (NLVR) dataset and see if the problem is solvable
with a general architecture. The NLVR dataset consists of pairs of synthetically generated images and their language descriptions, each annotated by a binary label indicating whether the relational statement, describing the image, is true or false. The
images are simple and contain few two-dimensional shapes, however the sentences

© Springer Nature Switzerland AG 2018
S. Kalajdziski and N. Ackovska (Eds.): ICT 2018, CCIS 940, pp. 208–216, 2018.
https://doi.org/10.1007/978-3-030-00825-3_18

associated with them are natural and unstructured. The simplicity of the verbal and image content allows us to rely on their structured representation and thus separating the linguistic problem from the computer vision related part.

A number of approaches have been explored for visual question answering [5, 11, 14, 15]. In particular, an end-to-end architecture consisting of convolutional neural network (CNN) for processing the image and a long short-term memory network (LSTM) that is trained jointly on the encoded question and image to produce the output have been put forth [11]. We extend this monolithic architecture approach and generate question embeddings in vector space in order to tackle on the linguistic problem, and use several approaches for generating image feature vectors utilizing convolutional neural networks (CNN) [10] and capsule networks [12]. In this research, two models, inspired by the RN-based architecture introduced by Santoro et al. in [13] were evaluated. The structure and parameters of the general RN model needed to be adjusted to account for the constraints of this particular task. On the raw image version of the Cornel NLVR published test set, our RN-based models have achieved an accuracy of 89.1%, outperforming the previous state-of-art models.

In what follows, we present a brief discussion of the related research and give a description of the dataset used for our study. Then, we highlight the primary findings of our empirical research in evaluating the performances of two end-to-end RN-based architectures.

2 Related Research

In recent years, a range of benchmarks for visual reasoning have been proposed, including the datasets for image captioning [4], relational graph prediction [9] and visual question answering [7, 11]. Corpora, containing complex real-world images [7] or simple synthetic images [11] paired with natural language statements have been specifically created for the tasks that pertain to answering visual reasoning questions.

The majority of the approaches for visual reasoning follow the same pipeline – recurrent neural networks (RNN) are used to encode the questions and convolutional neural networks are used as feature extractors for the images [11]. The findings that extend across several studies is that attention mechanism has been successful in learning the alignments between different objects by locating important image regions as well as the joint representation of the visual objects [3, 5, 11, 13–15].

The authors of the NLVR dataset have provided multiple baseline models [14] against which our proposed models will be compared. Two single modality models, image-only and text-only, were used to illuminate how the particularities of a domain might influence the results. Accuracies slightly better than chance were reported. Three methods, namely, MAXENT classifier, a single layer perceptron and a concatenation of extracted features and RNN have been evaluated on the structured representation test set. In contrast, two alternative models were used for the raw image test set, namely, the well-established monolithic architecture CNN+RNN and neural module network (NMN) proposed by Andreas et al. [2]. Out of all eight baselines, the NMN architecture yielded the best accuracy on the NLVR test sets.

An end-to-end bidirectional attention architecture [15] and a semantic parsing architecture [5] evaluated on the same NLVR QA dataset are two research works closely related to the objective of our own. Our much simpler end-to-end architecture that uses a convolutional neural network as a feature extraction layer for the images, and a feed-forward network that learns the relations between the objects outperforms the results of the reported studies on the same visual reasoning challenge.

3 Dataset and Problem Statement

Data in the form of synthetic images paired with simple natural language statements publicly available as the Cornell NLVR dataset [14] has been targeted in this research. Each image consists of three boxes, containing non-overlapping objects (1–8) that may vary in size, position, color and shape is accompanied by a language statement as shown in Fig. 1. A binary class is assigned to each pair denoting whether the statement is true or false in the context of the image. In general, any combination of language statement and image is allowed and could be associated with true or false class label. Language statements and class labels have been collected through crowdsourcing.

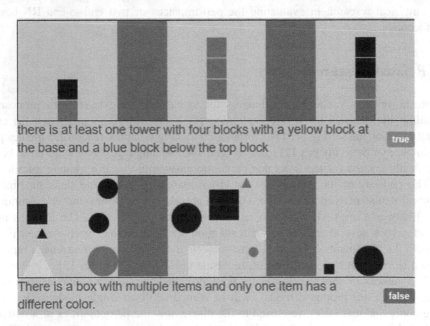

Fig. 1. Cornell NLVR task - Given an image and a statement, the model needs to predict whether the statement correctly describes the image or not.

The visual QA task associated with the NLVR dataset is a true or false answer to the question whether a language description matches or not the compositional structure of the scene presented in the corresponding image that requires some relational reasoning. The context (image) and the question being addressed become the ultimate arbiters.

The total amount of language statements paired with synthetic images is 92,244, are divided between a training (74,460), development (5,940) and two test sets, one published (5,934) and the other unreleased (5,910). Structured representation of images, the JSON representation from which the images were synthetically rendered, are also available, making the task of visual QA possible on both, unstructured and structured visual input.

4 End-to-End Models

This section provides a broadly-based discussion of the related concepts, means and concerns in current deep learning research pertaining to visual reasoning with the objective to articulate the rationale behind the research choices we made. A number of prominent deep learning architectures and practices have been considered and experimented with to determine the most suitable choice that match the task and context under investigation. We identified and followed trends that have proven their effectiveness in the field and explored those that appeared to be promising and yet not enough empirically-supported by research.

Two conventional neural network modules have dominated deep learning models for computer vision and NLP tasks, namely CNN and LSTM, respectively. We give a short briefing since they were components for generating image and question embeddings in our models as well.

Convolutional Neural Networks. Convolutional neural networks (CNNs or ConvNets) [10] are very similar to ordinary neural networks: they are made up of neurons that have learnable weights and biases. Each neuron receives some input, performs a dot product and follows it with a function that introduces non-linearity. The entire network expresses a single score function: from raw image pixels on one end, to class scoring at the other.

Three main types of layers are used to build ConvNet architectures: *convolutional*, *pooling* and *fully-connected* layer. Stacking of these layers allows the ConvNet to transform the image layer-by-layer from the original pixel values to the final class score.

Long Short-Term Memory Networks. Long short-term memory networks (LSTMs) [6] are a special kind of recurrent neural network that are capable of learning long-term dependencies. They are explicitly designed to remember information for long periods of time – they preserve the error that can be backpropagated through time and layers. Information can be stored in the LSTM cells, written to or read from, much like data is stored in computer's memory. This property allows them to model the spatial dependencies in sequential data, making it suitable for language processing, which was utilized in our deep learning model.

4.1 Relational Neural Network

In this research, two models, inspired by the RN-based architecture proposed in [13] were evaluated against a background of belief that having a network specifically designed for relational reasoning could be effective for answering the NLVR questions that are explicitly relational in nature. The idea underlying relational networks is to have a compact module with built-in capabilities for reasoning rationally with objects, that has no need for knowledge about the particular objects or relations present in the training data (e.g., type, existence). The kind of problems that this model would be able to address are pervasive, from spatial reasoning and language understanding to making complex inferences. The role of the RN module would be analogous to the critical role that relational reasoning plays in higher-level cognitive processes.

A relational network is a composite function $f\varphi$, operating on a set of objects $O = \{o_1, o_2, \ldots, o_n\}$

$$RN(O) = f\varphi\left(\sum_{i,j} g_\theta(o_i, o_j)\right) \tag{1}$$

The inner sum represent the step of pairing objects and inferring the type of relations g_θ that exists or do not exist between the objects in each pair. The parameters φ and θ of the functions $f\varphi$ and g_θ represent the synaptic weights learned by the $f\varphi$- and g_θ- MLPs. Different tasks and datasets are likely to afford different MLP structures.

The general architecture of the models we have evaluated is shown in Fig. 2. Convolutional networks were used as a feature extraction layer for the images, generating vector representations of objects. We would like to point out that the broad term "object" does not necessarily imply physical object, it can also represent a background or feature characteristic that might be used later for comparisons with other objects. LSTM was used for generating embeddings of the statements that serve the purpose of conditioning the g_θ function - a relation between two objects conditioned by the question. Then, the information from the two modalities is fused allowing the model to learn the correct relationships between them. As a final step of the pipeline processing, we train a classifier using the output of these networks in order to make a binary prediction.

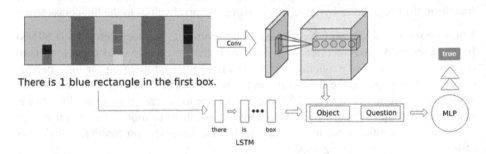

There is 1 blue rectangle in the first box.

Fig. 2. Visual representation of the general RN-based architecture.

4.2 Capsule Networks

Capsule networks (CapsNets) [12], in comparison with CNNs, are composed of capsules instead of neurons. A capsule is a small group of neurons that learns to detect a particular object within a given region of the image and outputs a vector, whose length represents the estimated probability that a given object is present. The object's orientation and position is encoded in the object's parameters, and if it is changed slightly the capsule outputs the same vector length with different probability distribution.

CapsNets are organized in multiple layers as shown in Fig. 3. The capsules in the lowest layer are called primary capsules, and each of them receives a small region of the image as an input. Higher-layers capsules are called routing capsules, and their job is to detect larger and more complex objects. Both, ConvNet and CapsNet were used in the proposed RN-based architecture for image processing in our task at hand – their performances evaluated and compared. It is worth noting that capsule networks have not been previously evaluated on visual QA tasks.

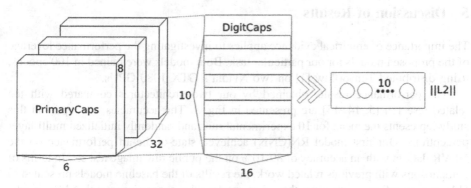

Fig. 3. Dynamic routing between capsules [12].

4.3   Experimental Setup

To address the research questions, an exploratory experiment was carried out investigating the performance of the two different end-to-end architectures on the NLVR raw image test set. Suitable adjustments of the structure and parameters of the general RN-based architecture proposed in [13] were made regarding our particular task and dataset.

RN(CNN). In the first experiment, we use CNNs to extract features from the raw images and thus converting it to a set of objects represented by vectors. This is done by convolving the images of size 200×500 through three convolutional layers with kernel size of 32 to feature maps of size 3×3. After the convolutions, each of the 32-dimensional cells in the feature map was tagged with a coordinate indicating its spatial position and fed as input to the relation network.

A bidirectional LSTM neural network creates the question embeddings - 50 unit word look-up embeddings are created for each word in a question and fed to the LSTM, a single word embedding at a time.

The final module of the RN is MLP consisting of three-layers, each with 50 neurons per layer, activated by a rectified linear unit (ReLU). The output of the network is a linear layer, producing binary classification. The loss is calculated using binary cross-entropy function, and the network is trained using Adam optimizer with a learning rate of $3e^{-4}$. To avoid overfitting, we apply 50% dropout to all layers.

RN(CapsNet). In our second experiment, we have tested a CapsNet encoded architecture, which extends the previous CNN model by including 32 primary capsules layer. The output of this network is a 16-dimensional vector of instantiation parameters – this is the point at which the capsules do their job. The output of this sub-network are two vectors of NLVRCaps – which has 2 capsules, one for each class label. Our intuition was that by defining the probability distributions of the relations detected within the capsules could lead to better performance results. Finally, we use a fully-connected layer to get the vector embedding of the image.

5   Discussion of Results

The importance of empirical evidence applies to investigating the performance leverage of the proposed models for our particular task. Both models were trained in 100 epochs, using distributed Tensorflow [1] on two Nvidia's GTX 1080 GPUs.

The performance results obtained by our two architectures compared with the related research [5, 14, 15] are presented in Fig. 4. The accuracies, we report in this study represents the mean for 10 experimental runs and randomly initialized multi-layer perceptrons. Our first model RN(CNN) achieved state-of-the-art performance on the NLVR dataset with an accuracy of 89.10% on the public raw image test set. In terms of comparisons with previous related work, the results of the baseline models presented in [14] are significantly weaker than our model, the more successful NMN model obtaining 66.1% accuracy. Our model have also outperformed the results of two subsequent research studies [5, 15] evaluating different deep learning architectures on the same dataset. The best result reported on the same raw image test set that our architectures were tested is the research by Goldman et al. [15]. Their end-to-end model with joint bidirectional attention and object-ordering pointer networks has achieved an accuracy of 69.7%. It should be noted that the results of the semantic parser [5], their most successful method W.+Disc yielding 84% accuracy, were performed only on the structured representation task. The results confirm the claims that relational neural networks are a powerful tool for relational inference on a single modality and multi-modal inputs [13].

Our evidence has shown that by utilizing the capsule networks, similar results could be obtained in terms of accuracy compared to our previous model results. The CapsNet encoded architecture yielded 0.47% improvement, however the training time of the second network model was considerably longer – therefore, the tradeoff between accuracy gains and computational complexity should be taken into account when dwelling on the choice of architecture. While the expectations attributed to the use of capsule networks were not met, to an extent, this result is indicative of the fact that there are many more degrees of freedom with respect to potential changes in the RN-structure that are worth exploring. In particular, one line of work that deserves attention should be augmenting the component for generating embeddings of the language statements.

We conclude that relational neural network in combination with the CNN network provides the best results in terms of both, classification accuracy and computational complexity. A notable advantage that we ascribe to the proposed end-to-end RN-based model is that we can feed the network with raw data without requiring any kind of pre-processing on the data itself. Our core insight is that the RN-based models exhibit the ability to recognize objects and spatial relationships between them, as well as perform higher-level skills, such as counting, logical inference and comparisons.

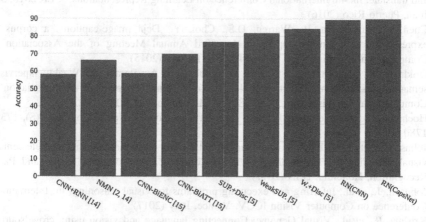

Fig. 4. Accuracies obtained by the proposed RN-based models compared with previously reported research conducted on the NLVR published test set: baseline models by Suhr et al. [14], and related research by Goldman et al. [5] and Tan et al. [15].

6 Conclusions

The key objective of this paper is to demonstrate how relational-based neural networks could be used on relatively unstructured and structured input from the NLVR dataset to solve visual question answering tasks. We have presented two end-to-end models with built-in mechanism that is capable of learning relations between objects in an image and assigns a truth value to the natural language statements describing the relations between the objects in a particular image. We have evaluated the models on the unstructured version of the NLVR test set (raw images), obtaining increased performance compared with the related research using the same dataset. The performance advantage afforded by the RN-based models highlight the power of relational neural networks when applied on visual reasoning tasks.

References

1. Abadi, M., et al.: TensorFlow: a system for large scale machine learning. In: Proceedings of the 12th USENIX Symposium on Operating Systems Design and Implementation, Savannah, GA, USA, pp. 265–283 (2016)
2. Andreas, J., Rohrbach, M., Darrell, T., Klein, D.: Neural module networks. In: IEEE Conference on Computer Vision and Patter Recognition CVPR (2015)
3. Bahdanau, D., Cho, K., Bengio, Y.: Neural machine translation by jointly learning to align and translate. In: 4th International Conference on Learning Representations, ICLR 2016, San Juan, Puerto Rico (2016)
4. Chen, J., Kuznetsova, P., Warren, D.S., Choi, Y.: Deja image-captions: a corpus of expressive descriptions in repetition. In: 53rd Annual Meeting of the Association of Computational Linguistic, pp. 504–514. ACL, Denver (2015)
5. Goldman, O., Latcinnik, V., Naveh, U., Globerson, A., Berant, J.: Weakly-supervised semantic parsing with abstract examples. In: 56th Annual Meeting of the Association of Computational Linguistic, pp. 1809–1819. ACL, Melbourne (2018)
6. Hochreiter, S., Schmidhuber, J.: Long short-term memory. J. Neural Comput. 9(8), 1735–1780 (1997)
7. Johnson, J., et al.: CLEVR: a diagnostic dataset for compositional language and elementary visual reasoning. In: Proceedings of IEEE Conference on Computer Vision and Patter Recognition, Honolulu, USA (2017)
8. Johnson, J., et al.: Inferring and executing programs for visual reasoning. In: International Conference on Computer Vision ICCV, Venice, Italy (2017)
9. Krishna, R., et al.: Visual Genome: Connecting language and vision using crowdsourced dense image annotations. Int. J. Comput. Vis. 123(1), 32–73 (2017)
10. Krizhevsky, A., Sutskever, I., Hinton, G.E.: ImageNet classification with deep convolutional neural networks. In: Pereira, F., et al. (eds.) Advances in Neural Information Processing System, vol. 25. NIPS, Lake Tahoe (2012)
11. Malinowski, M., Rohrbach, M., Fritz, M.: Ask your neurons: a neural-based approach to answering questions about images. In: International Conference on Computer Vision ICCV, Santiago, Chile (2015)
12. Sabour, S., Rosst, N., Hinton, G.E.: Dynamic routing between capsules. In: Guyon, I., et al. (eds.) Advances in Neural Information Processing System, vol. 30. NIPS, Long Beach (2017)
13. Santoro, A., et al.: A simple neural network module for relational reasoning. In: Guyon, I., et al. (eds.) Advances in Neural Information Processing System, vol. 30. NIPS, Long Beach (2017)
14. Suhr, A., Lewis, M., Yeh, J., Artzi, Y.: A corpus of natural language for visual reasoning. In: 55th Annual Meeting of the Association of Computational Linguistic, pp. 217–223. ACL, Vancouver (2017)
15. Tan, H., Bansal, M.: Object ordering with bidirectional matchings for visual reasoning, In: Proceedings of NAACL-HLT 2018. ACL, New Orleans (2018)

Explorations into Deep Neural Models for Emotion Recognition

Frosina Stojanovska[✉], Martina Toshevska, and Sonja Gievska

Faculty of Computer Science and Engineering, Ss. Cyril and Methodius University,
Skopje, Macedonia
stojanovska.frose@gmail.com, martina.tosevska.95@gmail.com,
sonja.gievska@finki.ukim.mk

Abstract. Deep emotion recognition is the central objective of our recent research efforts. This study examines the capability of several deep learning architectures and word embeddings to classify emotions on two Twitter datasets. We have identified several aspects worth investigating that appeared to challenge and contrast previously established notion that semantic information is captured by distributional word representations. Our evidence has shown that extending the word embeddings to account for the use of emojis and incorporating a suitable lexicon of emotional words can lead to a better classification of the emotional content carried by Twitter messages.

Keywords: Emotion detection · Deep learning
Deep neural networks · Word embeddings · Lexicon embeddings
Emoji embeddings

1 Introduction

Human behavior is guided by our emotions that affect our reasoning, decisions making, disposition and judgment of others. Decoding emotions is challenging not only because of their elusive nature but also due to the complexity and ambiguity of their manifestation. Yet, human nature betrays us. Not without reason do we view emotions as a phenomenon, for the perception of emotions lies in the eye of the beholder. The sound of someone's voice, the look on their face, the posture, body language, the phrases and words choices they use communicate not only what they want to convey or try to conceal, but offer insights into constructs, such as affective states, personality traits and interpersonal sensitivity. A large body of evidence documents the impact someone's inner emotional state has on language choices people make. New theories of emotion point out to the importance language has in emotion perception, by providing context that helps us correctly interpret what is shown on someone's face [4].

Studying human behavior is recurrently challenged by the huge gaps in the knowledge and models of the cognitive processes underlying it. Existing theoretical models of emotions vary in their expressiveness, dimensionality and level of

© Springer Nature Switzerland AG 2018
S. Kalajdziski and N. Ackovska (Eds.): ICT 2018, CCIS 940, pp. 217–232, 2018.
https://doi.org/10.1007/978-3-030-00825-3_19

complexity. The debate on their veracity still continues. From Ekman's categorical model of six basic emotions [11] to Robert Plutchik's Wheel of Emotions [26] with eight bipolar emotions to the Circumplex model [29], a two-dimensional Valence-Arousal space – the models representing emotional space consist of "man-made categories imposed on, rather than discovered in the world" [4].

A notable body of work exists revealing insights into the language of emotions. The predictive power of linguistic cues such as word counts and frequencies [3], language modeling [14], part of speech tags (POS) [6,32], Probabilistic Context-Free Grammar [8,17] and their combinations [13,22] have proved to be successful with varying performance and generalization power, especially when tested on cross-domain datasets. The more statistically inclined models tend to originate from machine learning (ML). The use of deep learning (DL) has complemented the list of traditional machine learning algorithms to tease out the hidden connections between words [1,5,12,35,37,38]. The problems with recognizing emotions in ambiguous text still persist. In particular, inconclusive situations for detecting emotional connotations arise when contradicting and similar affective cues make it difficult to discriminate between two "close" emotional categories, or when neutrally expressed text or more complex sentences, such as those with multi-entity topics, speculations, irony or humor are involved.

In what follows, we highlight the primary findings of our empirical research as they relate to our central commitment of detecting emotions in text. This study examines the capability of several deep learning architectures and word embeddings to classify 11 and 13 emotional categories on two Twitter datasets. We point to several interesting areas for future investigation and conclude the paper.

2 Research Objective

A broad research and development front in the field of deep learning affective research exists. Recurrently, new deep learning structures with a higher level of sophistication are being proposed and explored. A number of architecture types and parameters have been experimented within this study, but we choose to discuss the most successful ones for the emotion detection task on Twitter datasets.

Contrary to the heightened interest it has attracted, specific discussions regarding the understanding of the intricacies of the deep neural structures are notably absent. The importance of research in this area is not to be judged in terms of their performance only, but rather according to the extent to which it elucidates the nature of the phenomenon. We argue that the ways in which we doubt and investigate established research practices serves the purpose of better "explainability" of deep learning architectures.

A central commitment of our research to investigate subtle deep learning elements and their ability to leverage our research agenda in practice, matters largely for how they reveal hidden patterns in data. Studies have put forth the idea that some aspects that deep neural models are built upon may need to be reexamined, word embeddings in particular. We have conducted several exploratory studies to investigate the concerns and issues raised by related research [10,31].

3   Dataset

In this study, we utilize two Twitter datasets to explore the conditions under which the proposed deep learning models for emotion detection obtain better performance results.

1. SemEval-2018 Task 1 multi-label classification[1] - The data used for this multi-label classification task were annotated in 11 emotional categories: *anger* (2,859), *anticipation* (1,102), *disgust* (2921), *fear* (1,363), *joy* (2,877), *love* (832), *optimism* (2,291), *pessimism* (895), *sadness* (2,273), *surprise* (396) and *trust* (400). In addition, a neutral class was added. The data were divided into a training set of 6,838, a development set consisting of 886, and a testing set of 3,259 tweets.
2. Crowdflower Twitter emotion[2] - 40,000 tweets annotated with one emotional label for the following 13 emotional categories: *sadness* (5,165), *neutral* (8,638), *anger* (110), *worry* (8,459), *surprise* (2,187), *love* (3,842), *fun* (1,776), *hate* (1,323), *happiness* (5,209), *relief* (1,526), *boredom* (179), *empty* (827) and *enthusiasm* (759). The dataset is separated into a training set consisting of 30,000 instances and a test set of 10,000 tweets. This dataset, contrary to the previous one, does not contain emojis.

3.1   Data Preprocessing

At the onset of the pre-processing pipeline, the Unicode text of the tweets was repaired and tokenization of each tweet into sentences was performed. Subsequently, a number of context-appropriate corrections of the language variations of English e-dialect were performed:

- Each username is replaced with the token "@user".
- Letters occurring more than two times consecutively are replaced with one occurrence. For example, the word *gooooooood* would be changed to *good*.
- The # sign from each hashtag is removed and the text of the hashtag is kept.
- Apostrophes are removed from the contractions of the negation of auxiliary verbs (e.g., don't, isn't) to preserve consistency with the representation of these words in the word embeddings.

---

[1] https://competitions.codalab.org/competitions/17751#learn_the_details-datasets, last accessed: May 2018.
[2] https://data.world/crowdflower/sentiment-analysis-in-text, last accessed: May 2018.

Part-of-speech tagging and lemmatization i.e., extracting the base form of words conclude the preparation of data for the next stage of transforming the text into suitable vector representation to be fed into the deep neural network models. Standard functions offered in NLTK package[3] have been utilized for performing all of the above pre-processing functions.

4   Investigating the Suitability of Word Embeddings

4.1   Word Embeddings

Of primary importance to this research is to investigate the appropriateness of traditional word embeddings for the task of detecting emotions in tweets. Two word vector spaces, GloVe and word2vec, were explored. We underline some of their shortcomings revealed in our exploratory studies. Addressing these concerns have led us to experiment with some extensions of these word embeddings that were deemed to be beneficial for the task at hand.

Word embeddings are vector representations of words that map a word into a common space preserving word meanings, their relationships and the semantic information. Each word in the training vocabulary is represented as a dense real-valued vector. The dimensions of the vector space represent the latent features of a word. Pre-training embeddings on larger corpora, not necessarily (even preferably) from the same context is a well-established practice that has been proven to lead to better performances. Of particular interest for our study were corpora built from Twitter data.

GloVe. GloVe (Global Vectors for Word Representation) [24] is a log-bilinear regression model for unsupervised learning of word representations that combines the advantages of two model families: global matrix factorization and local context window methods. It captures meaningful linear substructures, efficiently leveraging global word-word co-occurrence matrix statistics. There are several pre-trained GloVe word embeddings; a 200-dimensional pre-trained GloVe word embeddings trained on 2 billion tweets with 1.2 million word vocabulary[4] were used in this study.

Word2vec. Word2vec [20,21] is a predictive model that captures the context of a given word by mapping it into a fixed dimensional real-valued vector representation where words with similar meaning have similar representation. These vectors capture syntactic and semantic regularities in language, such that the operation "King - Man + Woman" results in a vector very close to "Queen". We have used the 400-dimensional word2vec vectors[5] that were trained on a corpus of ten million tweets [7].

---

[3] https://www.nltk.org/, last accessed: May 2018.

[4] https://nlp.stanford.edu/projects/glove/, last accessed: May 2018.

[5] https://github.com/felipebravom/AffectiveTweets/releases/tag/1.0.0, last accessed: May 2018.

4.2   Exploratory Studies on Word Embeddings

As previously stated, embedding vector space preserves the semantic information and the relationships between words. Hence, words that are similar appear closer in the space, unlike dissimilar words that are distant. When words need to be separated according to the emotion they express and its intensity, it is important that the embeddings capture their emotional sense. The analysis presented in [31] casts doubts on researchers' claims that distributional word embeddings produce a representation of their semantic meaning. In contrast to previous claims, the results of their experiments investigating the semantic similarity of emotions carried by GloVe and word2vec vectors point out to entirely opposite conclusions.

Drawing on these assertions, we have conducted exploratory experiments to investigate the ability of the word embeddings in our model to represent the semantic similarity between emotions present in our selected corpus data. The first experiment is to test the embedding similarity of the opposite emotions derived from the Plutchik's wheel of emotions [25, 27]. Table 1 presents the results of the cosine similarity between an emotion and the opposite emotion. These emotional words are opposite, so the expectation is that they should be separated, i.e. have small similarity value close to zero. The table reveals results that are in-line with [31], with the exception of surprise and anticipation, with an implication that both types of pre-trained embeddings failed to differentiate between opposite emotions in their corresponding vector spaces.

Table 1. Cosine similarity between an emotion and the opposite emotion (smaller value is better).

Emotion	Opposite emotion	Glove	word2vec
Surprise	Anticipation	0.257	0.114
Fear	Anger	0.568	0.545
Trust	Disgust	0.327	0.505
Joy	Sadness	0.592	0.571
Average		0.436	0.434

It is also well worth investigating the questions raised in [21] on the potential of the embedding space to represent meaningful and interesting linear algebraic relationships between emotional words. For example, the pairing of the primary emotions joy and trust result in the emotion of love, represented by a dyad ($Joy + Trust = Love$). Our second exploratory experiment tests the capability of the embeddings of our choice to preserve the relationships between an emotion and its emotional constituents for a selected list of twenty-four dyads derived from Plutchik's wheel of emotions [2, 25]. The expectation is that the similarity between the combination of emotions and the resulting dyad is high, close to one. However, Table 2 reveals that in general both types of word embeddings are unable to capture these relationships. The GloVe embeddings have an average similarity of 0.416, whereas the average similarity of word2vec embeddings is 0.325. The conclusion derived from the experiments is that commonly used distributional vector spaces do not possess the capability to map the relationships between opposite emotions or between an emotion and its counterparts.

4.3 Augmenting the Word Representations

Emoji Embeddings. In general, pre-trained word-embeddings do not include emojis in the vocabulary. However, the use of emojis in the context of our Twitter dataset is abundant. Inspired by the idea proposed in [10], we have appropriately modified and extended the traditional GloVe and word2vec vector space.

Table 2. Cosine similarity of dyads - combination of emotions (bigger value is better).

Emotions	Result of emotions	Glove	word2vec
Joy + Trust	Love	0.679	0.415
Joy + Fear	Guilt	0.562	0.523
Joy + Surprise	Delight	0.503	0.382
Trust + Fear	Submission	0.281	0.052
Trust + Surprise	Curiosity	0.226	0.156
Trust + Sadness	Sentimentality	0.113	0.128
Fear + Surprise	Awe	0.358	0.169
Fear + Sadness	Despair	0.634	0.704
Fear + Disgust	Shame	0.610	0.366
Surprise + Sadness	Disappointment	0.581	0.596
Surprise + Disgust	Unbelief	0.046	0.152
Surprise + Anger	Outage	0.249	0.180
Sadness + Disgust	Remorse	0.514	0.310
Sadness + Anger	Envy	0.374	0.206
Sadness + Anticipation	Pessimism	0.331	0.318
Disgust + Anger	Contempt	0.423	0.368
Disgust + Anticipation	Cynicism	0.360	0.326
Disgust + Joy	Morbidness	/	/
Anger + Anticipation	Aggression	0.373	0.445
Anger + Joy	Pride	0.553	0.442
Anger + Trust	Dominance	0.213	0.249
Anticipation + Joy	Optimism	0.425	0.360
Anticipation + Trust	Hope	0.576	0.233
Anticipation + Fear	Anxiety	0.583	0.391
Average		0.416	0.325

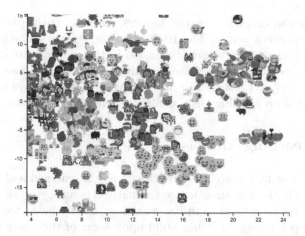

Fig. 1. Visual representation of 400-dimensional emoji embeddings derived from word2vec projected to two-dimensional space using the t-SNE algorithm.

The emoji embeddings are formed using the Unicode emojis with their text description. Each emoji embedding is generated by optimizing the probability of a match between the emoji representational vector and the vector of the emoji description constructed by summing the individual existing word embeddings of the words in its description. Figure 1 gives a visual representation of the 400-dimensional emoji embeddings derived from wor2vec and projected into a two-dimensional space created by using the t-SNE algorithm presented in the original paper [10–19]. Our research hypothesis was that incorporating emoji-related information within the word embeddings would yield some performance advantage on the emotion detection task on Twitter data.

Lexicon Embeddings. Lexicons, frequently used in the field of affective analysis, are lists of affective words and their associated metadata tags (e.g., category, valence, intensity, ...). To address the question of whether or not, and to what extent, the use of a lexicon could impact the performance of our deep learning model, we needed to investigate and choose the most suitable one.

The selection of a suitable lexicon is as important as the selection of the word embedding space. The manner of its creation, as well as its vocabulary, are determined by the context and problem under investigation. The construction of lexicons is costly because of the human expertise they rely upon. They often suffer from small vocabulary coverage. Current approaches to remedy these shortcomings include automatic creation of context-sensitive lexicons and expansion of the existing small human-annotated lexicons using appropriate algorithms.

Our lexicon of choice was W2V-DP-CC-Lex[6] [7], a multi-label emotion lexicon, which is an extension of the NRC word-emotion association lexicon [36]. This lexicon associates each word with a value of association for 8 emotions:

---

[6] https://www.cs.waikato.ac.nz/ml/sa/files/, last accessed: May 2018.

anger, anticipation, disgust, fear, joy, sadness, surprise, trust, plus the negative and positive sentiment values. The 10-dimensional lexicon embeddings are created with this lexicon, where each dimension in the vector space is respectively one of the eight emotions or positive/negative sentiment. We have followed the research by Shin et al. presented in [33], suggesting several ways to incorporate a lexicon into a deep learning CNN model for sentiment analysis.

5   Deep Learning Models

Over the past decade, research on deep learning has emerged and a remarkable number of tasks in the domain of image and natural language processing has been accomplished. We have tried to utilize what is currently being recommended in the field of deep learning NLP and build upon some of the many findings both consistent and inconsistent.

5.1   CNN

Convolutional Neural Networks (CNNs) also referred to as *ConvNets* are neural networks that analyze higher-order features in the data via convolutions. They are specially well suited for object recognition in images [23] since their creation was inspired by the inner workings of the visual cortex of animals [9]. CNNs can arrange the neurons in a two- or three-dimensional structure. Two main layers used to build CNN are:

- **Convolutional Layer** - transforms the input data by using a patch of locally connecting neurons from the previous layer. The convolutional operation is known as feature detector of the CNN. Major components of this layer are: *filters, activation maps, parameter sharing* and *layer-specific hyperparameters.*
- **Pooling Layer** - operates independently on every depth slice of the input and resizes it spatially, using the *max* operation (alternative operations are average pooling, L2-norm pooling, etc.).

5.2   LSTM

Long Short-Term Memory networks (LSTMs) are a specific type of recurrent neural network (RNN), able of learning long-term dependencies [15]. LSTMs preserve the error that is backpropagated through time and layers. LSTM can maintain a more constant error, and recurrent nets can continue to learn over many time steps (over 1000). LSTM networks overcome the vanishing and exploding gradient problems in standard RNNs, where the error gradients would otherwise decrease or increase at an exponential rate.

A LSTM block is composed of four main components: a **cell**, an **input gate**, an **output gate** and a **forget gate**, given with the following equations of each particular component:

$$blockinput \rightarrow z_t = g(W_z x_t + R_z y_{t-1} + b_z) \qquad (1)$$

$$inputgate \rightarrow i_t = \sigma(W_i x_t + R_i y_{t-1} + p_i \odot c_{t-1} + b_i) \qquad (2)$$

$$forgetgate \rightarrow f_t = \sigma(W_f x_t + R_f y_{t-1} + p_f \odot c_{t-1} + b_f) \qquad (3)$$

$$cellstate \rightarrow c_t = i_t \odot z_t + f_t \odot c_{t-1} \qquad (4)$$

$$outputgate \rightarrow o_t = \sigma(W_o x_t + R_o y_{t-1} + p_o \odot c_t + b_o) \qquad (5)$$

$$blockoutput \rightarrow b_t = o_t \odot h(c_t) \qquad (6)$$

where g, h, and σ are activation functions, $\odot$ is element-wise multiplication. x_t is input vector at time t, W are input weight matrices, R represents the recurrent weight matrices, p are peephole weight vectors and b are the bias vectors. The cell remembers values over arbitrary time intervals.

5.3   BiLSTM

Bidirectional recurrent neural networks [30] are combining two independent RNNs together. The first network is feed with the input sequence in normal time order, while the time-reversed input is supplied to the second network. The features of the two separated networks are combined usually with output concatenation at each time step, or other approaches, such as summarization can be used.

5.4   Baseline Model

The first layer of the model is the construction of a matrix representation of the tweet W_e, formed with the word embeddings for each word. The matrix has the same dimension for every tweet, i.e. the maximum number of words is set to 150. If the real number of words is smaller than 150, then the matrix is constructed with padded rows. So, the dimension of the matrix is $150 \times n$, where n is the word embedding dimension. The next layer performs one-dimensional convolution over the embedded word matrix using kernel size 3 and 32 output filters in the convolution. The activation function for this convolution layer is the ReLU activation function given in Eq. 7. Additionally, the layer weights are normalized using batch normalization.

$$f(x) = \begin{cases} 0 \ for \ x \leq 0 \\ x \ for \ x \geq 0 \end{cases} \qquad (7)$$

The next layer is the pooling layer that uses the max function with a window size of dimension 2, over the result of the convolutional layer. These two parts are added subsequently with 64 output filters. The dropout regularization layer with a rate of 0.1 is added before the Bidirectional LSTM sub-network. The BiLSTM sub-model is constructed with two LSTM networks, each with 128 units. The outputs of the forward and backward LSTMs are combined by concatenation.

The output is classified using a fully connected layer with sigmoid activation function. The formula of the function is represented in Eq. 8. This activation function gives the probability for each of the possible classes for the tweet. Figure 2 shows a visual representation of the model.

$$f(x) = \frac{1}{1 + e^{-x}} \tag{8}$$

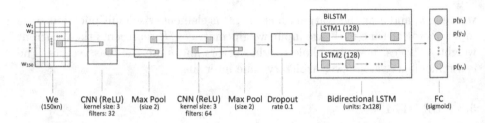

Fig. 2. Visual representation of the CNN-BiLSTM model.

5.5   Extended Model

The second model presents a modification of the baseline model that incorporates previously described emoji embeddings, while the third model extends the second by adding the lexicon embeddings. Figure 3 visualizes the new model that consists of two CNN-BiLSTM sub-models, each one represents the baseline model. The CNN-BiLSTM sub-models are executed in parallel with different matrix representations of the tweets W_e and W_l. The W_e matrix represents the word embedding matrix constructed in the same way as in the initial model, and W_l is the matrix with lexicon embeddings of the words that is fed to the

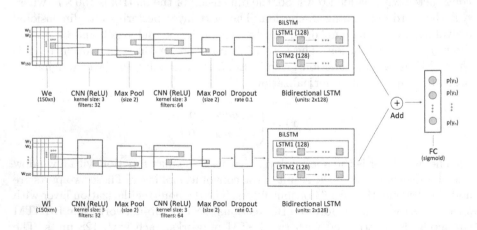

Fig. 3. Visual representation of the CNN-BiLSTM model with incorporated lexicon.

second parallel sub-model. The output feature vectors of the two separate sub-models are joined with addition and then the class is determined through the fully connected layer with sigmoid activation function.

5.6 Model and Training Parameters

A short description of the training parameters for the baseline and the extended model presented in Sects. 4.4 and 4.6, respectively follows.

The kernel weights matrix is initialized with Xavier uniform initializer, while random orthogonal matrix is an initialization for the recurrent kernel weights matrix used for the linear transformation of the recurrent state. Batch normalization [16] and Dropout layer [34] are the two applied methods for regularization of the model. The learning was performed using the AMSGrad variant [18] of the Adam optimizer from [28] with a learning rate set to 0.001. The models were trained for 100 epochs and the most appropriate epoch model was chosen to be the model with lowest validation loss; the most successful model version was used for evaluating the performance. The loss function for the emotion detection model is the categorical cross-entropy between predictions and targets for each class and binary cross-entropy for the multi-label classification.

6 Experiments: Discussion of Results

In this paper, we explore the extent to which augmented word representations could improve the performance of the proposed deep neural architecture. This section presents the comparative evaluation results obtained by the three models: the baseline CNN-BiLSTM architecture, the extended model that uses emoji2vec embeddings and the third model, which incorporates both, emoji2vec and lexicon embeddings. Two word vector representations, GloVe and word2vec, were used in the baseline model and as a basis for the emoji and lexicon embeddings. The third model yielded the best performance results on the problem under our investigation for both datasets.

The results obtained by our three models trained and tested on the SemEval-2018 dataset are summarized in Table 3. We report on three evaluation metrics following the recommendation of the SemEval task. Accuracy was defined using the Jaccard index i.e., Jaccard similarity between the set of predicted labels and the corresponding set of true labels.

The best performing model achieved accuracy rates of 0.444 with GloVe embeddings and 0.468 for word2vec embeddings, placing us at the 10th position out of 37 participants at SemEval 2018 Task1: Affect in tweets. We should point out that the small size of the SemEval dataset, consisted of only 6838 instances, was always considered to be a challenging task for deep learning models. Notwithstanding our primary concern, we have also observed that the model using GloVe vectors in terms of its F1-micro and F1-macro results. We could hypothesize that the reason may be attributed to the higher dimensionality of the word2vec word embeddings.

Table 3. Results for multi-label emotion detection of the models with the SemEval-2018 dataset (Model1 = CNN-BiLSTM, Model2 = CNN-BiLSTM + emoji2vec, Model3 = CNN-BiLSTM + emoji2vec + lexicon).

Model	Acc	F1 micro	F1 macro
GloVe embeddings			
Model1	0.350	0.464	0.263
Model2	0.390	0.508	0.299
Model3	0.444	0.554	0.378
word2vec embeddings			
Model1	0.406	0.504	0.320
Model2	0.432	0.542	0.361
Model3	0.468	0.577	0.435

Pre-trained word-embeddings do not include emojis, although their use on social media people is frequent. We extended the word embedding vectors with emoji2vec and evaluated the model. The performance advantage obtained with the emoji vectors points out to the added value of their integration in the process of prediction and separation of emotions. To further investigate the ways in which word vector space could be improved on the task of emotion separation, we have reviewed the behavior of adding lexicon features that associate words with emotions. The results indicate that the inclusion of a lexicon could further improve the performance of our model on the task.

Classical confusion matrix cannot be plotted for a multi-label emotion detection task, so we have computed confusion matrices, one for each of the 11 emotions. In other words, TPs, FPs, TNs and FPs are calculated for each emotion. The confusion matrices for the best model (the third model), presented in Fig. 4, have shown that the proposed deep model performs well in distinguishing whether the particular emotion is present in a tweet or not, with the exception of two emotions *surprise* and *trust*. We could hypothesize that the reason for such underperformance could be the small number of tweets labeled in these categories. If *trust* and *surprise* classes were ignored, the third model obtained accuracy of 0.481, F1-micro of 0.590 and F1-macro measure of 0.501.

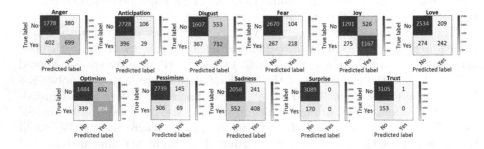

Fig. 4. Confusion matrices for the SemEval-2018 dataset.

We performed the same experiments with the Crowdflower Twitter emotion dataset. We wanted to see if the pattern of results replicated with the second dataset even though there were significant differences between the datasets. It should be noted that the tweets of the second dataset do not contain emojis, hence no need for consideration of emoji2vec. In addition, each tweet is associated with one emotional class, so the tasks of the two experiments differ slightly. Table 4 displays the obtained performance results. The models yielded accuracy rates of approximately 32% with both word embedding vectors, although the expectations in terms of better performance with lexicon embeddings incorporated in Model 3 were not met. The top-5 accuracy was 0.79 for GloVe embeddings and 0.78 for word2vec.

The effects and implications presented in this study have noteworthy implications for the performance leverage of a modest size deep architectures on a challenging task. Considering the obvious challenges pertaining to the limited performance achievable by deep architectures of modest size, as well as the considerable timing and processing requirements that need to be met, we have identified the positive trends in accomplishing the emotion detection task when extensions to the commonly used word embeddings were included.

Although the findings of the studies pertaining to distributional word representations might be contradictory, they actually are demonstrations of the shortcoming of word embeddings, which indicates that their effectiveness varies with tasks and contexts. Our exploratory study indicates a need for another perspective on what is conveyed by the word vector space. The results of our initial explorations have provided insights regarding future solutions that should involve augmenting word representations tailored to increase the performance on emotion detection task, in the context of interest.

Table 4. Results for emotion detection of the models with the Crowdflower Twitter dataset (Model1 = CNN-BiLSTM, Model3 = CNN-BiLSTM + emoji2vec + lexicon).

Model	Acc	Top 3 acc	Top 5 acc	$F1_{micro}$	$F1_{macro}$
GloVe embeddings					
Model1	0.324	0.620	0.782	0.324	0.107
Model3	0.327	0.632	0.786	0.327	0.106
word2vec embeddings					
Model1	0.326	0.623	0.782	0.326	0.105
Model3	0.321	0.618	0.774	0.321	0.119

7   Conclusions

The ability of deep neural networks to classify the emotional content in the verbal message has been the question posed in this research. Two Twitter datasets were utilized in this study in order to evaluate the behavior of the proposed models.

We have presented several deep learning models, based on a combination of CNN and Bidirectional LSTM. The baseline model is a sequence of convolutional and pooling layers followed by a bidirectional LSTM. The extensions to the baseline model incorporated augmented word representations, emoji and lexicon embeddings in particular.

By experimenting with different deep architecture models and investigating the expressiveness and capabilities of word representations to capture emotional connotations in text, this article contributes to the conversation of how deep learning architectures could support the efforts. Implications include the relevance of empirical evidence and rigorous testing to point towards future progression opportunities.

Acknowledgements. This work was partially financed by the Faculty of Computer Science and Engineering at the "Ss. Cyril and Methodius" University.

References

1. Abdul-Mageed, M., Ungar, L.: Emonet: fine-grained emotion detection with gated recurrent neural networks. In: Proceedings of the 55th Annual Meeting of the Association for Computational Linguistics (Volume 1: Long Papers), vol. 1, pp. 718–728 (2017)
2. Adliterate.com: Robert plutchik's psychoevolutionary theory of basic emotions. http://www.adliterate.com/archives/Plutchik.emotion.theorie.POSTER.pdf. Accessed 09 May 2018
3. Alm, C.O., Roth, D., Sproat, R.: Emotions from text: machine learning for text-based emotion prediction. In: Proceedings of the Conference on Human Language Technology and Empirical Methods in Natural Language Processing, pp. 579–586. Association for Computational Linguistics (2005)
4. Barrett, L.F., Lindquist, K.A., Gendron, M.: Language as context for the perception of emotion. Trends Cogn. Sci. **11**(8), 327–332 (2007)
5. Bertero, D., Siddique, F.B., Wu, C.S., Wan, Y., Chan, R.H.Y., Fung, P.: Real-time speech emotion and sentiment recognition for interactive dialogue systems. In: Proceedings of the 2016 Conference on Empirical Methods in Natural Language Processing (EMNLP), pp. 1042–1047 (2016)
6. Binali, H., Wu, C., Potdar, V.: Computational approaches for emotion detection in text. In: Proceedings of the 4th IEEE International Conference on Digital Ecosystems and Technologies, pp. 172–177. IEEE (2010)
7. Bravo-Marquez, F., Frank, E., Mohammad, S., Pfahringer, B.: Determining word-emotion associations from tweets by multi-label classification. In: Proceedings of the 2016 IEEE/WIC/ACM International Conference on Web Intelligence (WI), pp. 536–539 (2016)
8. Charniak, E.: Statistical parsing with a context-free grammar and word statistics. In: Proceedings of the Fourteenth National Conference on Artificial Intelligence and Ninth Conference on Innovative Applications of Artificial Intelligence, pp. 598–603. AAAI Press (1997)
9. Eickenberg, M., Gramfort, A., Varoquaux, G., Thirion, B.: Seeing it all: convolutional network layers map the function of the human visual system. NeuroImage **152**, 184–194 (2017)

10. Eisner, B., Rocktäschel, T., Augenstein, I., Bosnjak, M., Riedel, S.: emoji2vec: Learning emoji representations from their description. In: Proceedings of The Fourth International Workshop on Natural Language Processing for Social Media, pp. 48–54 (2016)
11. Ekman, P.: An argument for basic emotions. Cogn. Emot. **6**(3–4), 169–200 (1992)
12. Ghosh, A., Veale, T.: Magnets for sarcasm: making sarcasm detection timely, contextual and very personal. In: Proceedings of the 2017 Conference on Empirical Methods in Natural Language Processing (EMNLP), pp. 482–491 (2017)
13. Gievska, S., Koroveshovski, K., Chavdarova, T.: A hybrid approach for emotion detection in support of affective interaction. In: Proceedings of the 2014 IEEE International Conference on Data Mining Workshop, pp. 352–359. IEEE (2014)
14. Hasan, M., Rundensteiner, E., Agu, E.: EMOTEX: detecting emotions in twitter messages. In: Proceedings of the Sixth ASE International Conference on Social Computing (SocialCom 2014). Academy of Science and Engineering (ASE) (2014)
15. Hochreiter, S., Schmidhuber, J.: Long short-term memory. Neural Comput. **9**(8), 1735–1780 (1997)
16. Ioffe, S., Szegedy, C.: Batch normalization: accelerating deep network training by reducing internal covariate shift. In: Proceedings of the 32nd International Conference on Machine Learning, vol. 37, pp. 448–456 (2015)
17. Jelinek, F., Lafferty, J.D., Mercer, R.L.: Basic methods of probabilistic context free grammars. In: Laface, P., De Mori, R. (eds.) Speech Recognition and Understanding. NATO ASI Series, vol. 75, pp. 345–360. Springer, Heidelberg (1992). https://doi.org/10.1007/978-3-642-76626-8_35
18. Kingma, D.P., Ba, J.: Adam: a method for stochastic optimization. CoRR abs/1412.6980 (2014)
19. van der Maaten, L., Hinton, G.: Visualizing data using t-SNE. J. Mach. Learn. Res. **9**, 2579–2605 (2008)
20. Mikolov, T., Chen, K., Corrado, G.S., Dean, J.: Efficient estimation of word representations in vector space. CoRR abs/1301.3781 (2013)
21. Mikolov, T., Yih, W., Zweig, G.: Linguistic regularities in continuous space word representations. In: Proceedings of the 2013 Conference of the North American Chapter of the Association for Computational Linguistics: Human Language Technologies, pp. 746–751 (2013)
22. Najdenkoska, I., Stojanovska, F., Gievska, S.: Detecting emotions in tweets based on hybrid approach. In: Proceedings of the 15th International Conference on Informatics and Information Technologies (2018)
23. Patterson, J., Gibson, A.: Deep Learning: A Practitioner's Approach. O'Reilly Media Inc., Newton (2017)
24. Pennington, J., Socher, R., Manning, C.: Glove: global vectors for word representation. In: Proceedings of the 2014 Conference on Empirical Methods in Natural Language Processing (EMNLP), pp. 1532–1543 (2014)
25. Plutchik, R.: Emotion: A Psychoevolutionary Synthesis. Harpercollins College Division, New York (1980)
26. Plutchik, R.: Emotions: a general psychoevolutionary theory. In: Scherer, K.R., Ekman, P. (eds.) Approaches to Emotion, pp. 197–219. Erlbaum, Hillsdale (1984)
27. Plutchik, R.: Integration, differentiation, and derivatives of emotion. Evol. Cogn. **7**(2), 114–125 (2001)
28. Reddi, S.J., Kale, S., Kumar, S.: On the convergence of adam and beyond. In: Proceedings of the International Conference on Learning Representations (2018)
29. Russell, J.A.: A circumplex model of affect. J. Pers. Soc. Psychol. **39**(6), 1161 (1980)

30. Schuster, M., Paliwal, K.K.: Bidirectional recurrent neural networks. IEEE Trans. Sig. Process. **45**(11), 2673–2681 (1997)
31. Seyeditabari, A., Zadrozny, W.: Can word embeddings help find latent emotions in text? Preliminary results. In: Proceedings of the Thirtieth International Florida Artificial Intelligence Research Society Conference, pp. 206–209 (2017)
32. Shaheen, S., El-Hajj, W., Hajj, H., Elbassuoni, S.: Emotion recognition from text based on automatically generated rules. In: 2014 IEEE International Conference on Data Mining Workshop (ICDMW), pp. 383–392. IEEE (2014)
33. Shin, B., Lee, T., Choi, J.D.: Lexicon integrated CNN models with attention for sentiment analysis. In: Proceedings of the 8th Workshop on Computational Approaches to Subjectivity, Sentiment and Social Media Analysis, pp. 149–158 (2017)
34. Srivastava, N., Hinton, G., Krizhevsky, A., Sutskever, I., Salakhutdinov, R.: Dropout: a simple way to prevent neural networks from overfitting. J. Mach. Learn. Res. **15**(1), 1929–1958 (2014)
35. Wang, Z., Zhang, Y., Lee, S., Li, S., Zhou, G.: A bilingual attention network for code-switched emotion prediction. In: Proceedings of COLING 2016, the 26th International Conference on Computational Linguistics: Technical Papers, pp. 1624–1634 (2016)
36. Warriner, A.B., Kuperman, V., Brysbaert, M.: Norms of valence, arousal, and dominance for 13,915 English lemmas. Behav. Res. Methods **45**(4), 1191–1207 (2013)
37. Zahiri, S.M., Choi, J.D.: Emotion detection on TV show transcripts with sequence-based convolutional neural networks. In: The Workshops of the Thirty-Second AAAI Conference on Artificial Intelligence, New Orleans, Louisiana, USA, 2–7 February 2018, pp. 44–52 (2018)
38. Zhou, C., Sun, C., Liu, Z., Lau, F.C.M.: A C-LSTM neural network for text classification. CoRR abs/1511.08630 (2015)

Medical Real-Time Data Analytics System Design Aspects, Reference Architecture and Evaluation

Magdalena Kostoska(✉), Monika Simjanoska, Bojana Koteska,
and Ana Madevska Bogdanova

Faculty of Computer Science and Engineering, University Ss. Cyril and Methodius,
Rugjer Boskovikj 16, 1000 Skopje, Macedonia
{magdalena.kostoska,monika.simjanoska,
bojana.koteska,ana.madevska.bogdanova}@finki.ukim.mk

Abstract. This paper presents a medical distributed system whose main goal is to enable real-time triage processing and further propagation of the data and patient history. The system uses various biosensors and mobile devices, as well as additional communication equipment. The system aims to ease the decision of priority in treatment, as well as to further monitor and alarm state change.

We propose a novel approach for military and civil real-time data analytics. We introduce our data analytics platform and application model and discuss their main design requirements and challenges, based on use case scenarios. We analyze the system in the described scenarios, we test the limitations of the system given the system architecture and evaluate it. We present the obtained results.

Keywords: Real-time data analytics · Cloud computing · IoT
Health care system

1 Introduction

Time is critical regarding the survival rate of severely injured patients [4]. The faster the response, the better chances are for survival and shorter aftercare. One of the most important aspects in this situation is evaluating the state of the patient, especially in situations where multiple victims require help simultaneously, i.e. mass casualty incidents (MCI). This is the task of triage - to determine priority for treatment and represents a standard classification task based on vital data parameters of the injured. The concept of mass casualty triage originates from the need of militaries to efficiently and effectively treat multiple battle casualties. Many of the strategies used to triage and treat wounded soldiers have been promoted for the civilian setting. The skill to prioritize patients for treatment and transport during a mass casualty incident is considered essential [10], particularly since we have been witnesses to many MCI situations in the last years. Natural disasters (i.e. earthquakes, fires, floods etc.) or human induced events

© Springer Nature Switzerland AG 2018
S. Kalajdziski and N. Ackovska (Eds.): ICT 2018, CCIS 940, pp. 233–246, 2018.
https://doi.org/10.1007/978-3-030-00825-3_20

as terrorist attacks happen every day and the ability to select and help the most critically wounded has great impact in final casualty number outcome.

Important aspect in the military battle field, as well as in case of MCI, is that medical personnel is limited and valuable resource. They cannot commit to every patient and have very limited time to assess the state of the patient and treat them. Because triage requires rapid (often measured in seconds) assessment and decision-making, [6] any additional information on the patient state or visible state monitoring helps tremendously. Our system augments the process of triage by using biosensors, includes mobile applications with the triage algorithms and allows real-time assessing and monitoring the state of the injured persons. The proposed system is adaptable to many environments, including military, as well as civilian.

The paper is organized as follows: in the next Sect. 2, we give an overview of the existing solutions such as electronic health monitoring and biosensors usage in massive disasters. In Sect. 3 we describe two distinct situation where the system can be used, then in Sect. 4, we describe the triage proccess. In Sect. 5, we describe the overall system architecture, as well as the software architecture and potential hardware usage. We give system dissemination in Sect. 6. Conclusive remarks are given in the last Sect. 7.

2   Related Work

The challenges in designing an interactive systems for emergency responses are analyzed and defined since 2006 by Kyng et al. [8]. Since then the biosensors improved, became more available and started to be used in every-day life. They are often proposed for monitoring patients' condition.

A framework to collect patients' data in real time, perform monitoring, and propose medical and/or life style engagements is proposed by Benharref and Serhani [2]. Their proposed framework integrates mobile technologies to collect and communicate vital data from a patient's wearable biosensors. The data are stored in the Cloud and are available to physicians, paramedics, etc. Yang et al. propose intelligent home-based platform [16] that includes an open-platform-based intelligent medicine box, intelligent pharmaceutical packaging with communication capability enabled by passive radio-frequency identification and a flexible and wearable bio-medical sensor device. Lee et al. propose health-care service that uses clothing with embedded biosensors and bio-information framework and service platform for mobile systems that will enable constant analyzes and presentation of the data [9]. A interesting concept of unobtrusive observing in the homes of elderly people offers the U-Health Smart Home [1]. It represents self-managed intelligent system connected via broadband Internet access to back-end health-care providers. The U-Health smart home uses small-sized medical body sensors and actuators that are capable of collecting physiological data from the body of the monitored inhabitant and possibly delivering some drugs.

One of the important researches in this area is establishing an ad hoc sensor network infrastructure. Example of this is the CodeBlue, developed by

Malan et al., where the network integrates low-power, wireless vital sign sensors, PDAs, and PC-class systems [12]. Usage of biosensors in massive disasters is researched by Gao et al. [5] where Advanced Health and Disaster Aid Network (AID-N) is used to replace the manual recording of vital signs by usage of electronic triage tag. Similar approach is used by Lui et al. [11] where an analysis of six key issues in employing the disaster aid system is performed, as well as challenges. Kramp et al. suggest usage of wireless and mobile biomonitoring system, called BlueBio [7], intended to monitor injured persons, especially in the prehospital work.

3   Use Cases

The designed system is adaptable and can be used in military, as well in civilian environment. We futher describe both such scenarios.

3.1   Use Case 1: Civil MCI Environment

In case of a major disaster, prompt paramedic attention is crucial in order to save people's lives. Many countries have created emergency protocols (e.g., MRMI) to deal with such situations. The protocol that is described in this scenario relies and extends the existing protocol for major disasters in Macedonia. The protocol for major disasters involves several existing structures and organizations, like general hospitals, the army and volunteers. An extension of major disasters protocol includes wearable biosensors that can be attached to injured people by on-site medics. Attaching a sensor only takes a few seconds, allowing a medic to rapidly deploy sensors to multiple patients in the area he/she is responsible for. These sensors measure basic vital signs, as well as more complex vital signals (e.g. ECG), in order to provide critical information about the patient's vital medical state.

As soon as the sensors are attached, they start emitting data to nearby portable Edge devices, e.g., a smartphone. Such devices perform only basic, lightweight analytics, such as triage decision tree, in order to determine the overall stability of the injured person and inform on-site medics. The device screen shows the latest values of all vital signs for the selected patient, and according to that, medics can give first aid. The primary triage is performed on the site. The primary decision is based on the respiratory and heart rate. This parameters are obtained by sensor equipped with Bluetooth and are passed to the mobile device (via the Bluetooth connection). The mobile phone constantly performs the decision tree algorithm and in case of red or yellow classification alarms. The data for these critically injured persons are transmitted via ad-hoc network to a medical personnel on the site. Persons with critical injuries are relocated to the on-site dedicated area for immediate treatment. In order to perform secondary triage additional biosensors must be attached to the injured (SPO2 and blood pressure sensors). All data are further sent to the medic. The medic inputs manually part of the data required for the secondary triage (e.g. Glascow coma scale) and the next classification is performed. The overview of the scenario is depicted in Fig. 1.

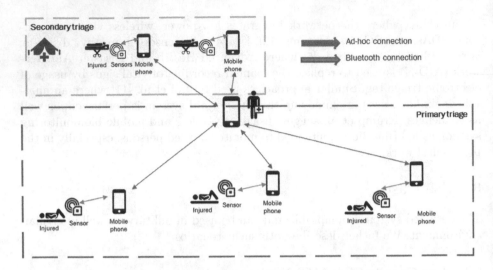

Fig. 1. Civil MCI usage system overview (Color figure online)

3.2   Use Case 2: Military Environment

In case of injury on battlefield, the military operations follow protocol described by the mission, country or alliance. In general case there is a First-Aid responder (FAR) in the team which is responsible for the team welfare and immediate medical response. Depending of the type of action involved the FAR may, or may not be, near all the team members. This may render more difficult the assessment of the medical state of the soldiers. By equipping the army team members with wearable biosensors before going into battle, we ensure continuous monitoring of the state of each soldier and thus enable the FAR to overview and react in case of need. The data of the soldiers are further distributed to all higher instances included in the action or battle. Depending of the equipment, Internet may not available on the battlefield, and as a redundancy the data can be transmitted via Bluetooth to the mobile terminals and send further using soldiers radios and HF radios. The temporary stored data are synced to the cloud as soon as there is Internet connection established. FAR is responsible only for fast medical procedures and inputs data for the injured body parts and medications. The entities involved in the saving protocol are the military transport such as ambulance/helicopter further referred to as Role 1, the improvised hospital tent and the battlefield remote hospital further referred to as Role 2. All of them can access the central database, retrieve the patients' data collected from the battlefield, during the transport and from the hospital tent. As an alternative/option the data can be transmitted from one instance to another using Bluetooth protocol. Figure 2 presents the overview of the system usage in military environments.

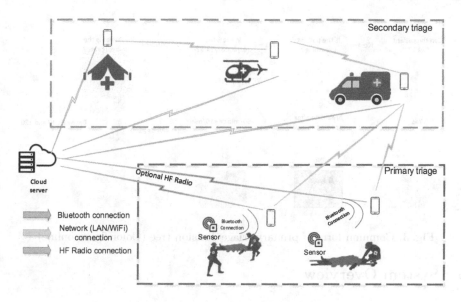

Fig. 2. Military usage system overview (Color figure online)

4 Triage

Numerous variations of the triage classification process exist. The process and assigned states vary between countries and systems used. One of the most common procedure is to assign color that represents the class/severity of injury. Almost all the systems use the same colors and have similar or equal meaning:

- Black - Deceased
- Red - Life-threatening injuries that requires immediate treatment
- Yellow - Non-life-threatening injuries. The injured should be monitored and require treatment when possible
- Green - Minor injurie, does not require immediate attention

Depending on where the triage is performed there is primary triage - on the scene of the accident and secondary triage - medical point at the site of a major incident. The primary triage is fast and on-site. The state should be reassessed frequently. The secondary triage is an ongoing process that takes place after the patient has been moved to a dedicated area on the site for initial treatment. This process is more detailed. Based on the secondary triage the injured are prioritized and sent to further treatments in hospitals. Figure 3 depicts common decision tree for most primary triage protocols as START [3] or Triage sieve [15]. Secondary triage is performed when injured are placed/evacuated to dedicated medical facility on the site. A medical personnel performs this triage and more physiological variables are required for this triaging process which uses the triage sort algorithm, depicted in Table 1 [15]. The process sums the values of each variable and categorizes the injured in three possible categories: T1 (red-immediate), T2 (yellow-urgent) and T3 (green-delayed).

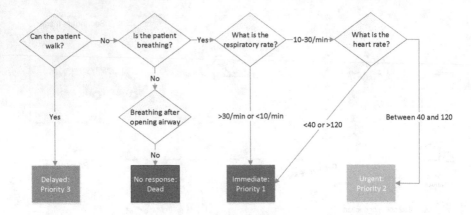

Fig. 3. Common form of primary triage decision tree (Color figure online)

5   System Overview

The overall system consist of hardware elements such as biosensors and communication equipment, software modules and describes the interaction between the elements.

5.1   Hardware System Elements

Suitable selection of hardware elements is an important factor since the system enables collaboration and data exchange between many different devices. It is recommendable for the hardware elements to have open-source or API in order to program the desired behavior and to support the intended communication protocols. The smart devices selected in this system depend on Android OS as most commonly used mobile OS and offers more control over desired behavior. Since most of the biosensors available on the market support Bluetooth connection, as an important parameter for selecting devices is the availability of API for the sensor. Further we describe the selected biosensors.

Zephyr Bioharness 3. The Zephyr Bioharness 3[1] sensor monitors basic vital parameters as heart rate, heart rate variability, and respiratory rate and includes 3-axis accelerometer that determines body orientation and activity. It is Bluetooth low energy capable and thus supports Bluetooth data exchange. It streams data at a frequency of 250 Hz. Besides the API that the manufacturer offers, another advantage is the possibility to use it either with chest strap or patch. The chest strap is more suitable for military environment of the system since it offers more stability, but it has to be put beforehand since it is difficult to attach it to injured person. On the other hand, while the patch is not as stable, it is easily applicable to wounded person. The sensor is shown in Fig. 4.

---

[1] https://www.zephyranywhere.com/.

Table 1. Triage sort evaluation

Physiological variable	Value	Score
Respiratory rate	10–29	4
	>29	3
	6–9	2
	1–5	1
	0	0
Systolic blood pressure	>90	4
	76–89	3
	50–75	2
	1–49	1
	0	0
Glascow coma scale	13–15	4
	9–12	3
	6–8	2
	4–5	1
	3	0
Total	12	**T3**
	11	**T2**
	10–1	**T1**

MyTech Wrist Cuff Blood Pressure Monitor. The MyTech[2] sensor measures systolic/diastolic blood pressure and pulse rate from the wrist and measured results can be transferred to mobile device through bluetooth. It uses oscillometric method of measurement. The sensor is shown in Fig. 5.

Nonin. The Nonin[3] sensor monitors blood oxygen saturation levels and pulse rates and measured results can be transferred to mobile device through bluetooth. The sensor is shown in Fig. 6.

Concurrent Sensor Usage. As mentioned before, in primary triage only the Zephyr sensor is used, while in secondary triage all biosensors are used. In order to decrease the concurrency for the Bluetooth connection, the SPO2 and Blood pressure sensors are directly connected to the Zephyr sensor (as depicted in Fig. 7) and only the Zephyr sensor is connected via Bluetooth to the medic's tablet or mobile phone.

---

[2] http://www.mytech-tel.com/.
[3] http://www.nonin.com/.

Fig. 4. Zephyr biosensor   **Fig. 5.** MyTech biosensor   **Fig. 6.** Nonin biosensor

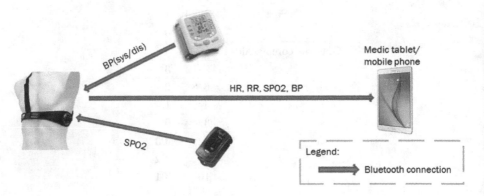

Fig. 7. Secondary triage

5.2   Software Modules and Interaction

The main software architecture is modular. It consists of two main android mobile modules: one for primary triage and other for secondary triage. The primary triage module is hosted in the FAR device, whether civil or military. In military environment the team commander also has the module in order to be to date with the situation and make strategic decisions. The data from the primary triage application for the critically injured persons (labels as red or yellow by the triage decision tree) are further transferred to the medic device and processed for secondary triage with the additional sensors. Depending of the use cases the secondary triage is hosted either in Role 1 (military) or ER ambulance reception or vehicle (civil). Figure 8 depicts the main software elements of the solution (modules), the devices they use, as well as the communication protocols used in the exercise in military environment.

The identification of injured persons is based on the unique MAC address of the sensor assigned to the person. All assignments of sensors to persons are logged. At real-time monitoring, based on the MAC address of the sensor sending data, the person is identified. If the sensor is removed from one person, it can

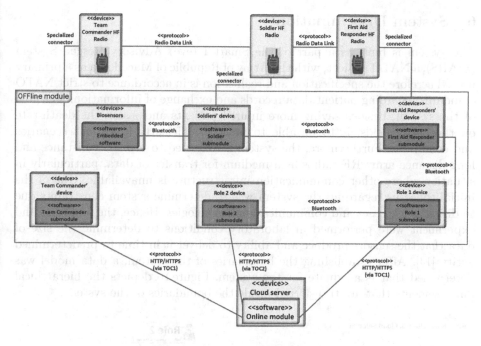

Fig. 8. Modules communication

be assigned to new person and log should be updated. All the readings from the sensors are kept and can be interchanged between modules or delivered to Cloud.

The medic can monitor all the injured persons that has been once labeled as critical and can see details, history or perform additional triage. Figure 9 shows some of the application interfaces available to the medic.

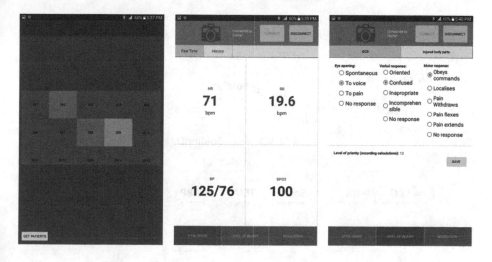

Fig. 9. Medic application interfaces

6   System Dissemination

The system is built as a part of the Smart I (eye) Advisory Rescue System
(SIARS), a NATO project, with the Army of Republic of Macedonia as a primary
user. Therefore the specification and realization is in accordance to strict NATO
standards regarding patient data records and exchange of information. The goal
of the system, besides saving more injured patients and lessen the death-rate
on the battle fields, is to be able to function efficiently in various scenarios
and conditions. Furthermore, the system is extended to use short-distance and
long-distance army RF radios as a medium for transfer of data, particularly in
situation where other communication infrastructure is unavailable. One of the
initial concerns regarding the system was to determine system constraints due
to different use cases and communication technologies. Hence, the first tests and
experiment were performed in laboratory conditions to determine the size of
data that the sensors produce and ability to deliver it in time to predetermined
entity [14]. After establishing the boundaries of the system, a data model was
determined that is adequate for the system. Figure 10 depicts the hierarchical
data elements that are transferred inside the boundaries of the system.

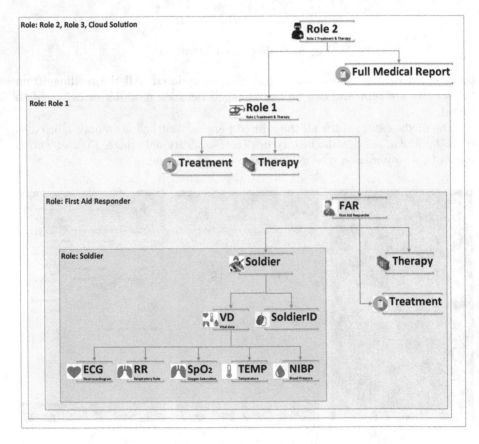

Fig. 10. Hierarchical data elements depicted by role

Further the system was tested in accordance to manufacturer specification, data size growth regarding time and battery life of the rechargeable sensors included. The most relevant system dissemination, outside controlled laboratory conditions, occurred in 2017 in the **17th EADRCC consequence management field exercise** called "Bosnia and Hercegovina 2017", jointly organized by the Euro-Atlantic Disaster Response Coordination Centre (EADRCC) and Ministry of Security of Bosnia. In the exercise the system was directly included and tested and even further extend to exchange information with others country participant standardized health facilities using web services.

6.1   Data Size Growth and Transfer

We have conducted several tests regarding the data size growth in primary triage, as well as transfer in situations where only direct Bluetooth transfers was available. The results are shown in Table 2. **While the data shows linear growth, the data transfer speed doesn't show the same growth.** This is mainly due the usage of Bluetooth protocol, which handles the given file sizes effortlessly. Figure 11 shows data chart, where is clearly depicted that data transfer speed is directly affected by the data size growth. We have also tested the system for 13 h continuous monitoring and the data size at the end of the period was

Table 2. Primary triage data size growth and transfer time

Time of measurement	Total data size	Time of transfer
2.5 min	39.16 KB	31 s
5 min	53.17 KB	34 s
10 min	78.75 KB	35 s

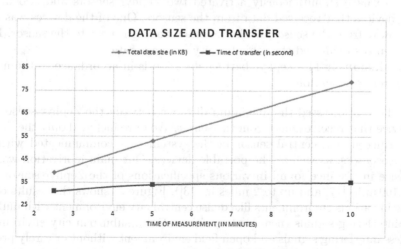

Fig. 11. Primary triage data size growth and transfer time

around 2700 KB, which supports the linear data growth. We have also tested the additional data size for secondary triage, shown in Table 3. Since blood pressure and SPO2 are not continuously measured (with much less frequency and upon demand), the additional data does not affect significantly. We have also measured the complete timeline (time required for transfer from FAR to Role 1 + Measurements (1 SPO2 +1BP) + Transfer to Role 2) which in average required 3.5 min.

Table 3. Measuring BP and SPO2 during secondary triage

Time of measurement	Number of measurements (SPO2 + BP)
2 min 16 s	3 SPO2 + 2 BP
4 min 30 s	6 SPO2 + 5 BP
10 min	10 SPO2 + 8 BP

6.2   Battery Life and Distance Between Sensors and Devices

We have conducted various scenarios to test the battery life expectancy of the sensors, as well as to establish is there correlation between battery duration and distance between the sensors and the device that receives data from the sensors. Since only the Zephyr sensor has embedded rechargeable batteries, while the other sensors use replaceable batteries, we have concentrated on the Zephyr sensor. After continuous monitoring we have come to the following conclusions:

- **Conclusion 1:** The battery duration is not correlated to the distance between the Zephyr sensor and the device that receives the data.
  - We have simultaneously activated two Zephyr sensors and two mobile devices that received data from the sensors. One of the devices was 20 m away from the sensor, while the other was right next to the sensor. Both sensors exhibited the same battery drainage rate.
- **Conclusion 2:** The Zephyr battery duration is in accordance to the manufacturer specification.

We have also tested the maximum distance between the Zephyr sensor and the device that receives data from the sensor. We have included only the Zephyr sensor since it is a central sensor in the system that communicated with the other biosensors, as well as the portable devices. Our main motivation was the difference in distance found in various specifications of the Zephyr sensor (e.g. up to 100 m in [17] and up to 2 miles in [13]). Figure 12 shows the results of the testing in various environments like dense central city environment with multiple other interfering signals (near central city station), suburban city environment with less interfering signals and open field environment without or rarely present interfering signals (like 17th EADRCC field exercise conducted in and around abandoned mine).

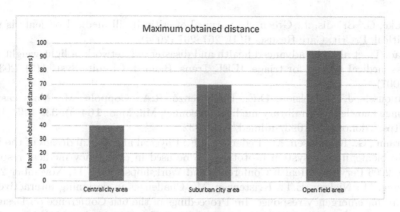

Fig. 12. Maximum obtained distance in various environments

7 Conclusion

In this paper we propose adaptable real-time distributed analytics system based on mobile devices and biosensors that can replace manual vital data collection with automatic one and thus improve, speed-up and maintain the triage decision process, which is crucial in situation with large number of casualties, where minutes or even seconds can make difference in numbers of saved lives. The system is adaptable and can be used in military and civil situations. Due to military usage the system follows and implements all required army and NATO standards and in according to these standards the hardware is selected. The system can be easily adapted to exchange data with other systems.

We have tested the model in laboratory and real-life simulated conditions in order to verify it. We performed numerous measuring activities using the chosen biosensors and the system uninterruptedly performs within the initial determined boundaries. We have also observed stability of the system in the required timespan as well as scalability of the model. We have successfully integrated three different sensors and reprogramed them in order to collaborate, synchronize and exchange data.

Acknowledgment. This paper is supported by SIARS, NATO multi-year project NATO.EAP.SFPP 984753.

References

1. Agoulmine, N., Deen, M.J., Lee, J.S., Meyyappan, M.: U-health smart home. IEEE Nanotechnol. Mag. **5**(3), 6–11 (2011)
2. Benharref, A., Serhani, M.A.: Novel cloud and SOA-based framework for e-health monitoring using wireless biosensors. IEEE J. Biomed. Health Informat. **18**(1), 46–55 (2014)
3. Benson, M., Koenig, K.L., Schultz, C.H.: Disaster triage: START, then SAVE - a new method of dynamic triage for victims of a catastrophic earthquake. Prehospital Disaster Med. **11**(2), 117–124 (1996)

4. Duke, G., Briedis, J., Green, J.: Survival of critically ill medical patients is time-critical. PrCrit. Care Resusc. **6**(4), 261–267 (2004)
5. Gao, T., et al.: The advanced health and disaster aid network: a light-weight wireless medical system for triage. IEEE Trans. Biomed. Circuits Syst. **1**(3), 203–216 (2007)
6. Janousek, J.T., Jackson, D.E., De Lorenzo, R.A., Coppola, M.: Mass casualty triage knowledge of military medical personnel. Mil. Med. **164**(5), 332–335 (1999). https://doi.org/10.1093/milmed/164.5.332
7. Kramp, G., Kristensen, M., Pedersen, J.F.: Physical and digital design of the Blue-Bio biomonitoring system prototype, to be used in emergency medical response. In: 2006 Pervasive Health Conference and Workshops, pp. 1–11, November 2006
8. Kyng, M., Nielsen, E.T., Kristensen, M.: Challenges in designing interactive systems for emergency response. In: Proceedings of the 6th Conference on Designing Interactive Systems, DIS 2006, pp. 301–310. ACM, New York (2006). https://doi.org/10.1145/1142405.1142450
9. Lee, T., Lee, S.H.: Dynamic bio-sensing process design in mobile wellness information system for smart healthcare. Wirel. Pers. Commun. **86**(1), 201–215 (2016)
10. Lerner, E.B., et al.: Mass casualty triage: an evaluation of the data and development of a proposed national guideline. Disaster Med. Public Health Preparedness **2**(S1), S25–S34 (2008)
11. Liu, J., Wang, Q., Wan, J., Xiong, J., Zeng, B.: Towards key issues of disaster aid based on wireless body area networks. KSII Trans. Internet Inf. Syst. **7**(5), 1014–1035 (2013)
12. Malan, D., Fulford-jones, T., Welsh, M., Moulton, S.: CodeBlue: an ad hoc sensor network infrastructure for emergency medical care. In: International Workshop on Wearable and Implantable Body Sensor Networks (2004)
13. Rae systems by Honeywell: BioHarness 3 Datasheet (2014). https://www.raesystems.com/sites/default/files/content/resources/Datasheet_BioHarness3_DS-1066-03.pdf. Accessed 10 May 2018
14. Simjanoska, M., Koteska, B., Kostoska, M., Bogdanova, A.M., Ackovska, N., Trajkovikj, V.: Information system for biosensors data exchange in healthcare. In: Stojanov, G., Kulakov, A. (eds.) ICT Innovations 2016. AISC, vol. 665, pp. 230–239. Springer, Cham (2018). https://doi.org/10.1007/978-3-319-68855-8_23
15. Smith, W.: Triage in mass casualty situations. Contin. Med. Educ. **30**(11) (2012). http://www.cmej.org.za/index.php/cmej/article/view/2585
16. Yang, G., et al.: A health-iot platform based on the integration of intelligent packaging, unobtrusive bio-sensor, and intelligent medicine box. IEEE Trans. Ind. Informat. **10**(4), 2180–2191 (2014)
17. Zephyr Technology: BioHarness 3.0 User Manual (2012). https://www.zephyranywhere.com/media/download/bioharness3-user-manual.pdf. Accessed 10 May 2018

Character Traits in Online Education: Case Study

Ermira Idrizi[(✉)], Sonja Filiposka, and Vladimir Trajkovik

Faculty of Computer Science and Engineering,
Ss. Cyril and Methodius University, Skopje, Republic of Macedonia
ermiraidrizi@gmail.com,
{sonja.filiposka,trvlado}@finki.ukim.mk

Abstract. In this paper we seek to better understand the outcomes of online education by observing the role of learner's personality traits. Under the assumption that the behaviors that maximize learning are dependent on the delivery method, we compared learning outcomes of students participating in two classes set up on an interactive e-learning platform. Our results confirm that personality traits are independent variables worthy of consideration in online settings. The preliminary results reveal that Extraversion, Openness and Conscientiousness personality traits are the most influential factors on the student's perceived quality of experience. Specifically, we argue that online education demands a particular set of behavioral patterns (i.e., low companionship, achievement orientation) necessary to navigate the eccentricity of online environment (e.g., social isolation, schedule flexibility). We discuss the theoretical implications of our results in the context of online education and offer practical suggestions for online teaching design.

Keywords: Quality of experience · Learning style · Big Five
Online education

1 Introduction

Online education is commonly considered a form of distance education because students are physically separated from each other and the instructor. This teaching approach features electronic learning or e-learning, which relies on computer network technology, often via the Internet, to transfer information from instructors to participants and vice versa and is a widely spread teaching approach in higher education institutions [1].

Significant research has followed the increasing academic interest in online education, with particular attention to understanding the efficiency of the online teaching approach compared to classroom teaching [2–4]. The student experience in online classes is different compared to traditional face-to-face classes, and patterns of engagement seem to differ between the two. For example, in online classes students felt more detached from their peers and professors, more compelled to be self-sufficient in their studies, and less assisted by their professor, than their professors believe them to be. Students can also feel overwhelmed by the technological assumption of online

© Springer Nature Switzerland AG 2018
S. Kalajdziski and N. Ackovska (Eds.): ICT 2018, CCIS 940, pp. 247–258, 2018.
https://doi.org/10.1007/978-3-030-00825-3_21

study, particularly if they start off without enough technical knowledge or support. Researchers have recommended that unlike the faster, real-time pace of face-to-face classes, the additional time available for online activities might allow students to think about course material more critically and reflectively, leading to deeper understanding of the course content. Others have suggested that the less challenging or personal nature of e-learning might give confidence to shyer students to engage more, or to feel less pressure than in face-to-face interactions [5, 6].

The perceived usefulness and the user's attitude are used to predict the intention of the students to use the system. The relationship between Quality of Service (QoS), Quality of Experience (QoE) and online learning tools, have been investigated in [7, 8]. Deeper understanding of factors influencing QoE in higher education should be investigated in order to create better utilization of the resources in the distance learning environment [9, 10].

A better understanding of how student's personality and character affect their academic performance during online classes would lead to better adapting, designing and evaluating online classes and so the students grasp of QoE would increase together with their satisfaction with online education. The focus of this paper is to determine how personality traits affect student's academic performance, and how this differs when comparing traditional classes with online classes. The method of investigating how character traits influence the quality of achieved learning results using different media presentations and delivery styles is presented. We performed a quality of experience (QoE) study including 70 students from the Faculty of Computer Science and Engineering that enrolled two distance learning courses from the computer science study program, both set up on the faculty's Moodle interactive e-learning platform. In the study we introduced the "Big Five" personality traits (Neuroticism, Extraversion, Openness, Agreeableness and Conscientiousness) as subjective input variables. The results analysis shows how personality traits correlate to the delivery method offered for following the online course.

This paper is organized as follows: in Sect. 2 we discuss the characteristics of online education and relate to the personality types. Section 3 describes the methodology used for the research, the participants and the procedure employed. In Sect. 4 we present the results of the experiment we have performed using the methodology together with the discussion of the outcomes of the experiment. Finally, the last section concludes the paper.

2   Characteristics of Online Education

Online education offers a variety of advantages for students and education institutions, while changing the scheme of education. Flexible schedules seem to be one of the most appealing attributes of online education. The broad accessibility to technology enables online students to do class work anytime anywhere. Consequently, the measure of learning in online courses heavily relies on the students, who can choose convenient times to concentrate on learning [11]. This feature has proven to be of great value, especially to students facing irregular schedules. Other advantages ascribed to online learning include reduced travel time and expenses.

Online learning mostly consists of blended learning and fully online courses. Blended learning primarily employs face-to-face sessions, including distance learning/lecturing sessions, and online materials that are also provided to students. Fully online learning has no face-to-face sessions, and most learning processes are provided through an online environment. Therefore, this type of instruction can present students with freedom from learning restrictions. The new concept for online education known as Massive Open Online Courses (MOOCs) is available for both blended and fully online courses and is attracting the interest of both educators and students. Though institutes of higher education recognize some potential benefits, the impact on teaching and learning is still being discussed. On the other hand, it has often been suggested that a great deal of these participants have difficulty with continuing their education online, leading to aggravating drop-out rates. While the world-wide use of MOOCs as fully online courses has increased rapidly, the course completion rate is still one of the most serious problems impacting their success. Based on these features of online learning in addition to the quality of the online program, personality and learning styles play a tremendous aspect on student's academic performance since online learning methods differ from traditional classrooms [12].

2.1   Personality in Online Education

The character of students' thought process is critical to learning and could potentially determine their academic achievement. Students differentiate in how they process, encode, memorize, organize, and implement the information they learn; some are thoughtful learners and others process information more seemingly [13]. Personality Traits have been seen for a long time as important factors of students' academic achievement, numerous studies undermine that personality traits as Conscientiousness and Agreeableness play a significant role in their success. Most studies have been carried out behalf of the relationship of Personality traits and traditional education [14–16], and the newest studies focus on how online learning can contribute on adapting to specific character traits and the learning environment to be more personalized [17]. Learning strategies mediate the connection among personality traits and academic achievement, by examining the relationships between personality traits, learning styles, and academic achievement among college students. Formal education is multidimensional in nature in the sense that numerous factors may influence learning results. This multidimensionality advocates that each educational event is unique; it is a particular combination of the various dimensions that gather in learning. To maximize results, learners should recognize and adapt to the particularities of the event. They need to ensure that their behavioral efforts match the eccentricity of a central course. This assumption determines studies in education where learning is partially seen as the fit between learners' behavioral patterns and the characteristics of the learning event. Personality refers to an individual's social reputation; behavioral patterns that govern individuals' responses across situations where these behavioral tendencies have been instrumental in explaining learning outcomes in a variety of ways.

2.2   The Big Five Character Traits

The Big Five personality ranges have long been used as a predictor of individual preferences and performance in various educational contexts in academic research [18]. The vast majority of work in this area has focused on the relationship between personality and student performance focusing on traditional methods of delivery and teaching. In this paper, we will try to identify if those specifics of students' performance also apply to online classes. The Big Five structure of personality traits is appearing as a powerful and avaricious model for understanding the connection between personality and different academic behaviors [19]:

- **Conscientiousness** is depicting students being disciplined, organized, and achievement-oriented. Students who are high on the conscientiousness range also tend to: spend time preparing, finish important tasks right away, pay attention to details, enjoy having a set schedule. Students who are low in this trait tend to: dislike structure and schedules, make messes and not take care of things, fail to return things or put them back where they belong, postpone important tasks.
- **Neuroticism** refers to degree of emotional stability, impulse control, and anxiety. This trait is characterized by sadness, moodiness, and emotional instability. Individuals who are high in this trait tend to experience mood swings, anxiety, irritability and sadness. Those who are low in this trait are typically: emotionally stable, deal well with stress, rarely feel sad or depressed, don't worry much.
- **Extraversion** is shown through a higher degree of sociability, assertiveness, and talkativeness. Students who are high in extraversion are outgoing and tend to gain energy in social situations, they tend to: enjoy being the center of attention, like to start conversations, enjoy meeting new people, have a wide social circle of friends and acquaintances. Students who rate low on extraversion tend to: prefer solitude, feel exhausted when they have to socialize a lot, find it difficult to start conversations, carefully think things through before they speak, dislike being the center of attention.
- **Openness** is reflected in a strong intellectual curiosity and a preference for novelty and variety. Students who are high in this trait tend to be more adventurous and creative, they are open to trying new things and focused on tackling new challenges. Those who are in the lower range on this trait: dislike change, resist new ideas and dislikes abstract or theoretical concepts.
- **Agreeableness** refers to being helpful, cooperative, and sympathetic towards others. This personality dimension includes attributes such as trust, altruism, kindness, affection, and other prosocial behaviors. Students who are higher in this trait tend to be more cooperative, have a great deal of interest in other people, feel empathy and concern for other people. Those who are in the lower range of this trait in this trait tend to: take little interest in others.

Personality and motivation are associated with particular differences in learning styles, and it is recommended that educators go beyond the current persistence on cognition and include these variables in understanding academic behavior [9]. For instance, the dependability and achievement orientations that characterize individuals high in conscientiousness are associated with a strong motivation to learn; a drive that, in turn, exhibits substantial connections with learning outcomes.

It is worth noting that the ability of personality traits to explain phenomena is expected to increase in environments wherein individual behaviors are less constrained by contextual clues. This notion is particularly relevant in online contexts where learners' discretion over learning behaviors amplifies. Compared to face-to-face classes, online students are less restrained to fulfill social expectations. They possess more freedom over the nature and frequency of social interactions. Consequently, we anticipate personality to be an outstanding independent variable of learning in online settings.

3   Methodology

The main goal of this paper is to determine if personality traits are influential in the participation of students who are enrolled in online or more traditional courses. This study used a field experiment to empirically test our belief that different types of education materials delivery combined with learning styles and character traits affect student's academic performance [20]. The current physical environment and equipment at our university campus gives a good basis to experiment with the proposed educational contents and scenarios, while our University uses the MOODLE platform for organizing courses and tracking their achievements is also enables us to use it for our purpose one establishing two online courses, deliver them to students and track their accomplishment.

3.1   Participants

For this study the sample populations were students enrolled in two courses Search Engines (C1) and Designing Dynamic Websites (C2) at the Faculty of Computer Science and Engineering, Ss. Cyril and Methodius University in Skopje, R. Macedonia. The two distance learning courses "Search Engines" and "Designing Dynamic Web Sites" were set up on the faculty's MOODLE interactive e-learning platform.

The students were informed about the experiment before the beginning of the course. A few of them decided not to participate in this experiment, some discarded the experiment. Thus, the total number of participants which took part in this experiment was 70 (40 females and 30 males). The average age of participants was 21 ranging from 19 to 22. All participants were randomly divided into two groups of 35 students (group A and B). For the duration of one semester the selected students attended both courses. Their learning results were tested at the final exam. For motivating students to participate seriously they gained an extra credit for their course grade based on their individual performance. This study tried to determine how character traits and learning styles affect student's academic performance especially while taking online classes.

3.2   Course Delivery

As introduced earlier, for the purpose of this research we have used two experimental courses (C1 and C2) with two groups of students (group A and group B). The first course (C1 course) can be considered as slightly less advanced course on the level of introduction to computer science, while the second (C2 course) is a more advanced course that requires some previous knowledge in computer science. The group's participants were randomly chosen from students enrolled on both courses. In order to experiment with the character traits and preferred student learning styles, we used different presentation types for delivering the educational content of each course:

- **Offline document content** - PDF documents, presentations and url links with related content were designed and spread to students. This makes it possible for students to independently manage their time and learn at their own selected pace.
- **Offline video content** - video presentations were recorded and delivered to the students in the form of a streaming video. This gives the opportunity for students to watch the material presented in a more animated fashion but still create their own learning schedule.
- **Online video conferencing** - live video conferences were prepared with the professor of each course. The lectures were scheduled at fixed time, and students needed to be enrolled for the appropriate course in order to be able to participate in the video conference. This delivery method requires that students attend classes at fixed times, so it differs from the previous delivery methods were students had the freedom of organizing their time at their own. But, at the end of the lecture, students have the opportunity to cooperate with the professor and among themselves.

For each course, students were divided into two groups (A and B). The A group of students that attended the C1 course, were asked to choose their preferred content delivery type (one of the three educational contents described above). According to their choice, they were divided into three stereotypes, and to each stereotype the lectures were presented according to their preferences. The B group of students that attended the C1 course had no possibility to choose a preferred content delivery type. The choice of the type of education materials delivery (one of the three types described above) was made by the professor, without taking into consideration the student's preferences. For the C2 course, students from B group choose their preferred content delivery type; while students from A group were given the content delivery type chosen by the professor. At the end of both courses (C1 and C2), a survey was carried out with the participants in order to gather feedback results about the students' observation for the quality of experience during these two courses.

3.3 Procedures

This experiment was conducted at the Faculty of Computer Science and Engineering, Ss. Cyril and Methodius University in Skopje. Two significant groups of students (A and B), each containing three subgroups, as described in the previous section were organized for comparison purposes. The Moodle interactive interface was used for management of the student-content during the experiment, as well as for the teacher-content interaction. None of the students had accessed the material previous of the experiment. All participants attended both courses during one semester. At the beginning of the semester the participants received a short explanation about the way the experiment will be carried out, and their required duties. During the experiment students were asked to complete questionnaires: personality questionnaire (character traits), questionnaires about their preferred learning styles, questionnaire indicating their intentions to continuously use various educational content delivery types as well as questionnaire for assessing the students' QoE. The experiment was completed with a final exam, through which students' learning outcomes were tested. Students took the exam as a regular final exam, for both courses and earned an extra credit for taking part in this study case.

For the purposes of this study we processed and analyzed the data collected from the final exam, their character traits, and also the data from QoE survey. We compared the final test results in terms of correlation coefficient of test results and character traits for both groups, A and B. We analyzed the character traits preference influences on the test outcomes for both groups of students.

4 Result Analysis and Discussion

Learning outcome relates to the degree of knowledge gathered by a person after studying certain material. We analyzed how character traits affect the learning results during those courses and their exam scores of the two groups of students on both courses, after study sessions. And at the end we analyzed the Quality of Experience students experienced during this experiment.

4.1 Correlating Test Results and Character Traits

In order to provide corresponding simulation for the different character traits, the students were offered to follow the two courses via three different types of interaction: Listening to the lectures via interactive video conferencing, receiving video streaming lectures, and using PDF-based materials.

Student's personality characteristics were measured while introducing the "Big Five" personality traits (Neuroticism, Extraversion, Openness, Agreeableness and Conscientiousness) as subjective input variables. Character traits play a significant role in student's academic performance, in view of the fact that while they derive from the character of the student and so determine how a student is willing to learn, organize their learning environment and adapt to the particular method of teaching.

Conscientiousness as a character trait favored also in traditional teaching, shows the highest average in this study (Table 1):

Table 1. Conscientiousness character trait average for the two courses

Course_C1	Course_C2
Group A = 5.26	Group A = 5.3
Group B = 5.20	Group B = 5.26

Students with this trait are more organized and are more willing to succeed, so they exhibit these characteristics while attending online classes. Students with this trait prefer to have PDF materials in addition to listening to online materials in order to ensure the process of learning.

Extraversion is the second significant trait measured in this study which indicates that it is still important for students to be social and have the alternative of assertiveness while attending online classes (Table 2).

Table 2. Extraversion character trait average for the two courses

Course_C1	Course_C2
Group A = 5	Group A = 5.1
Group B = 5.07	Group B = 5.0

Even though **Agreeableness** is typically a character trait related positively with good academic performance, in our case the averages are lower, probably because in online classes being helpful and sympathetic toward others does not have a big implication with the performance during those classes (Table 3).

Table 3. Agreeableness character trait average for the two courses

Course_C1	Course_C2
Group A = 3.89	Group A = 4.1
Group B = 4.11	Group B = 3.89

As the results show Conscientiousness and Extraversion are more significant for a student's academic performance, where agreeableness does not play a significant role during online classes.

We have compared the two sample groups, based on the correlation coefficient of how test results and character traits correlate. It is notable that the correlation coefficient for the first sample Group A, see Fig. 1, which attended the first course while choosing the preferred way of learning, shows a positive correlation, while Group B shows a slightly negative correlation of the test results and character traits (Tables 4 and 5).

Table 4. Correlation coefficient for Group A for the first course

Correlation coefficient Course_1	Group A values
Emotional	−0.19
Extraversion	0.15
Conscientiousness	0
Agreeableness	0.39
Openness	0.22

For the second course C2 we also calculated the correlation coefficient, see Fig. 2, and the examination shows that for the second sample group B on which students were allowed to choose the preferred way of learning the correlation coefficient is in the

Table 5. Correlation coefficient for Group B for the second course

Correlation coefficient Course_1	Group B values
Emotional	0.02
Extraversion	−0.13
Conscientiousness	0.16
Agreeableness	0.03
Openness	−0.13

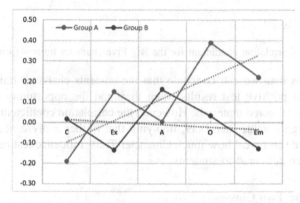

Fig. 1. Correlation coefficient for the Big Five character traits – Course 1

positive axes, while for the second sample Group Λ; the correlation coefficient has a slightly more negative tendency (Tables 6 and 7).

Table 6. Correlation coefficient for the Big Five character traits – Course 2; Group B

Correlation coefficient Course_2	Group B values
Emotional	0.08
Extraversion	−0.1
Conscientiousness	0.07
Agreeableness	0.25
Openness	0.10

Table 7. Correlation coefficient for the Big Five character traits – Course 2; Group A

Correlation coefficient Course_2	Group A values
Emotional	−0.027
Extraversion	0.018
Conscientiousness	0.29
Agreeableness	0.17
Openness	−0.24

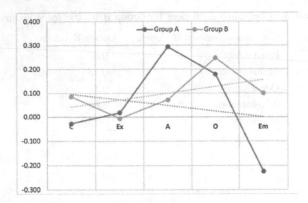

Fig. 2. Correlation coefficient for the Big Five character traits – Course 2

Based on this data we can conclude that the students character traits are slightly more exposed to positive test results when they had the opportunity to choose the preferred way of delivered materials. However, the correlation coefficients in overall do not achieve a higher value than 0.39, implying that further analyze needs to be conducted so that we can reveal how much Character Traits influence test results based on how materials are delivered to students.

4.2 QoE of the Two Classes

Qualities of Experience are the measurements of how students felt and were satisfied during the online classes. Based on the study flow throughout the semester students had to declare and fill out questionnaires how they agreed with the overall experience they underwent during the two online courses (6 absolutely agree to 1 absolutely disagree). Students had to fill out a questionnaire based on their experience they endured during the two online courses they attended, not based on their final exam success but based on their overall experience, the satisfaction with the service, delivered teaching materials. Qualities of Experience of the two class do not explain well how found these results. It is important for this section that the authors mention how the results are interpreted and what's is the most important of the results found.

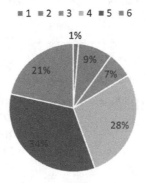

Fig. 3. Combined QoE distribution for the two online courses

The percentage presented in Fig. 3 shows that students had an overall good experience during this experiment. We counted and summed each response in categories from 1 to 6 which associate with the corresponding scale of satisfaction, and then converted those results to percentage. Unrelated to the test results, the students overall experience with the two online courses was satisfying, with 34% of the students giving a mark of 5 that expresses high level of agreeing with the way the course was delivered.

5 Discussion and Conclusion

This paper is a pursuit to better understand the online teaching approach through the inquiry of personality traits. Our results indicate that personality traits are worthy of research in online education since they do have impact on how students attend and finalize online courses. More exactly we found that: conscientiousness and achievement orientation are significant predictors of learning in online environment.

The role of personality in education should be considered in the aspect of teaching delivery preferences. The correlation coefficients for almost all our analyses were positively related with the sample groups which had the opportunity to choose the preferred way of delivering. According to these findings, we suggest that more attention to learners' personality traits needs to be given in online educational environments in order to allow students to adopt the online courses based on their learning styles. In our view, as technology matures, online education will experience significant changes with respect to the type of electronic interactions between learners and instructors, further rising the difficulty of this delivery method. Such advances in complexity propose different sets of behaviors from learners to exploit results. In other words, different association patterns between personality traits and learning outcomes. We view this research attempt as a potential area for more investigations and as a crucial part to make online education a more impressionable environment for the significant growth of student's shifting to these settings.

Finally, these results could help to better adapt and consolidate online classes, so that teachers are more aware of what students are looking in for while attending online classes, and so adjust to their needs and preferences to achieve the highest results. Based on the results of the QoE it proves that students do like online classes and their experience is positively related with the quality of those classes attended.

References

1. Loya, A., Gopal, A., Shukla, I., Jermann, P., Tormey, R.: Conscientious behaviour, flexibility and learning massive open on-line courses. Procedia – Soc. Behav. Sci. **191**, 519–525 (2014)
2. Aharony, N., Bar-Ilan, J.: Students perceptions on MOOCs: an exploratory study. Interdiscip. Study J. e-Skill Lifelong Learn. **12**, 145–162 (2016)
3. Aharony, N., Shonfeld, M.: ICT use: educational technology and library and information science students perspectives - an exploratory study. Interdiscip. J. e-Skills Lifelong Learn. **11**, 191–207 (2015)

4. Barrick, M.R., Mount, M.K.: The Big Five personality dimensions and job performance: a meta analysis. Pers. Psychol. **44**, 1–26 (1991)
5. Komarraju, M., Schmeck, R.R., Karau, S.J.: The Big Five personality traits, learning styles, and academic achievement. Pers. Individ. Differ. **51**, 472–477 (2011)
6. Nakayama, M., Mutsuura, K., Yamamoto, H.: Impact of learner's characteristics and learning behaviour on learning performance during a fully online course. Electron. J. e-Learn. **12**, 394–408 (2014)
7. Malinovski, T., Vasileva-Stojanovska, T., Jovevski, D., Vasileva, M., Trajkovik, V.: Adult students' perceptions in distance education learning environments based on a videoconferencing platform–QoE analysis. J. Inf. Technol. Educ. **14**, 1–19 (2015)
8. Malinovski, T., Vasileva, M., Vasileva-Stojanovska, T., Trajkovik, V.: Considering high school students' experience in asynchronous and synchronous distance learning environments: QoE prediction model. Int. Rev. Res. Open Distrib. Learn. **15**(4), 91–112 (2014)
9. Kimovski, G., Trajkovic, V., Davcev, D.: Resource manager for distance education systems. In: Proceedings of IEEE International Conference on Advanced Learning Technologies, pp. 387–390 (2001)
10. Koceska, N., Trajkovik, V.: Quality of experience using different media presentation types. In: Information Technology Based Higher (2017)
11. Allen, E., Seaman, J.: Changing Course: Ten Years of Tracking Online Education in the United States. Babson Survey Research Group and Quahog Research (2013)
12. Singh, R.N., Hurley, D.C.: The effectiveness of teaching-learning process in online education as perceived by university faculty and instructional technology professionals. J. Teach. Learn. Technol. **6**, 65–75 (2017)
13. Tsai, C.-W.: Applications of social networking for universal access in online learning environments. Univ. Access Inf. Soc. **16**, 269–272 (2017)
14. Busatto, V.V., Prins, F.J., Elshaut, J.L., Hamaker, C.: The relation between learning styles, the Big Five personality traits and achievement motivation in higher education. Pers. Individ. Differ. **26**, 129–140 (1998)
15. de Feyter, T., Caers, R., Vinga, C.: Unraveling the impact of the Big Five personality traits on academic performance: the moderating and mediating effects of self-efficacy and academic motivation. Learn. Individ. Differ. **22**, 439–443 (2012)
16. Hakimi, S., Hejazi, E., Lavasani, M.G.: The relationships between personality traits and students' academic achievement. Procedia – Soc. Behav. Sci. **29**, 836–845 (2011)
17. Kratky, P., Tvarozek, J., Cuda, D.: Big Five personality in online learning and games: analysis of student activity. Int. J. Hum. Cap. Inf. Technol. Prof. **7**, 33–46 (2016)
18. Varela, O.E., Cater III, J.J., Michel, N.: Online learning in management education: an empirical study of the role of personality traits. J. Comput. High Educ. **24**, 1–17 (2013)
19. Wingo, N.P., Ivankova, N.V., Moss, J.A.: Faculty perceptions about teaching online: exploring the literature using the technology acceptance model as an organizing framework. Online Learn. **21**, 15–35 (2017)
20. Vasileva-Stojanovska, T., Vasileva, M., Malinovski, T., Trajkovik, V.: An ANFIS model of quality of experience prediction in education. Appl. Soft Comput. **34**, 129–138 (2014)

Amplitude Rescaling Influence on QRS Detection

Ervin Domazet and Marjan Gusev[✉]

FCSE, Ss. Cyril and Methodius University, 1000 Skopje, Macedonia
ervin_domazet@hotmail.com, marjan.gushev@finki.ukim.mk

Abstract. When we record the electrical activity of the heart we generate a signal called an electrocardiogram. Within the electrocardiogram, the information that explains the heart's health is based on the detection of QRS complexes. The focus of this paper is on a wearable ECG sensor that uses a low sampling frequency and bit resolution while it converts the analog signal to digital data. The overall goal is to see if an efficient industrial QRS detector can be developed within these constraints. In particular, we set a research question to investigate how amplitude rescaling affects sensitivity and positive predictive rate of the Hamilton algorithm for QRS detection and improved it by optimizing it based on amplitude ranges. We used the MIT-BIH Arrhythmia ECG database to evaluate performance. The original recordings are sampled with a sampling frequency of 360 Hz with a 11-bit resolution over a 10 mV range. Our experiments include testing rescaled signals on a sampling frequency of 360 Hz using different maximum amplitudes. We found that rescaling impacts performance and that the optimization parameters need to tuned to obtain the expected performance. However, the performance decreases when the maximum amplitude is lower than 9 bits.

Keywords: ECG · QRS detection · Hamilton · Sampling frequency
Amplitude · AD conversion bit resolution · Beat detection

1 Introduction

An electrocardiogram (ECG) shows the electrical activity of the patient's heart over time. This information is considered as an indicator of the patient's state of health, yet not all features of the ECG can be extracted. The persistence of heart characteristics over time in the ECG was revealed by Lugovaya [12]. This means that if the hidden information is extracted from the ECG signal, most of the cardiac disorders can be detected before they happen.

Algorithms specialized in the extraction of heartbeats from signals streamed from wearable devices, such as the one described in [7], face a lot of challenges. Different kinds of noise is one of the challenges, which can stem from internal and external sources.

An ECG can be described as a repeating pattern of characteristic P, QRS and T waves, as illustrated in Fig. 1. The QRS wave is constructed with the signal

© Springer Nature Switzerland AG 2018
S. Kalajdziski and N. Ackovska (Eds.): ICT 2018, CCIS 940, pp. 259–272, 2018.
https://doi.org/10.1007/978-3-030-00825-3_22

starting from the local minimum Q, up to the local maximum R, and finishing at the local minimum S.

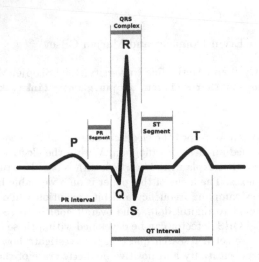

Fig. 1. Periodical nature of an ECG signal [3].

The process of QRS wave detection starts with applying Digital Signal Processing (DSP) algorithms, which consists of at least four stages. Subsequent to *Data Preprocessing*, the *QRS detection* stage plays a crucial role especially for further processing, including *Classification* and *Rhythm identification*.

Developing an industrial QRS detector has been a hot research topic since the late 80's. As advances in IoT rapidly continues, so do the trends in ECG processing. That is why, our primary focus in this research, is the *QRS detection* stage.

Hamilton has published a relatively good QRS detector [8]. In this research, we aim to optimize QRS detection performance especially on lower sample rates and amplitudes and to improve the original algorithm.

The paper is organized in the following way. Section 2 gives an overview of the current approaches for QRS Detection and discusses existing analysis of the influence of amplitude rescaling on QRS detection performance. Details for Hamilton's algorithm are given in Sect. 3. Section 4 is an elaboration on the testing methodology and Sect. 5 evaluates the results. The obtained results are discussed and compared to other available QRS detectors in Sect. 6. Finally, Sect. 7 summarizes the obtained results and gives directions for the future research.

2   Related Work

Kohler et al. [10] give an overview of existing QRS detection methods.

Undoubtedly, there are many published approaches, and they can be considered as by-products of the previous approaches, or their combination.

Pahlm, and Sörnmo [17] comment that algorithms sustain a standard procedure even though their approaches differ. The steps they follow are: *Reducing noise* to an acceptable level, then applying a *Thresholding* and *Deciding* whether the peak is a beat.

The Pan and Tomkins algorithm [18] and Hamilton's algorithm [8] are among the most cited papers in the category of *Differentiation (Derivation) based* approaches. Their algorithms are fast and can be treated as real time algorithms. Another feature is that they can quickly adapt to changes in signal quality.

They differ in the way they apply filters and detections rules. We have to confirm that Hamilton's approach introduces more detection rules increasing the complexity and performance of the algorithm.

Among the most cited DSP based filtering approaches is Afonso's algorithm [1]. His study is based on using filter banks, and the reported results at a 360 Hz sampling rate suggest that these types of algorithms can run very fast with promising performances.

Gusev et al. [6] have proposed a pattern matching algorithm in order to match the QRS pattern with default patterns.

An increasing amount of papers address Neural networks. Xue et al. [21] have presented such an approach with excellent performance. QRS detection is based on an Artificial Neural Network (ANN). An adaptive whitening filter is used to filter low frequencies. Whereas the QRS complex is detected with a linear matched filter, which compares the output against the high frequency signal input. High frequency signals are compared against.

Poli et al. [19] presents a solution for calculating a threshold based on genetic algorithms. QRS complexes are detected with a linear and nonlinear polynomial filter, with an applied adaptive local maxima threshold. Parameters are optimized via genertic algorithms and, are successful in decreasing detection errors.

The performance of any QRS detection algorithm, can be further increased with improved noise elimination. DWT algorithms are extensively used as the best filtering method. The philosophy behind DWT algorithms lies in decomposing the signal into two different signals, called details and approximation, as well as reconstructing the signal by creating an inverse transformation of these two signal segments. This operation eliminates certain levels of noise, though there is a limit on the frequency range. Generally 250 Hz is suitable for such algorithms. Li et al. [11], Bahoura [4], Shambi [20] and Martinez [14] have reported successful implementations of DWT-based QRS detectors. Although one might think that DWT-based algorithms are complex and require a lot of processing time, Milchevski and Gusev [15] have reported a fast algorithm that achieves DWT filtering very fast.

Martinez [13], has proposed a phasor-transform based algorithm for eliminating baseline drift, on a 360 Hz sampling frequency. This algorithm converts each sample into a phasor and correctly identifies feature points.

Ajdaraga and Gusev [2] have analyzed how the sampling frequency and resolution impacts ECG signals and their QRS detection. They report that even the rescaled signals obtain good performance when fine tuning the threshold parameters.

3   Hamilton's QRS Detection Algorithm

Algorithm details on Hamilton's QRS detector are already presented in [8]. Generally, algorithms for QRS detection publish their conceptual work, but lack implementational details.

For this particular case though, EP Limited [9] has released an open source implementation of the algorithm, which we will use to improve it. They provide a complete C-implementation for Hamilton's algorithm with different variations. It includes three different detectors and a fundamental beat classification unit.

Figure 2 presents the processing steps to detect a peak and classify it as a beat. The processing is executed in two different phases, the first is *DSP Filtering* and the second *Peak Detection*. The primary aim of the first phase is to eliminate noise stemming from different sources, such as breathing, muscle or skin movements. This is important for proper beat detecting. Whereas, the latter phase aims to detect a peak, and classify it as a real beat or an artifact.

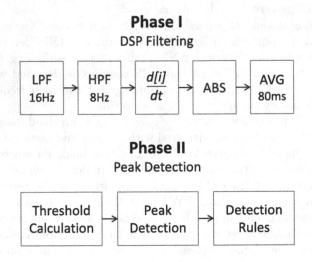

Fig. 2. Architectural representation of Hamilton's QRS detection algorithm.

Phase I implements a total of five sequential steps: a 16 Hz low pass filter (LPF), an 8 Hz high pass filter (HPF), a differentiation filter ($\frac{d[i]}{dt}$), a filter that calculates an absolute value (ABS), and a filter that calculates an average over an 80 ms moving window of samples.

The first two filters suppress environmental and internal noise, including baseline drift and act as a bandpass filter (BPF). The essence of the detection is the differentiation filter that aims to calculate a slope of differences. Since the slope can be negative, the ABS stage calculates only positive portions, and the AVG filter generates energy peaks that can be detected using rather simple rules in Phase II.

Phase II is responsible for peak detection by calculating a local maximum and comparing to two thresholds: *static threshold* (MIN_AMP_PEAK) with a fixed value, and *dynamic threshold* (DT) which is affected by the amplitudes of the last eight real and noise peaks. Any peak value lower than the static threshold will classify as a artifact, and any value over both the static and dynamic peaks will classify as a real peak. Otherwise it will classify as a noise peak.

The main optimization method used in this research is to choose an optimal value of the static threshold. In addition to this, we have explored the dependence of threshold values on different sampling frequencies and bit resolutions.

4   Testing Methodology

The experiments conducted during this study aim to find the dependence of sampling frequency and bit resolution on QRS detection performance. Particularly, we aim to find the optimal value and correct adjustment of the static threshold with scaled amplitudes to boost maximum performance. This optimization will improve the existing algorithm and reach higher performance.

4.1   Testing Environment

The experiments were conducted on the MIT-BIH Arrhythmia database [16], which is considered as the most important benchmark for ECG monitoring. It contains two-channels of 48 half-hour ECG recordings. These recordings are publicly available on Physionet.org [5]. The sampling frequency is 360 Hz per channel, with a 11-bit resolution over a 10 mV range. Each record is accompanied by an annotation file made by physicians, and therefore enables a good-quality evaluation of the QRS detection algorithm.

Although the database has a total of 48 records, there are four records that contain paced beats, which may introduce a higher rate of errors (a paced beat followed by a QRS to be misinterpreted). Therefore, our test cases use only 44 records.

Each record contains two channels of data representation, and we used the first one which represents the ML II signal in most cases.

4.2   Test Cases

Two experiments were conducted for the purpose of this research:

- *Exp. 1*: Impact of rescaled amplitudes on the performance
- *Exp. 2*: Optimal static threshold value to boost the performance

The first experiment *Exp. 1* aims to determine the impact of different amplitudes on a 360 Hz sampling frequency. Test cases within this experiment start with the initial data records from the default MIT BIH Arrhythmia ECG database using an 11-bit resolution. The following test cases gradually scale the amplitudes by a factor varying from 1.00 to 0.25 with decremental steps of 0.05.

Note that if a scaling factor 1 is used, then it corresponds to the original signal records with a 11-bit resolution, and the factor 0.5 corresponds to signals with half of the amplitude, i.e. to a 10-bit resolution. Consequently, the test case with scaling factor 0.25 corresponds to a quarter of the original signal, and represents a 9-bit resolution.

The second experiment *Exp. 2* addresses the optimal value of the static threshold. A crossed dependence check is performed to check the dependence of various amplitudes and the static threshold parameter on performance.

The test cases include measurements where the input is a pair of static threshold and the scaling factor of the amplitude. Static threshold values change from 2 to 10 with step 1, and amplitude scaling factor from 0.25 to 1 with step 0.05. Similar to the previous experiment, the test cases were conducted on the first channel of signals with an original sampling frequency of 360 Hz.

4.3   Test Data

Performance evaluation of proposed algorithms are done with existing metrics based on measured correctness. Here we use the usual correctness classification:

- *Correctly detected beats* denoted as true positives TP;
- *Extra detected peaks* are considered as false positives FP meaning that the algorithm has detected a peak that is not a real beat;
- *Missed beats* identified as false negatives FN meaning that the algorithm has not detected real beats.

To measure the performance we will use number of errors as an indicator. The smaller the number of errors - the better the algorithm. Following the previous definition, the total number of (false) errors is equal to the sum of FP and FN, calculated by (1).

$$False\ Errors = FP + FN \tag{1}$$

Although the number of errors can be efficiently used to compare two different algorithms, we still have to compare this number with total number of QRS beats. The total number of QRS beats can be calculated as a sum of correctly detected beats and those that were missed according to (2)

$$Total\ QRS = TP + FN \tag{2}$$

So instead of number of errors one can use the performance measure called *Relative Error* (RE), that explains the relative magnitude of false errors compared to the total number of beats, which is calculated by (3).

$$Relative\ Error = \frac{False\ Errors}{Total\ QRS} = \frac{FP + FN}{TP + FN} \tag{3}$$

5   Evaluation of Results

This section evaluates the results from the experiments.

5.1   Performance Achieved on Rescaled Amplitudes

Figure 3 presents the optimal performance of the execution of Hamilton's algorithm over signals sampled with 360 samples per second and 11-bit resolution.
The x-axis portrays the amplitude scaling factor represented as a percentage,
and the y-axis the number of false errors.

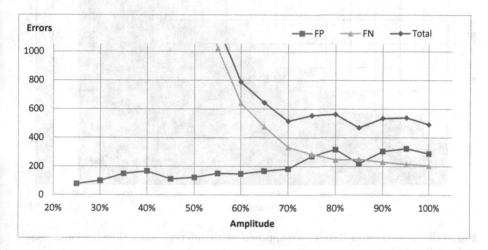

Fig. 3. False error detections of Hamilton's approach for MIT BIH Arrhythmia
database signals

The presented results are based on measurements where Hamilton's original
algorithm uses the default static threshold value MIN_AMP_PEAK = 7. It is
obvious that the threshold parameter can not follow the downscaled amplitudes,
and the number of generated false positives (extra generated peaks that are not
real beats). Note that rescaled amplitudes with a scaling factor higher than 65%
obtain a relatively good performance for the fixed static threshold parameter
(the number of errors was lower than 600).

5.2   Optimal Static Threshold Value to Boost the Performance

Presented results for the default static threshold $MIN\_AMP\_PEAK = 7$ show
big discrepancies and performance fluctuations when the amplitudes are rescaled.

Second experiment defines test cases on different pairs of a static threshold
parameter and amplitude scaling factor. The figures show charts where x-axis
represents the amplitude scaling factor measured in %, and y-axis the static

threshold parameter. The presented values are three dimensional, where the third dimension is a colored scale of relative error. The darkest squares are those with the highest performance and smallest relative error presented in %, and the lightest color is the worst performance and highest relative error.

Figure 4 presents the Relative Error obtained by executing Hamilton's algorithm on different pairs of static threshold values and amplitude scaling factors.

The x-axis scale starts from 25% to 100%, and the y-axis from static threshold value 2 up to 10. The colored relative error scale starts from 0.4% (darkest) up to over 1.0% (lightest).

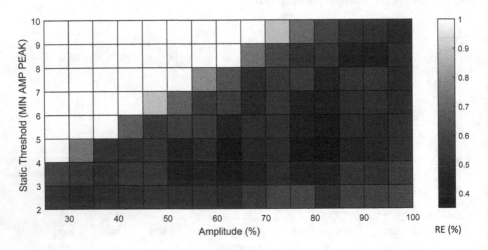

Fig. 4. Relative error of Hamilton's approach with different static threshold and amplitude scaling factors.

We observe that the best configuration for original signals with a sampling frequency of 360 Hz is when the static threshold is 4 and amplitude multiplier is 65%, with a relative error of 0.402% (total false errors 405).

6   Discussion

6.1   Performance Impact of Rescaled Amplitudes

Table 1 summarizes the evaluation of best and worst performance for the fixed static threshold, and gives the behavior expressed by the average number of errors and fluctuations when different amplitudes are used with the same algorithm and fixed static threshold parameter.

The fixed value of the static parameter with value equal to 7 showed fluctuation between 400 and 580 false errors in a dataset with 100733 beats. This is a relatively a high fluctuation, since it is equal to 45% of the false detections. However, the relative error in comparison to the total number of beats is between 0.578% and 0.397%.

Table 1. The performance behavior for the fixed static threshold parameter equal to 7

Performance	Amplitude	Errors	Relative error
Best	85%	470	0.467%
Worst	25%	7085	7.033%
Average	-	1726	1.713%
Fluctuation	-	6615	6.566%

When analyzing the fluctuations of the algorithm performance, one can conclude that the fixed static threshold parameter will only obtain good performance on selected amplitudes.

Next we will analyze which static threshold parameter achieves the best performance.

6.2   Selecting an Optimal Static Threshold

Table 2 presents the best performances from the default threshold and the best performance from different thresholds.

Table 2. Optimal static threshold parameters that reach the highest performance using the Hamilton's approach

Amplitude (%)	STHR	Errors	RE (%)
100	9	474	0.471
95	8	454	0.451
90	8	451	0.448
85	4	407	0.404
80	4	423	0.420
75	4	503	0.499
70	3	457	0.454
65	4	405	0.402
60	3	462	0.459
55	3	436	0.433
50	4	517	0.513
45	4	498	0.494
40	2	487	0.483
35	2	478	0.475
30	2	508	0.504
25	2	580	0.576

When we analyze results, we firstly observe that the default fixed static threshold does not yield the best performance. One can observe that the best static threshold is 9 instead of 7.

Interestingly, the default amplitude is not the best option either. We found that the best performance is achieved when rescaling by a scaling factor of 65% and a static threshold of 4 achieved. In this case, the relative error is 0.402%, and the QRS sensitivity and positive predictivity rate are high (99.81% and 99.79% respectively).

Figure 5 presents false error detections for the best chosen static threshold parameter from Table 2.

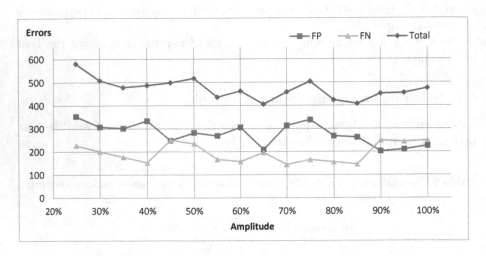

Fig. 5. False error detections of improved Hamilton's approach with optimal selection of threshold parameters for MIT BIH Arrhythmia database signals sampled on **360 Hz** and **11-bit** resolution.

Note, that the performance achieved with optimized static parameter (Fig. 5) is much better than the one obtained with a fixed static parameter (Fig. 5). Interestingly, signals using 65% of the amplitude reached the minimal number of errors is achieved for an amplitude scaled by a scaling factor of 65%.

In addition, we have also tested the influence rescaling had on different sampling frequencies. Figure 6 presents the optimal performance of the improved Hamilton algorithm using the optimal static threshold parameter over resampled signals to a 125 Hz sampling frequency with the original 11-bit resolution.

On can conclude that the performance of the optimal chosen threshold parameter is improved even on different sampling frequencies.

Figure 7 compares false error detections for typical values of bit resolutions by applying the optimal static threshold parameter from Table 2. Note that the performance difference of the algorithm for amplitudes using 9, 10 or 11 bits is negligible, whereas we observe big increase of the number of errors up to 2.246% for 8-bits.

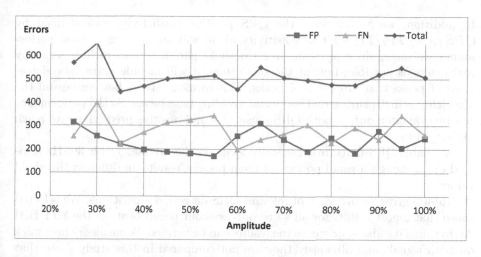

Fig. 6. False error detections of improved Hamilton's approach with optimal selection of threshold parameters for MIT BIH Arrhythmia database signals sampled on **125 Hz** and **11-bit** resolution.

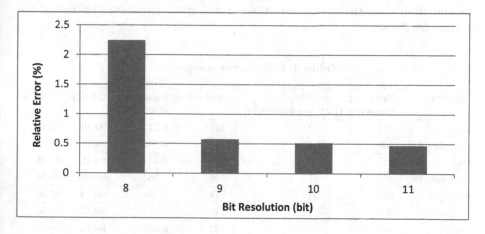

Fig. 7. False error detections comparison for different bit resolutions.

Bit resolution is sometimes called sampling resolution. Note that low number of bits used by the AD converter or rescaling to an amplitude with a small scaling factor can yield an increased signal-to-quantization-noise ratio. Thus, in this case, it will produce an increased number of errors in QRS detection algorithms. Our experimental research has proven that 8-bit resolution will increase the number of errors significantly when compared to 9, 10 or 11-bit resolution.

6.3   Comparison to Other Studies

Performance comparison with other approaches will be realized through conventional performance metrics for QRS sensitivity $QRS_{SE} = TP/(TP + FN)$.

In addition, we also include the QRS positive predictivity rate defined by $QRS_{+P} = TP/(TP + FP)$. Sensitivity alone will not give us a good performance estimate, since one can tune threshold parameters to include as many real peaks as possible, but, at the same time, it will include extra false detections of peaks that are not real peaks. This measure shows how successful the algorithm is in capturing real beats and avoiding false peaks. So the performance can be evaluated only if both QRS sensitivity and positive predictive rate reach higher values.

Available QRS detectors, generally focus on the original 11 bit 360 Hz sampled data, though a small portion of them present results on different threshold values.

Table 3 gives an overview of relevant QRS detection algorithms. We selected those that report results for all 44 records without paced beats in the MIT BIH Arrhythmia database, or where the results can be verified. Some algorithms work on both signals and obviously they are not compared in this study. Note that most of the algorithms do not publish information on their performance was measured on one or two channels, and we present the algorithms that use one channel for QRS detection. In our algorithm, we apply the improved version of the Hamilton's algorithm that uses the optimal static parameter instead of the original fixed value.

Table 3. Performance comparison

Algorithm	Sampling frequency (Hz)	Bit resolution (bit)	Total errors	Relative error[b] (%)	QRS_{SE}	QRS_{+P}
Bahoura[a,b] [4]	250	11	291	0.271	99.89	99.84
Our algorithm	360	11	405	0.402	99.81	99.78
Our algorithm	125	11	444	0.441	99.78	99.78
Hamilton [8]	360	11	569	0.564	99.68	99.76
Pan[a] [18]	200	11	771	0.768	99.73	99.50
Afonso [1]	360	11	732	0.872	99.57	99.56
GQRS [5]	360	11	562	0.558	99.72	99.72
WQRS [5]	360	11	1411	1.401	99.79	98.82
SQRS [5]	360	11	1899	1.885	98.73	99.38
SQRS125 [5]	125	11	3951	3.922	96.19	99.88

[a] Values are computed according to the record-by-record tables in the referred works.
[b] QRS_{SE} and QRS_{+P} calculated by (3).

The most important observation is that the optimized Hamilton approach (our algorithm) produces better results than those reported in the original algorithm. More importantly, even downscaled signals will obtain good performance results if the static threshold parameter is tuned according to the provided results.

It is even more interesting to compare our results with the Pan Tompkins algorithm. Our algorithm running at 360 Hz correctly finds 375 more peaks, and makes 366 less errors.

7   Conclusion

In this study we primarily focused on finding the influence of the signal amplitude on QRS detection performance.

We have observed that rescaling the amplitude directly affects the performance and increases the number of false detections. The reason why is the selection of the static threshold.

The experiment showed that choosing a fixed value of the static threshold will not yield the best performance. Therefore, one needs to tune its value in order to obtain a good performance.

Our research showed which optimal values of the static threshold result with best performance, and one needs to update this value according to the input amplitude. This optimizes Hamilton's algorithm and makes for a better solution, when is compared to other approaches.

This study is useful for future designers, gives the optimal value of the threshold parameter to be used by an AD converter for satisfactory QRS detection performance. For example, we concluded that 8-bit resolution will not yield performances over 98%, while using a 9, 10 or 11-bit resolution may achieve performances of QRS_{SE} and QRS_{+P} over 99.80%.

This research motivates further research and modeling regarding the optimal value of the threshold parameter.

References

1. Afonso, V.X., Tompkins, W.J., Nguyen, T.Q., Luo, S.: ECG beat detection using filter banks. IEEE Trans. Biomed. Eng. **46**(2), 192–202 (1999)
2. Ajdaraga, E., Gusev, M.: Analysis of sampling frequency and resolution in ECG signals. In: 2017 25th conference on Telecommunication Forum (TELFOR), pp. 1–4. IEEE (2017)
3. Atkielski., A.: Schematic diagram of normal sinus rhythm for a human heart as seen on ECG, January 2007. https://commons.wikimedia.org/wiki/File:SinusRhythmLabels.svg
4. Bahoura, M., Hassani, M., Hubin, M.: DSP implementation of wavelet transform for real time ECG wave forms detection and heart rate analysis. Comput. Methods Programs Biomed. **52**(1), 35–44 (1997)
5. Goldberger, A.L., et al.: Physiobank, physiotoolkit, and physionet. Circulation **101**(23), e215–e220 (2000)
6. Gusev, M., Ristovski, A., Guseva, A.: Pattern recognition of a digital ECG. In: Stojanov, G., Kulakov, A. (eds.) International Conference on ICT Innovations, pp. 93–102. Springer, Heidelberg (2016). https://doi.org/10.1007/978-3-319-68855-8_9
7. Gusev, M., Stojmenski, A., Guseva, A.: ECGalert: a heart attack alerting system. In: Trajanov, D., Bakeva, V. (eds.) ICT Innovations 2017. CCIS, vol. 778, pp. 27–36. Springer, Cham (2017). https://doi.org/10.1007/978-3-319-67597-8_3

8. Hamilton, P.S., Tompkins, W.J.: Quantitative investigation of QRS detection rules using the MIT/BIH arrhythmia database. IEEE Trans. Biomed. Eng. **12**, 1157–1165 (1986)
9. Hamilton, P.: Open source ECG analysis software documentation (2002)
10. Kohler, B.U., Hennig, C., Orglmeister, R.: The principles of software QRS detection. IEEE Eng. Med. Biol. Mag. **21**(1), 42–57 (2002)
11. Li, C., Zheng, C., Tai, C.: Detection of ECG characteristic points using wavelet transforms. IEEE Trans. Biomed. Eng. **42**(1), 21–28 (1995)
12. Lugovaya, T.S.: Biometric human identification based on ECG (2005)
13. Martínez, A., Alcaraz, R., Rieta, J.J.: Application of the phasor transform for automatic delineation of single-lead ECG fiducial points. Physiol. Meas. **31**(11), 1467 (2010)
14. Martínez, J.P., Almeida, R., Olmos, S., Rocha, A.P., Laguna, P.: A wavelet-based ECG delineator: evaluation on standard databases. IEEE Trans. Biomed. Eng. **51**(4), 570–581 (2004)
15. Milchevski, A., Gusev, M.: Improved pipelined wavelet implementation for filtering ECG signals. Pattern Recognit. Lett. **95**, 85–90 (2017)
16. Moody, G.B., Mark, R.G.: The impact of the MIT-BIH arrhythmia database. IEEE Eng. Med. Biol. Mag. **20**(3), 45–50 (2001)
17. Pahlm, O., Sörnmo, L.: Software QRS detection in ambulatory monitoring: a review. Med. Biol. Eng. Comput. **22**(4), 289–297 (1984)
18. Pan, J., Tompkins, W.J.: A real-time QRS detection algorithm. IEEE Trans. Biomed. Eng. **3**, 230–236 (1985)
19. Poli, R., Cagnoni, S., Valli, G.: Genetic design of optimum linear and nonlinear QRS detectors. IEEE Trans. Biomed. Eng. **42**(11), 1137–1141 (1995)
20. Shambi, J., Tandon, S., Bhatt, R.: Using wavelet transforms for ECG characterization. IEEE Eng. Med. Biol. **16**, 77–83 (1997)
21. Xue, Q., Hu, Y.H., Tompkins, W.J.: Neural-network-based adaptive matched filtering for QRS detection. IEEE Trans. Biomed. Eng. **39**(4), 317–329 (1992)

Novel Data Processing Approach for Deriving Blood Pressure from ECG Only

Monika Simjanoska[1(✉)], Martin Gjoreski[2], Matjaž Gams[2],
and Ana Madevska Bogdanova[1]

[1] Faculty of Computer Science and Engineering, Ss. Cyril and Methodius University,
Rugjer Boshkovikj, 16, 1000 Skopje, Macedonia
{monika.simjanoska,ana.madevska.bogdanova}@finki.ukim.mk
[2] Department of Intelligent Systems, Jožef Stefan Institute,
Jožef Stefan International Postgraduate School,
Jamova cesta 39, 1000 Ljubljana, Slovenia
{martin.gjoreski,matjaz.gams}@ijs.si

Abstract. Blood pressure is one of the most valuable vital signs. Recently, the use of bio-sensors has expanded, however, the blood pressure estimation still requires additional devices. We proposed a method based on complexity analysis and machine learning techniques for blood pressure estimation using only ECG signals. Using ECG recordings from 51 different subjects by using three commercial bio-sensors and clinical equipment, we evaluated the proposed methodology by using leave-one-subject-out evaluation. The method achieves mean absolute error (MAE) of 8.2 mmHg for SBP, 8.7 mmHg for DBP and 7.9 mmHg for the MAP prediction. When models are calibrated using person-specific labelled data, the MAE decreases to 7.1 mmHg for SBP, 6.3 mmHg for DBP and 5.4 mmHg for MAP. The experimental results indicate that when a person-specific calibration data is used, the proposed method can achieve results close to a certified medical device for BP estimation.

Keywords: Blood pressure · ECG · Machine learning
Complexity analysis · Classification · Regression · Stacking

1 Introduction

Blood pressure (BP) increase, hypertension, is one of the key factors for cardiovascular diseases [19,22]. The recent advances in bio-sensors technology has brought the opportunity to continuously monitor physiological signals (e.g., ECG, PPG, EMG, etc.) and consequently calculate or estimate the vital parameters: heart rate, respiratory rate, peripheral capillary oxygen saturation (SpO2) and blood pressure. BP estimation is considered to be a great challenge since the methodologies reported in the literature [12,13,21,23,26] usually require multiple physiological signals and devices for its estimation. However, in our previous research we proved that the systolic BP (SBP), diastolic BP (DBP) and mean

© Springer Nature Switzerland AG 2018
S. Kalajdziski and N. Ackovska (Eds.): ICT 2018, CCIS 940, pp. 273–285, 2018.
https://doi.org/10.1007/978-3-030-00825-3_23

arterial pressure (MAP) can be estimated by using only the ECG signal as a single source of information [24]. The methodology proposed relied on a combination of complexity analysis and machine learning (ML) techniques to build regression models that are able to predict the actual SBP, DBP and MAP values. By using a train-validation-test evaluation, we achieved a mean absolute error (MAE) of 8.6 mmHg for SBP, 18.2 mmHg for DBP, and 13.5 mmHg for the MAP prediction. By applying a probability distribution-based calibration, the MAE decreases to 7.7 mmHg for SBP, 9.4 mmHg for DBP and 8.1 mmHg for MAP.

In this paper, we consider a different evaluation of the methodology by performing leave-one-subject-out instead of the traditional train-validation-test evaluation and allowing a person-specific calibration to adapt the models to a particular user. The results obtained show significant improvement, especially for the DBP and MAP, decreasing the MAE error ~ -10 for non-calibrated DBP case and ~ -6 for the non-calibrated MAP case. When using a person-specific calibration, the improvements obtained are ~ -3 for DBP and MAP.

The rest of the paper is organized as follows. The proposed method is briefly described in Sect. 2. The experimental results are presented in Sect. 3, followed by a discussion in Sect. 4 and the conclusions of the study given in Sect. 5.

2   Methods and Materials

2.1   Methods

The complete methodology is comprehensively explained in [24] and is briefly depicted in Fig. 1. Raw ECG signals are divided into 30-s segments, each accompanied with SBP and DBP values. Those values pass through the preprocessing method, labelling the segments into the appropriate BP class and applying the low-pass filter. Hereupon, the signals are forwarded to the module for complexity analysis and feature extraction. Having computed the complexity metrics (signal mobility, signal complexity, fractal dimension, entropy and autocorrelation), the feature vectors are inputted to the classification module, which implements a stacking ML approach. The output of the classification module, in a combination with the extracted features, is inputted to a regression module which outputs the actual SBP and DBP estimation. The last module is a calibration module which allows for person-specific calibration. The calibration is performed by considering the mean error of the predictions for five randomly selected instances (measurements) of each subject, compared to the actual absolute SBP and DBP values. The error is either added or subtracted from the predicted values, depending on the model's tendency to predict higher or lower values.

2.2   Materials

The database we created for this research (publicly available online [1]) is built by using three different commercial ECG sensors (whose reliability is proven in previous studies [2,3,5,6,11,14,15,18,20,27]) and the reference SBP and DBP values measured by using an electronic sphygmomanometer. The second database

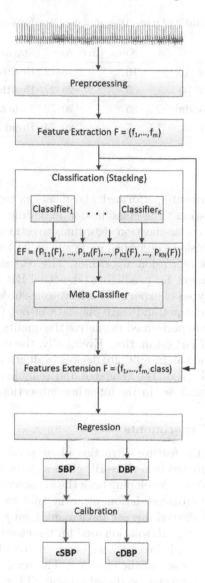

Fig. 1. Proposed methodology for blood pressure estimation.

that we use to additionally evaluate the methodology is obtained from the Physionet database and is created by using clinical equipment [9]. A summarized information of the datasets is provided in the following Table 1. Most of the participants are healthy (33). The rest 18 unhealthy participants were measured in hospital conditions, 11 of which are with cardiovascular problems and 7 are with brain injuries.

Table 1. Datasets summary information.

Source	Num. part	Age	Status
Cooking hacks sensor [10]	16	16–72	Healthy
180° eMotion FAROS [4]	3	25–27	Healthy
Zephyr Bioharness module [25]	25	20–73	14 healthy, 11 unhealthy
Charis Physionet database [17]	7	20–74	Brain injuries

3   Results

Four types of experiments are performed: classification, regression, feature analysis and devices evaluation. The classification experiments were performed to measure the ability of the classification algorithms to estimate BP class (hypotension, normal or hypertension - described in details in [24]) and were needed as an additional input for the regression algorithms. The regression experiments provide the error measurement for predicting the actual BP values of the method. Here we present three types of experiments, leave-one-subject-out (LOSO) interdataset, LOSO within dataset and leave-one-dataset-out (LODO). The Feature analysis experiments were performed to analyze the quality of the chosen features for the specific tasks of BP estimation. Eventually, the evaluation experiments were performed to provide an insight into devices evaluation, i.e. the performance of the wearable technology vs. the validated and reliable clinical monitors. The details for each experiment are in the following subsections.

3.1   Classification Experiments

The preprocessing and the feature extraction phase produced a total number of 3129 feature vectors mapped into three BP classes (hypotension - 0, normal - 1 and hypertension - 2). When developing both the classification and the regression models, we used LOSO cross-validation, meaning that we trained 51 models by including 50 subjects in the training set and leaving 1 subject out for testing. The performance of the stacking ML solution used for the classification was evaluated through the F-measure as a balanced mean between precision and recall for each class, and the overall accuracy of the classifier. The recall shows the proportion of the given class cases correctly predicted among all the instances that belong in the given class:

$$Recall = \frac{True\_positives}{Real\_positives} \tag{1}$$

Precision is a measure showing the proportion of the given class cases correctly predicted among all instances predicted to belong in the given class:

$$Precision = \frac{True\_positives}{Predicted\_positives} \tag{2}$$

$$F\text{-}measure = \frac{2 * Precision * Recall}{Precision + Recall} \tag{3}$$

The stacking design produced the results presented in Table 2.

Table 2. Stacking approach results.

Class/metric	Precision	Recall	F-measure	Accuracy
0	0.71	0.67	0.69	0.73
1	0.58	0.89	0.71	
2	0.94	0.63	0.76	

3.2 Regression Experiments

The predicted BP classes were used to extend the initial feature vectors and prepare the data for regression, as depicted in Fig. 1. Following the same principle for LOSO cross-validation, we evaluated three distinct models for predicting the SBP, DBP and MAP. To improve the prediction, we applied a calibration procedure as described in the Sect. 2. The results are sublimated in Table 3. The regression models were evaluated by using the Mean Absolute Error (MAE) and Root Mean Squared Error (RMSE). MAE is the average error obtained from the absolute differences between the actual, a_i, and the predicted values p_i, for $i = 1, ..., n$, where n is the number of instances within a subject. MAE weights all the differences equally and is calculated as:

$$MAE = \frac{\sum_1^n |p_i - a_i|}{n} \qquad (4)$$

To obtain higher weight for the large errors, which is important for the BP problem, the differences between the actual absolute and the predicted values are first squared, then averaged, and afterwards a square root of the average is performed. The RMSE is calculated according to the following equation:

$$RMSE = \sqrt{\frac{\sum_1^n |p_i - a_i|^2}{n}} \qquad (5)$$

Table 3 presents the MAE and RMSE evaluation for SBP, DBP and MAP. For each subject, $ID = 1, ..., 51$, and the errors from both the prediction and the calibration are presented for all three cases. The results show that the calibration goes in favor of the prediction by reducing the overall MAE from 8.24 ± 5.34 ($\mu \pm \sigma$) to 7.11 ± 5.29 for the SBP, from 8.75 ± 7.90 to 6.28 ± 5.02 for the DBP case, and from 7.92 ± 9.66 to 5.35 ± 4.16 for the MAP case.

Given the four datasets used in the experiments, two more experiments were performed using LOSO within a dataset and LODO evaluations. The results for the LOSO within dataset are presented in Table 4. The different datasets are labelled 1–4 (according to the device used for the measurements). Considering the obtained errors, for the SBP the MAE of 8.05 is close to the mean MAE obtained from the LOSO testing in Table 3. For the DBP case, the most critical

Table 3. MAE and RMSE evaluation for SBP, DBP and MAP.

ID	SBP				DBP				MAP			
	Prediction		Calibration		Prediction		Calibration		Prediction		Calibration	
	MAE	RMSE	MAE	RMSE	MAE	RMSE	MAE	RMSE	MAE	RMSE	MAE	RMSE
1	10.8	11.3	3.1	3.5	17.0	15.5	2.9	3.2	15.8	14.4	3.1	3.2
2	9.7	10.6	10.1	10.1	4.9	6.5	4.2	4.6	9.3	9.9	4.2	4.7
3	5.8	6.8	6.8	9.0	6.3	7.0	2.8	3.3	7.5	7.7	2.3	2.5
4	5.3	5.3	5.3	5.3	4.9	5.0	4.9	5.0	5.6	5.7	5.6	5.7
5	10.2	13.0	7.8	10.0	5.6	7.3	4.6	5.9	5.7	7.1	6.1	6.9
6	12.5	13.0	2.8	3.8	6.2	8.0	4.0	5.1	9.3	10.5	4.7	5.3
7	6.3	6.7	2.6	3.7	4.5	5.6	3.6	4.5	5.3	5.8	2.5	3.2
8	6.7	7.5	6.3	7.5	5.5	7.7	5.2	6.3	4.5	5.6	3.4	4.0
9	5.0	6.4	4.6	6.2	4.5	5.3	3.1	3.5	4.0	4.5	4.6	5.5
10	5.1	4.9	5.0	5.0	7.6	7.4	7.4	7.7	6.2	7.8	6.5	8.6
11	20.9	19.8	4.8	7.0	15.5	18.4	5.2	10.2	20.8	18.2	7.1	7.3
12	6.1	7.2	6.2	7.1	2.1	2.4	2.4	2.3	3.4	3.3	3.7	3.5
13	10.2	9.2	3.4	3.3	8.2	7.2	4.0	5.1	10.9	10.5	5.2	5.2
14	10.2	13.7	16.9	19.0	8.6	12.2	9.9	11.9	6.5	9.1	14.4	16.1
15	12.3	14.8	9.2	11.6	15.9	18.0	10.5	10.9	11.1	14.8	8.0	10.5
16	19.0	21.2	18.1	21.4	11.7	10.7	11.3	10.1	7.2	9.5	7.1	9.7
17	8.6	9.8	10.9	11.0	15.4	15.6	4.1	4.8	7.7	7.5	3.1	3.5
18	29.3	17.9	31.3	20.0	52.6	33.6	35.4	24.8	68.5	32.5	25.6	12.6
19	7.1	7.7	7.1	7.9	20.2	21.5	6.1	7.5	15.0	14.6	3.8	4.8
20	7.1	7.2	5.2	5.8	6.7	7.5	5.7	6.2	7.6	7.5	3.4	3.3
21	10.5	11.4	11.4	10.1	14.4	12.3	13.5	12.0	7.6	9.1	11.9	9.5
22	10.9	11.8	3.4	4.6	4.5	4.8	6.2	6.1	11.8	10.9	4.1	4.8
23	4.9	5.6	5.5	7.5	4.1	5.5	3.5	4.7	3.3	4.4	3.9	5.0
24	5.8	7.1	5.8	7.1	8.6	9.5	8.6	9.5	2.3	4.0	2.3	4.0
25	7.0	7.0	7.0	7.0	3.1	3.1	3.1	3.1	3.0	3.0	3.0	3.0
26	8.7	9.8	8.7	9.8	9.5	10.1	9.5	10.1	2.9	5.0	2.9	5.0
27	17.6	17.6	17.6	17.6	7.2	7.2	7.2	7.2	0.7	0.7	0.7	0.7
28	2.3	2.7	2.3	2.7	8.3	8.4	8.3	8.4	2.6	4.5	2.6	4.5
29	4.6	5.3	4.6	5.3	6.3	7.2	6.3	7.2	4.7	7.7	4.7	7.7
30	4.2	4.8	4.2	4.8	3.6	5.0	3.6	5.0	1.6	2.4	1.6	2.4
31	4.0	4.6	4.0	4.6	2.6	3.6	2.6	3.6	2.7	3.6	2.7	3.6
32	8.3	9.9	8.3	9.9	8.2	10.4	8.2	10.4	7.0	9.3	7.0	9.3
33	2.4	3.2	2.4	3.2	7.4	8.8	7.4	8.8	4.8	6.9	4.8	6.9
34	6.8	6.8	6.8	6.8	4.4	4.4	4.4	4.4	6.6	6.6	6.6	6.6
35	13.3	13.3	13.3	13.3	11.9	11.9	11.9	11.9	12.1	12.1	12.1	12.1
36	7.9	7.9	7.9	7.9	0.4	0.4	0.4	0.4	1.2	1.2	1.2	1.2
37	6.6	6.6	6.6	6.6	2.1	2.1	2.1	2.1	2.7	2.7	2.7	2.7
38	0.5	0.5	0.5	0.5	6.1	6.1	6.1	6.1	4.7	4.7	4.7	4.7
39	2.3	2.3	2.3	2.3	5.3	5.3	5.3	5.3	3.2	3.2	3.2	3.2
40	4.8	4.8	4.8	4.8	6.5	6.5	6.5	6.5	4.2	4.2	4.2	4.2
41	4.9	4.9	4.9	4.9	5.3	5.3	5.3	5.3	5.8	5.8	5.8	5.8
42	3.2	3.2	3.2	3.2	6.1	6.1	6.1	6.1	7.1	7.1	7.1	7.1
43	0.7	0.7	0.7	0.7	4.0	4.0	4.0	4.0	1.6	1.6	1.6	1.6
44	11.4	11.4	11.4	11.4	2.0	2.0	2.0	2.0	2.5	2.5	2.5	2.5
45	16.0	20.3	11.4	13.2	12.1	15.2	9.6	11.2	12.7	15.8	13.4	16.3
46	10.2	12.6	6.7	13.2	3.6	4.1	4.9	5.3	7.0	7.2	4.2	5.4
47	3.0	3.6	4.5	5.2	6.2	6.7	2.2	2.5	2.4	3.0	2.2	2.7
48	4.4	5.1	4.3	5.8	17.2	17.8	3.9	4.7	11.8	12.6	4.1	5.2
49	11.6	16.2	11.4	15.8	20.3	21.0	5.6	6.8	15.2	16.8	7.0	7.8
50	8.2	9.8	5.0	6.1	8.5	10.4	8.8	10.3	8.7	11.0	8.9	11.0
51	5.1	6.4	4.2	5.4	12.2	14.3	6.2	7.7	7.4	8.7	4.5	5.5
Mean	8.2	8.8	7.1	7.8	8.7	9.1	6.3	6.7	7.9	7.9	5.4	5.8
SD	5.3	5.0	5.3	4.8	7.9	6.1	5.0	3.9	9.7	5.5	4.2	3.5

datasets are number 2 and 4. Perhaps, this is due to the reduced number of participants available in those datasets. However, the calibration method is still able to perform well and provides a mean MAE of 6.5 ± 0.99.

The results for the LODO are presented in Table 5. Given the MAE and RMSE, it can be perceived that if the datasets are completely unknown to the classifier, then the model performs worse even in the calibration case.

Table 4. Leave-one-subject-out within dataset results.

BP	Dataset	Prediction		Calibration	
		MAE	RMSE	MAE	RMSE
SBP	1	8.6	9.8	5.8	7.1
	2	8.0	8.9	6.1	6.6
	3	7.5	7.8	7.3	7.7
	4	8.1	10.3	7.6	10.1
DBP	1	7.4	8.5	5.5	6.6
	2	14.9	15.5	6.5	7.4
	3	6.4	7.0	6.4	6.8
	4	11.8	13.4	7.9	9.7

Table 5. Leave-one-dataset-out results.

BP	Prediction		Calibration	
	MAE	RMSE	MAE	RMSE
SBP	13.0	15.7	10.2	12.6
DBP	12.3	14.9	14.3	17.0

Finally, Table 6 presents a summarization of the results for the three different evaluations: leave-one-subject-out inter-dataset, leave-one-subject-out within dataset and leave-one-dataset-out. The main three observations are:

1. For both the SBP and DBP prediction, the LODO evaluation results are worse compared to the LOSO inter-dataset and within dataset.
2. Regarding the LOSO evaluations, for the SBP prediction the within dataset models have slightly better MAE, but worse RMSE, meaning that the models (within dataset and inter dataset) perform similarly.
3. However, for the DBP prediction, the inter-dataset models yield the best results.

The differences between the LOSO and LODO are visualized in Figs. 2 and 3, correspondingly. Both figures present the SBP absolute values, rather than the

Table 6. Comparison of the evaluation metrics (MAE and RMSE) for the three evaluations: LOSO inter-dataset, LOSO within dataset and LODO.

BP	Evaluation type	Prediction		Calibration	
		μ(MAE)	μ(RMSE)	μ(MAE)	μ(RMSE)
SBP	LOSO inter-dataset	8.2	**8.8**	7.1	**7.8**
	LOSO within dataset	**8.0**	9.2	**6.7**	7.9
	LODO	13.0	9.2	6.7	7.9
DBP	LOSO inter-dataset	**8.7**	**9.1**	6.3	**6.7**
	LOSO within dataset	10.1	11.1	6.7	7.6
	LODO	12.3	14.9	6.6	7.6

absolute errors as provided in Table 3, of a patient referred to as patient X, from the Charis/Physionet database. The patient is chosen to be suitable since there is high variability in the BP values which is appropriate to visually represent whether the predictions follow the trend of the actual absolute values. The real BP are marked with a black line. The blue line represents the stacking approach predictions, and the green line represents the predictions after the calibration. The red line represents the performance of a simple classifier - the one that always predicts the mean value from the training set. The x-axis shows the continuous instances (samples) for the particular patient and the y-axis presents the absolute values of the SBP in mmHg. In Fig. 2 it can be perceived that in

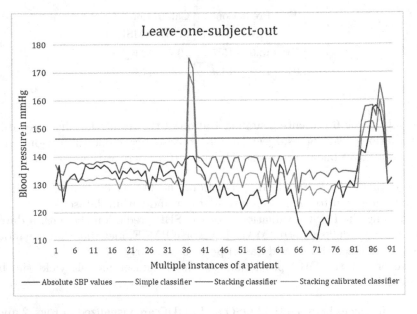

Fig. 2. Leave-one-subject-out for patient X. (Color figure online)

case of LOSO, the stacking classifier (before and after the calibration) follows the tendency of the actual BP values. On the contrary, when LODO, the stacking classifier (the blue line) is unable to predict better than the simple classifier predicting the mean BP values (the red line). Considering the results, it can be concluded that the stacking classifier needs to be provided with a training instance from the particular dataset before accurate predictions can be made.

3.3 Feature Analysis

Observing the experimental results, it can be noticed that the models are able to approximate the actual absolute BP values. In order to provide more insights into the usefulness of the complexity features, we provided additional analysis comparing each feature value with respect to the real BP values. Considering a sequence of actual BP values within 6 h period, in Fig. 4 we depicted the features values (y-axis) depending on the BP values sorted ascendingly. It can be seen that the increase in the BP values, influences the absolute value and the variability of the features.

Next, in Fig. 5 we present box-whisker plots to visualize the shape of the distributions, the mean value, and the variability of each complexity feature with respect to the three BP classes. It can be seen that for some of the features, e.g., Mobility, Complexity and Entropy, just the mean value itself has a discriminatory power for the three classes. In addition to the mean value, the variability of the feature values has some additional information. Even though, in some cases, the variability of the feature values may indicate noise in the data.

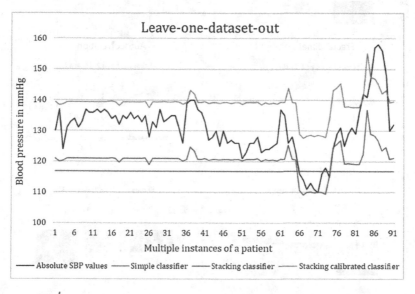

Fig. 3. Leave-one-dataset-out for patient X. (Color figure online)

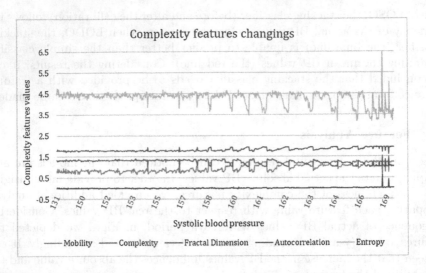

Fig. 4. BP changings tracking for patient X.

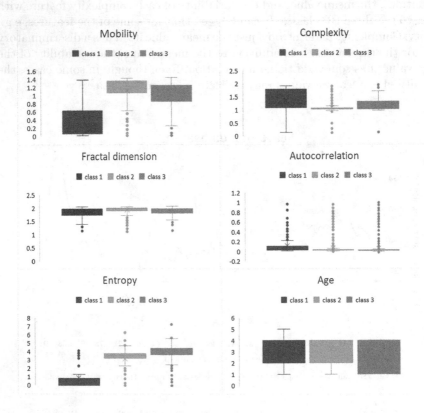

Fig. 5. Box-Whiskey plots per class for the complexity features.

3.4 Devices Evaluation

Even though all participant's groups are measured with different devices, we provide a comparison between the performance of the used wearable bio-sensors and the clinical monitors regularly used in medical practice, regarding our results. The comparison is made as a ratio between the prediction errors obtained from the experiments. Considering the gained measurements, the groups of participants (healthy, with brain injury and unhealthy), and the types of devices (bio-sensors and clinical monitors), we have compared the results for two main groups. The first group encompasses the healthy and unhealthy participants measured with wearable sensors technology (WS), and the second group comprise the participants with brain injuries measured with the regularly used clinical monitors (CM) in the hospitals. In Table 7 we have used the differences between the RMSE and MAE (R-M) to calculate the differences within the calibration and the prediction models in order to compute the ratios of errors between WS and CM. The differences between the RMSE and MAE were used to show the disparity of the actual SBP and DBP outputs and the predicted/calibrated outcome in the participant's readings. The calculations show that the errors are approximately the same for all the participants. Having computed the difference R - M for both the prediction and the calibration models, we calculated the ratios for both cases. Ratio P for SBP shows that the CM group suffers 2.6 times worse errors than the WS group; whereas in the DBP case the errors are the same. Ratio C is in regard of the condition after the calibration - the CM group suffers 3.4 times worse errors (even more than in the prediction case); whereas for DBP the errors are only 1.1 time worse.

Table 7. Concession difference.

BP	Group	Prediction (P)			Calibration (C)			Ratio P	Ratio C
		MAE (M)	RMSE (R)	R − M	MAE (M)	RMSE (R)	R − M		
SBP	WS	8.01	8.87	0.87	6.41	7.14	0.74	2.6	3.4
	CM	8.07	10.31	2.24	7.63	10.11	2.49		
DBP	WS	8.04	9.59	1.55	7.02	8.63	1.61	1.0	1.1
	CM	11.82	13.37	1.55	7.88	9.68	1.80		

4 Discussion

Our BP estimation system based on ECG sensor inputs enabled the reliable monitoring of various BP parameters on data obtained from 51 different subjects and 4 different ECG sensors. In the traditional train-validation-test evaluation of our method [24], we achieved MAE (non-calibrated/calibrated) measured in mmHg of 8.6/7.7 for SBP, 18.2/9.4 for DBP, and 13.5/8.13 for MAP. Performing a completely different approach to evaluate the methodology as presented in this paper, it turned out that the performance can be significantly improved.

The error on an unseen dataset, using another sensor, is 13 for the SBP and 12.3 for the DBP prediction (see no calibration "LODO" results in Table 5). If sensor-specific labelled dataset is provided, the MAE decreases to 8.2 for SBP and 8.7 for DBP prediction (see no calibration "LOSO inter-dataset" results in Table 3). Moreover, if person-specific labelled data is provided, the MAE decreases to 6.7 for SBP and 6.6 for DBP (see calibration results for "LODO" in Table 6). These results are close to a certified medical device for BP estimation (± 5 mmHg, and the SD within 8 mmHg according to BHS and AAMI standards [16]). Considering the time performance of the method, once the prediction model is built, the predictions can be considered real-time calculations.

5   Conclusions

Our method estimates SBP, DBP and the MAP from ECG sensor data. The method was tested on 51 different subjects and 4 different ECG sensors - part of which we obtained from online database and the rest from the database that we created. By performing leave-one-subject out evaluation, the method achieved results close to a certified medical device, especially when sensor-specific and person-specific labelled data is provided.

The proposed solution has promising real-world applications in civilian and military environments, however it should be tested with a dataset containing hundreds of diverse participants in variety of medical conditions. In the future work, the method could be enriched by an activity recognition module [7], or context-based BP estimation [8].

References

1. Blood pressure databases. http://www.webcitation.org/6ulZxAGP8
2. Ahonen, L., Cowley, B., Torniainen, J., Ukkonen, A., Vihavainen, A., Puolamäki, K.: Cognitive collaboration found in cardiac physiology: study in classroom environment. PloS One **11**(7), e0159178 (2016)
3. Bereksi-Reguig, M.A., Bereksi-Reguig, F., Ali, A.N.: A new system for measurement of the pulse transit time, the pulse wave velocity and its analysis. J. Mech. Med. Biol. **17**(01), 1750010 (2017)
4. Bittium Biosignals: Emotion faros (2016). http://www.megaemg.com/products/faros/
5. Cliff, D.P., et al.: The preschool activity, technology, health, adiposity, behaviour and cognition (PATH-ABC) cohort study: rationale and design. BMC Pediatr. **17**(1), 95 (2017)
6. Ding, H., Sarela, A., Helmer, R., Mestrovic, M., Karunanithi, M.: Evaluation of ambulatory ECG sensors for a clinical trial on outpatient cardiac rehabilitation. In: 2010 IEEE/ICME International Conference on Complex Medical Engineering (CME), pp. 240–243. IEEE (2010)
7. Gjoreski, M., Gjoreski, H., Luštrek, M., Gams, M.: How accurately can your wrist device recognize daily activities and detect falls? Sensors **16**(6), 800 (2016)
8. Gjoreski, M., Luštrek, M., Gams, M., Gjoreski, H.: Monitoring stress with a wrist device using context. J. Biomed. Inform. **73**, 159–170 (2017)
9. Goldberger, A.L., et al.: Physiobank, physiotoolkit, and physionet. Circulation **101**(23), e215–e220 (2000)

10. Hacks., C.: e-Health sensor platform V2.0 for Arduino and Raspberry Pi. https://www.cooking-hacks.com/documentation/tutorials/ehealth-biometric-sensor-platform-arduino-raspberry-pi-medical

11. Hailstone, J., Kilding, A.E.: Reliability and validity of the zephyr^{TM} bioharness^{TM} to measure respiratory responses to exercise. Meas. Phys. Educ. Exerc. Sci. **15**(4), 293–300 (2011)

12. Hsiu, H., Hsu, C.L., Wu, T.L.: A preliminary study on the correlation of frequency components between finger PPG and radial arterial BP waveforms. In: International Conference on Biomedical and Pharmaceutical Engineering, ICBPE 2009, pp. 1–4. IEEE (2009)

13. Ilango, S., Sridhar, P.: A non-invasive blood pressure measurement using android smart phones. IOSR J. Dent. Med. Sci. **13**(1), 28–31 (2014)

14. Johnstone, J.A., Ford, P.A., Hughes, G., Watson, T., Garrett, A.T.: Bioharness^{TM} multivariable monitoring device: part. i: validity. J. Sport. Sci. Med. **11**(3), 400 (2012)

15. Johnstone, J.A., Ford, P.A., Hughes, G., Watson, T., Mitchell, A.C., Garrett, A.T.: Field based reliability and validity of the bioharness^{TM} multivariable monitoring device. J. Sport. Sci. Med. **11**(4), 643 (2012)

16. Jones, D.W., Hall, J.E.: The national high blood pressure education program (2002)

17. Kim, N., et al.: Trending autoregulatory indices during treatment for traumatic brain injury. J. Clin. Monit. Comput. **30**(6), 821–831 (2016)

18. Miettinen, T., et al.: Success rate and technical quality of home polysomnography with self-applicable electrode set in subjects with possible sleep Bruxism. IEEE J. Biomed. Health Inform. (2017)

19. Mitchell, G.F.: Arterial stiffness and hypertension. Hypertension **64**(1), 13–18 (2014)

20. Morales, J.M., Díaz-Piedra, C., Di Stasi, L.L., Martínez-Cañada, P., Romero, S.: Low-cost remote monitoring of biomedical signals. In: Ferrández Vicente, J.M., Álvarez-Sánchez, J.R., de la Paz López, F., Toledo-Moreo, F.J., Adeli, H. (eds.) IWINAC 2015. LNCS, vol. 9107, pp. 288–295. Springer, Cham (2015). https://doi.org/10.1007/978-3-319-18914-7_30

21. Nitzan, M.: Automatic noninvasive measurement of arterial blood pressure. IEEE Instrum. Meas. Mag. **14**(1) (2011)

22. Rosendorff, C., et al.: Treatment of hypertension in patients with coronary artery disease. Hypertension **65**(6), 1372–1407 (2015)

23. Sahoo, A., Manimegalai, P., Thanushkodi, K.: Wavelet based pulse rate and blood pressure estimation system from ECG and PPG signals. In: 2011 International Conference on Computer, Communication and Electrical Technology (ICCCET), pp. 285–289. IEEE (2011)

24. Simjanoska, M., Gjoreski, M., Gams, M., Madevska Bogdanova, A.: Non-invasive blood pressure estimation from ECG using machine learning techniques. Sensors **18**(4), 1160 (2018)

25. Zephyr Technology: Zephyr BioHarness 3.0 user manual (2017). https://www.zephyranywhere.com/media/download/bioharness3-user-manual.pdf

26. Thomas, S.S., Nathan, V., Zong, C., Soundarapandian, K., Shi, X., Jafari, R.: BioWatch: a noninvasive wrist-based blood pressure monitor that incorporates training techniques for posture and subject variability. IEEE J. Biomed. Health Inform. **20**(5), 1291–1300 (2016)

27. Winderbank-Scott, P., Barnaghi, P.: A non-invasive wireless monitoring device for children and infants in pre-hospital and acute hospital environments (2017)

Performances of Fast Algorithms for Random Codes Based on Quasigroups for Transmission of Audio Files in Gaussian Channel

Daniela Mechkaroska[✉], Aleksandra Popovska-Mitrovikj, and Verica Bakeva Smiljkova

Faculty of Computer Science and Engineering, Ss. Cyril and Methodius University, Skopje, Macedonia
{daniela.mechkaroska,aleksandra.popovska.mitrovikj, verica.bakeva}@finki.ukim.mk

Abstract. RCBQ are cryptcodes proposed in 2007. Since then, for improving performance of these codes for data transmission through a binary-symmetric and Gaussian channels, several new algorithms have been proposed. The last proposed Fast-Cut-Decoding and Fast-4-Sets-Cut-Decoding algorithm give the best results for transmission of ordinary messages. In this paper we consider performances of these codes for transmission of audio files in Gaussian channel for different values of signal-to-noise ratio (SNR). We present and compare experimental results for bit-error and packet-error probabilities obtained using fast algorithms of RCBQ. In all experiments differences between sent and decoded audio files are considered. Also, a filter for enhancing the quality of the decoded audio files is proposed. The proposed filter makes additional cleaning of the noise in the decoded files, enabling clearer listening to the original sound even after transmission through a channel with lower values of SNR.

Keywords: Cryptcoding · Gaussian channel · Audio file · Filter Quasigroup

1 Introduction

In this paper we consider transmission of audio files through a Gaussian channel using cryptcodes based on quasigroups. In the coding algorithm of Random Codes based on Quasigroups (RCBQ) an encryption algorithm is used. Therefore these codes can correct some of the errors that occur during transmission through a noise channel and at the same time they encrypt the message. A few similar combinations of error-correcting codes and cryptographic algorithms are proposed for cryptographic proposes [4, 11, 12]. Here, we consider only error-correction capabilities of RCBQ.

© Springer Nature Switzerland AG 2018
S. Kalajdziski and N. Ackovska (Eds.): ICT 2018, CCIS 940, pp. 286–296, 2018.
https://doi.org/10.1007/978-3-030-00825-3_24

Random Codes Based on Quasigroups (RCBQ) are defined in [1]. Properties of these codes are investigated in [3,8–10]. Also, for improving the performances of these codes several new algorithms have been defined in [5–7]. The last proposed Fast-Cut-Decoding and Fast-4-Sets-Cut-Decoding algorithm [5] give the best results for transmission of ordinary messages. In this paper we compare performance of these Fast algorithms with Cut-Decoding and 4-Sets-Cut-Decoding algorithms proposed in [6,7] for transmission of audio files through a Gaussian channel. Also, we define a filter for enhancing the quality of the decoded audio files. In the decoding process of these codes, three types of errors appear: *more-candidate-error*, *null-error* and *undetected-error*. All of them make a noise in the audio files and the proposed median filter clear some damages made by first two kinds of errors. This filter cannot be applied for *undetected-errors* since we do not know where they appear.

The rest of the paper is organized on the following way. In Sect. 2, we briefly describe Cut-Decoding, 4-Sets-Cut-Decoding, Fast-Cut-Decoding and Fast-4-Sets-Cut-Decoding algorithms for RCBQ. Experimental results obtained with the four mentioned algorithms for different values of (SNR) are given in Sect. 3. In all experiments differences between sent and decoded audio files are considered. In Sect. 4, we define the filter for enhancing the decoded audio files and we analyze its ability to clean the noise. At the end, we give some conclusions for the presented results.

2   Description of Coding/Decoding Algorithms

RCBQs are designed using algorithms for encryption and decryption from the implementation of TASC (Totally Asynchronous Stream Ciphers) by quasigroup string transformation [2]. These cryptographic algorithms use the alphabet Q and a quasigroup operation $*$ on Q together with its parastrophe $\backslash$.

2.1   Description of Coding

At first, let describe Standard coding algorithm for RCBQs proposed in [1]. The message $M = m_1 m_2 \ldots m_l$ (of $N_{block} = 4l$ bits where $m_i \in Q$ and Q is an alphabet of 4-bit symbols (nibbles)) is extended to message $L = L^{(1)} L^{(2)} \ldots L^{(s)} = L_1 L_2 \ldots L_m$ by adding redundant zero symbols. The produced message L has $N = 4m$ bits $(m = rs)$, where $L_i \in Q$ and $L^{(i)}$ are sub-blocks of r symbols from Q. In this way we obtain (N_{block}, N) code with rate $R = N_{block}/N$. The codeword is produced after applying the encryption algorithm of TASC (given in Fig. 1) on the message L. For this aim, a key $k = k_1 k_2 \ldots k_n \in Q^n$ should be chosen. The obtained codeword of M is $C = C_1 C_2 \ldots C_m$, where $C_i \in Q$.

In Cut-Decoding and Fast-Cut-Decoding algorithms, instead of using (N_{block}, N) code with rate R, we use together two $(N_{block}, N/2)$ codes with rate $2R$ and for coding we apply the encryption algorithm (given in Fig. 1) two times, on the same redundant message L using different parameters (different keys or quasigroups). We obtain the codeword of the message as concatenation of two

Encryption	Decryption
Input: Key $k = k_1 k_2 \dots k_n$ and $L = L_1 L_2 \dots L_m$ **Output:** codeword $C = C_1 C_2 \dots C_m$	**Input:** The pair $(a_1 a_2 \dots a_r, k_1 k_2 \dots k_n)$ **Output:** The pair $(c_1 c_2 \dots c_r, K_1 K_2 \dots K_n)$
For $j = 1$ to m $\quad X \leftarrow L_j;$ $\quad T \leftarrow 0;$ $\quad$ For $i = 1$ to n $\quad\quad X \leftarrow k_i * X;$ $\quad\quad T \leftarrow T \oplus X;$ $\quad\quad k_i \leftarrow X;$ $\quad\quad k_n \leftarrow T$ $\quad$ **Output:** $C_j \leftarrow X$	For $i = 1$ to n $\quad K_i \leftarrow k_i;$ For $j = 0$ to $r - 1$ $\quad X, T \leftarrow a_{j+1};$ $\quad temp \leftarrow K_n;$ $\quad$ For $i = n$ to 2 $\quad\quad X \leftarrow temp \setminus X;$ $\quad\quad T \leftarrow T \oplus X;$ $\quad\quad temp \leftarrow K_{i-1};$ $\quad\quad K_{i-1} \leftarrow X;$ $\quad\quad X \leftarrow temp \setminus X;$ $\quad\quad K_n \leftarrow T;$ $\quad\quad c_{j+1} \leftarrow X;$ **Output:** $(c_1 c_2 \dots c_r, K_1 K_2 \dots K_n)$

Fig. 1. Algorithms for encryption and decryption

codewords of $N/2$ bits. In 4-Sets-Cut-Decoding and Fast-4-Sets-Cut-Decoding algorithms we use four $(N_{block}, N/4)$ codes with rate $4R$ and the codeword of the message is a concatenation of four codewords of $N/4$ bits.

2.2 Description of Decoding

The decoding in all algorithms for RCBQ is actually a list decoding and the speed of decoding process depends on the list size (a shorter list gives faster decoding).

In Standard decoding algorithm for RCBQs, after transmission through a noise channel, the codeword C will be received as message $D = D^{(1)} D^{(2)} \dots D^{(s)} = D_1 D_2 \dots D_m$ where $D^{(i)}$ are blocks of r symbols from Q and $D_i \in Q$. The decoding process consists of four steps: (i) procedure for generating the sets with predefined Hamming distance, (ii) inverse coding algorithm, (iii) procedure for generating decoding candidate sets and (iv) decoding rule.

Let B_{max} be a given integer which denotes the assumed maximum number of bit errors that occur in a block during transmission. We generate the sets $H_i = \{\alpha | \alpha \in Q^r, \quad H(D^{(i)}, \alpha) \leq B_{max}\}$, for $i = 1, 2, \dots, s$, where $H(D^{(i)}, \alpha)$ is Hamming distance between $D^{(i)}$ and α. The decoding candidate sets S_0, S_1, S_2, $\dots$, S_s are defined iteratively. Let $S_0 = (k_1 \dots k_n; \lambda)$, where λ is the empty sequence. Let S_{i-1} be defined for $i \geq 1$. Then S_i is the set of all pairs $(\delta, w_1 w_2 \dots w_{4ri})$ obtained by using the sets S_{i-1} and H_i as follows (w_j are bits). For each element $\alpha \in H_i$ and each $(\beta, w_1 w_2 \dots w_{4r(i-1)}) \in S_{i-1}$, we apply the inverse coding algorithm (i.e., algorithm for decryption given in Fig. 1) with input (α, β). If the output is the pair (γ, δ) and if both sequences γ and $L^{(i)}$ have the redundant zeros in the same positions, then the pair $(\delta, w_1 w_2 \dots w_{4r(i-1)} c_1 c_2 \dots c_r) \equiv (\delta, w_1 w_2 \dots w_{4ri})$ ($c_i \in Q$) is an element of S_i.

In Cut-Decoding algorithm, after transmission through a noisy channel, we divide the outgoing message $D = D^{(1)}D^{(2)} \ldots D^{(s)}$ in two messages $D_1 = D^{(1)}D^{(2)} \ldots D^{(s/2)}$ and $D_2 = D^{(s/2+1)} D^{(s/2+2)} \ldots D^{(s)}$ with equal lengths and we decode them parallel with the corresponding parameters. In this decoding algorithm we make modification in the procedure for generating decoding candidate sets. Let $S_i^{(1)}$ and $S_i^{(2)}$ be the decoding candidate sets obtained in the ith iteration of both parallel decoding processes, $i = 1, \ldots, s/2$. Then, before the next iteration we eliminate from $S_i^{(1)}$ all elements whose second part does not match with the second part of an element in $S_i^{(2)}$, and vice versa. In the $(i+1)^{th}$ iteration the both processes use the corresponding reduced sets $S_i^{(1)}$ and $S_i^{(2)}$.

In [7] authors proposed 4 different versions of decoding with 4-Sets-Cut-Decoding algorithm. The best results are obtained using 4-Sets-Cut-Decoding algorithm#3 and we use only this version. In this algorithm after transmitting through a noisy channel, we divide the outgoing message $D = D^{(1)}D^{(2)}...D^{(s)}$ in four messages D^1, D^2, D^3 and D^4 with equal lengths and we decode them parallel with the corresponding parameters. Similarly, as in Cut-Decoding algorithm, in each iteration of the decoding process we reduce the decoding candidate sets obtained in the four decoding processes, as follows. Let $S_i^{(1)}$, $S_i^{(2)}$, $S_i^{(3)}$ and $S_i^{(4)}$ be the decoding candidate sets obtained in the ith iteration of four parallel decoding processes, $i = 1, \ldots, s/4$. Let $V_1 = \{w_1 w_2 \ldots w_{r \cdot a \cdot i} | (\delta, w_1 w_2 \ldots w_{r \cdot a \cdot i}) \in S_i^{(1)}\}$, $\ldots, V_4 = \{w_1 w_2 \ldots w_{r \cdot a \cdot i} | (\delta, w_1 w_2 \ldots w_{r \cdot a \cdot i}) \in S_i^{(4)}\}$ and $V = V_1 \cap V_2 \cap V_3 \cap V_4$. If $V = \emptyset$ then $V = (V_1 \cap V_2 \cap V_3) \cup (V_1 \cap V_2 \cap V_4) \cup (V_1 \cap V_3 \cap V_4) \cup (V_2 \cap V_3 \cap V_4)$. Before the next iteration we eliminate from $S_i^{(j)}$ all elements whose second part is not in V, $j = 1, 2, 3, 4$.

The decoding rule is following. After the last iteration, if all reduced sets $S_{s/2}^{(1)}, S_{s/2}^{(2)}$ in Cut-Decoding (or $S_{s/4}^{(1)}, S_{s/4}^{(2)}, S_{s/4}^{(3)}, S_{s/4}^{(4)}$ in 4-Sets-Cut-Decoding) have only one element with a same second component then this component is the decoded message L. In this case, we say that we have a *successful decoding*. If the decoded message is not the correct one then we have an *undetected-error*. If the reduced sets obtained in the last iteration have more than one element then we have a *more-candidate-error*. If we obtain $S_i^{(1)} = S_i^{(2)} = \emptyset$ in some iteration of Cut-Decoding or $S_i^{(1)} = S_i^{(2)} = S_i^{(3)} = S_i^{(4)} = \emptyset$ in some iteration of 4-Sets-Cut-Decoding algorithm, then the process will finish (a *null-error* appears). But, if we obtain at least one nonempty decoding candidate set in an iteration then the decoding continues with the nonempty sets (the reduced sets are obtained by intersection of the non-empty sets only).

Also, for decreasing the number of unsuccessful decodings several modifications of decoding algorithms are defined in [6,9]. In RCBQ *null-error* occurs when more than predicted B_{max} bit errors appear in a block. Some of these errors can be eliminated if a few of iterations of the decoding process are canceled and all of them or part of them are reprocessed with a larger value of B_{max}. Therefore, in [9] authors have proposed a method for decreasing the number of *null-errors* by backtracking. Similar method with backtracking in the case of *more-candidate-error* is proposed in [6]. In the experiments with Cut-Decoding and 4-Sets-Cut-Decoding algorithms we use the following combination of these two methods with backtracking. If the decoding ends with *null-error*, then the last two iterations are canceled and the first of them is reprocessed with $B_{max}+2$. If the decoding ends with *more-candidate-error*, then the last two iterations of the decoding process are canceled and the penultimate iteration is reprocessed with $B_{max} - 1$. In the decoding of a message only one backtracking is made, except when after the backtracking for *null-error*, *more-candidate-error* appears and more than one decoding-candidate set is nonempty. In this case, we make one more backtracking for *more-candidate-error*.

As we mentioned previously, the speed of decoding process depends on the list size (a shorter list gives faster decoding). In all algorithms, the list size depends on B_{max}. For smaller values of B_{max}, we obtain shorter lists, but smaller number of corrected errors. On the other side, if B_{max} is too large, we have long lists and the process of decoding is too slow. Also, larger value of B_{max} can lead to ending of the decoding process with *more-candidate-error* (the correct message will be in the list of the last iteration, if there are no more than B_{max} errors during transmission). Therefore, with all decoding algorithms for RCBQ, *more-candidate-errors* can be obtained, although the bit-error probability of the channel is too small and the number of bit errors in a block is not greater than B_{max} (or no errors during transmission).

In order to solve this problem, in [5] authors propose modification of Cut-Decoding and 4-Sets-Cut-Decoding algorithms, called Fast-Cut-Decoding and Fast-4-Sets-Cut-Decoding algorithms. In these Fast algorithms instead of fixed value B_{max}, we start with $B_{max} = 1$. If we have successful decoding, the decoding is done. If not, we increase the value of B_{max} by 1 and repeat the decoding process with the new value of B_{max}, etc. The decoding finishes with $B_{max} = 4$ (for rate 1/4) or with $B_{max} = 5$ (for rate 1/8). Namely, in the new algorithms we try to decode the message using the shorter lists and in the case of successful decoding with small value of B_{max} ($B_{max} < 4$), we avoid long lists and slower decoding. Also, the number of *more-candidate-errors* is decreased.

3   Experimental Results

In this section, we present experimental results obtained by transmission of audio files. In our experiments we use Gaussian channel and RCBQ as an error-correcting code. We compare the results for code $(72, 576)$ obtained using four algorithms for RCBQ: Cut-Decoding, Fast-Cut-Decoding, 4-Sets-Cut-Decoding and Fast-4-Sets-Cut-Decoding for rate $R = 1/8$ and different values of SNR in Gaussian channel. In these experiments we use the audio that is consisted of one 16-bit channel with a sampling rate of 44100 Hz and it is a part of Beethoven's "Ode to joy" with a total length of approximately 4.3 s.

In the experiments we used the following parameters:

– In Cut-Decoding and Fast-Cut-Decoding algorithms - redundancy pattern: 1100 1100 1000 0000 1100 1000 1000 0000 1100 1100 1000 0000 1100 1000 1000 0000 0000 0000, for rate 1/4 and two different keys of 10 nibbles.

- In 4-Sets-Cut-Decoding and Fast-4-Sets-Cut-Decoding algorithms - redundancy pattern: 1100 1110 1100 1100 1110 1100 1100 1100 0000 for rate 1/2 and four different keys of 10 nibbles.
- In all experiments we used the same quasigroup on Q given in Table 1.

Table 1. Quasigroup of order 16 used in the experiments

*	0	1	2	3	4	5	6	7	8	9	a	b	c	d	e	f
0	3	c	2	5	f	7	6	1	0	b	d	e	8	4	9	a
1	0	3	9	d	8	1	7	b	6	5	2	a	c	f	e	4
2	1	0	e	c	4	5	f	9	d	3	6	7	a	8	b	2
3	6	b	f	1	9	4	e	a	3	7	8	0	2	c	d	5
4	4	5	0	7	6	b	9	3	f	2	a	8	d	e	c	1
5	f	a	1	0	e	2	4	c	7	d	3	b	5	9	8	6
6	2	f	a	3	c	8	d	0	b	e	9	4	6	1	5	7
7	e	9	c	a	1	d	8	6	5	f	b	2	4	0	7	3
8	c	7	6	2	a	f	b	5	1	0	4	9	e	d	3	8
9	b	e	4	9	d	3	1	f	8	c	5	6	7	a	2	0
a	9	4	d	8	0	6	5	7	e	1	f	3	b	2	a	c
b	7	8	5	e	2	a	3	4	c	6	0	d	f	b	1	9
c	5	2	b	6	7	9	0	e	a	8	c	f	1	3	4	d
d	a	6	8	4	3	e	c	d	2	9	1	5	0	7	f	b
e	d	1	3	f	b	0	2	8	4	a	7	c	9	5	6	e
f	8	d	7	b	5	c	a	2	9	4	e	1	3	6	0	f

In the experiments with Cut-Decoding and 4-Sets-Cut-Decoding algorithms we use $B_{max} = 5$ and for Fast algorithms the maximum value of B_{max} is 5.

In all decoding algorithms for RCBQ, when a null-error appears, the decoding process ends earlier and only a part of the message is decoded. Therefore in the experiments with audio files we use the following solution. In the cases when a null-error appears, i.e., all reduced sets are empty in some iteration, we take the strings without redundant symbols from all elements in the sets from the previous iteration and we find their maximal common prefix substring. If this substring has k symbols then in order to obtain decoded message of l symbols we take these k symbols and add $l - k$ zero symbols at the end of the message.

In Table 2, we present experimental results for bit-error probabilities BER_{cut}, BER_{f-cut}, BER_{4sets}, $BER_{f-4sets}$ obtained with Cut-Decoding algorithm, Fast-Cut-Decoding algorithm, 4-Sets-Cut-Decoding algorithm and Fast-4-Sets-Cut-Decoding algorithm, respectively and the corresponding packet-error probabilities PER_{cut}, PER_{f-cut}, PER_{4sets}, $PER_{f-4sets}$. For SNR smaller than -1, using of Cut-Decoding algorithm does not have sense since the values of bit-error probabilities are larger than the bit-error probability in the channel.

Table 2. Experimental results for BER and PER

SNR	BER_{cut}	BER_{f-cut}	BER_{4sets}	$BER_{f-4sets}$	PER_{cut}	PER_{f-cut}	PER_{4sets}	$PER_{f-4sets}$
-2	/	/	0.03658	0.04782	/	/	0.08451	0.08917
-1	0.04741	0.04019	0.01122	0.00825	0.08623	0.07589	0.02499	0.01644
0	0.01114	0.00713	0.00281	0.00081	0.02208	0.01418	0.00632	0.00165
1	0.00175	0.00086	0.00052	0.00003	0.00383	0.00177	0.00113	0.00012

From the experimental results given in Table 2 we can conclude that for all values of SNR the results for PER and BER obtained with Fast algorithms are better than the results obtained with Cut-Decoding and 4-Sets-Cut-Decoding algorithms, especially for larger values of SNR (low noise). Even more, for $SNR \geq 2$ bit-error and packet-error probabilities obtained with Fast algorithms are 0. Also, as it is concluded in [5], decoding with these new algorithms is much faster than with the old one.

For all experiments we also consider the differences between the sample values of the original and decoded signal. We present these analysis on graphs where the sample number in the sequence of samples consisting the audio signal is on the x-axes and the value of the sample is on the y-axis. In Fig. 2 the original audio samples are presented. Graphs for decoded audio files for considered values of SNR are given in Figs. 3, 4, 5 and 6. In Fig. 3 (for $SNR = -2$) we present only two graphs, the first one is for 4-Sets-Cut-Decoding algorithm and the second is for Fast-4-Sets-Cut-Decoding algorithm. For the other values of SNR, in Figs. 4, 5 and 6, we present 4 graphs (for decoded audio files using all 4 mentioned algorithms) in the following order: Cut-Decoding, Fast-Cut-Decoding, 4-Sets-Cut-Decoding and Fast-4-Sets-Cut-Decoding.

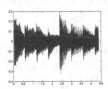

Fig. 2. Original audio samples

From these figures we can derive the same conclusions as from the tables for PER and BER. From the graphs it is evident that for all values of SNR, the results obtained using 4-Sets-Cut-Decoding algorithm are better than the results obtained with Cut-Decoding algorithm and the results obtained with the new Fast algorithms are better than correspnding results obtained with the old versions of these algorithms.

All audio files obtained in our experiments for transmission through a Gaussian channel with different SNR can be found on the following link: https:// www.dropbox.com/sh/5zq5ly6qtiho8d6/AACTQBgUDopFq9psdbaMb8BKa?dl=0.

If someone listened to these audio files, he/she would notice that as SNR decreases, the noise increases, but the original melody can be listened completely

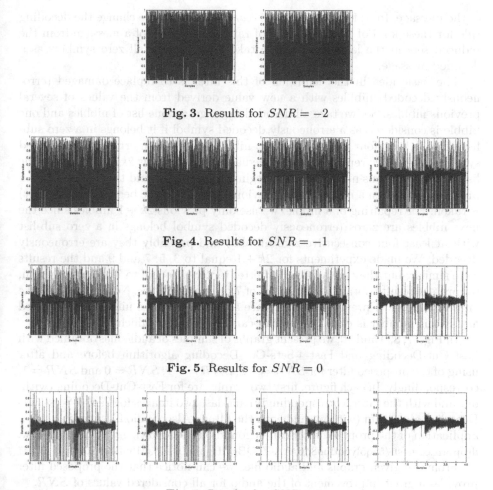

Fig. 3. Results for $SNR = -2$

Fig. 4. Results for $SNR = -1$

Fig. 5. Results for $SNR = 0$

Fig. 6. Results for $SNR = 1$

in background. This is also evident on the graphs. If we compare the graphs for files decoded with all 4 algorithms with the graph for original audio samples we can see that the samples of the original audio are contained on all graphs. This is the reason why we can still hear the original melody in the background intermixed with the noisy sounds. In order to clear some of these noise sounds, in the next subsection we propose a filter that can repair some of damages done by *null-errors* and *more-candidate-errors*.

4   Filter for Enhancing the Quality of Audio Decoded by Cryptcodes Based on Quasigroups

For repairing of a damage, a filter has to locate where it appears. Locating of the *null-errors* is easy since we add zero symbols in the place of the undecoded part of the message. In order to locate *more-candidate-errors* we change the decoding rule for these kind of errors. Instead of random selection of a message from the reduced sets in the last iteration, we take a message of all zero symbols as a decoded message.

The basic idea in the definition of this filter is to replace damaged (erroneously decoded) nibbles with a new value derived from the values of several previous nibbles. So, we take all decoded messages as one list of nibbles and one nibble is considered as a erroneously decoded symbol if it belongs in a zero sub-list with at least four consecutive zero nibbles. Then each erroneously decoded symbol (nibble) we replace with the median of the previous $2k+1$ nibbles in the list. If the erroneous nibble is at the beginning of the list and there are no $2k+1$ previous ones, then a median of all previous nibbles (to the beginning of the list) is taken. For repairing of a nibble, we use only previous $2k+1$ nibbles since the next nibbles are zeros (erroneously decoded symbol belongs in a zero sub-list with at least four consecutive zero nibbles) and probably they are erroneously decoded. We made experiments for $2k+1$ equal to 3, 5, 7 and 9 and the results were similar, but they are a little bit better for $2k+1$ equal to 7 or 9. Further on, we present results obtain with median of 7 previous nibbles. Notice that we take an odd number of previous nibbles since a median of these nibbles is computed and if this number is even the median can be a number which is not in Q.

In Figs. 7, 8, 9 and 10, we present graphs of samples of audio files obtained with Fast-Cut-Decoding and Fast-4-Sets-Cut-Decoding algorithm before and after using of the proposed filter for $SNR = -2$, $SNR = -1$, $SNR = 0$ and $SNR = 1$, correspondingly. In each figure, first two graphs are for Fast-Cut-Decoding (without and with the filter, correspondingly) and last two images for Fast-4-Sets-Cut-Decoding algorithm (without and with the filter). Also, audio files obtained after application of the proposed filter can be found on the following link: https://www. dropbox.com/sh/5zq5ly6qtiho8d6/AACTQBgUDopFq9psdbaMb8BKa?dl=0.

From the given graphs and audio files we can notice that the proposed filter provides a great improvement of the audio for all considered values of SNR.

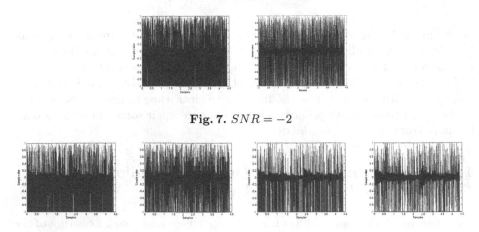

Fig. 7. $SNR = -2$

Fig. 8. $SNR = -1$

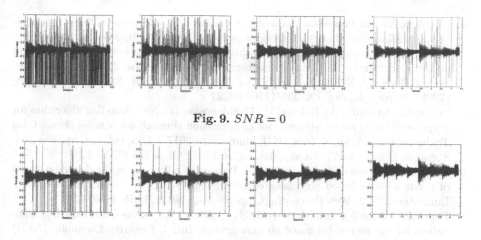

Fig. 9. $SNR = 0$

Fig. 10. $SNR = 1$

5 Conclusion

In this paper we present several experimental results obtained for transmission of audio files in Gaussian channel using four different coding/decoding algorithms for RCBQ. From all experiments made for investigation of the performances of Cut-Decoding, 4-Sets-Cut-Decoding, Fast-Cut-Decoding and Fast-4-Sets-Cut-Decoding algorithms for transmission of audio files we can conclude that all four algorithms show good performances. We can see that for all values of SNR a great part of the errors appeared during the transmission are corrected. Also, from the comparison we can conclude that Fast-4-Sets-Cut-Decoding and Fast-Cut-Decoding algorithms are better than the old versions of these algorithms. Moreover, Fast algorithms in all experiments are much faster than the old one. Also, we define a filter for improving the quality of decoded audio files. From the results obtained after application of this filter we can conclude that, for all values of SNR in a Gaussian channel, the proposed filter enables clearer listening to the original sound.

Acknowledgment. This research was partially supported by Faculty of Computer Science and Engineering at "Ss Cyril and Methodius" University in Skopje.

References

1. Gligoroski, D., Markovski, S., Kocarev, L.: Error-correcting codes based on quasi-groups. In: Proceedings of 16th International Conference on Computer Communications and Networks, pp. 165–172 (2007)
2. Gligoroski, D., Markovski, S., Kocarev, L.: Totally asynchronous stream ciphers + redundancy = Cryptcoding. In: Aissi, S., Arabnia, H.R. (eds.) Proceedings of the International Conference on Security and management, SAM 2007, pp. 446–451. CSREA Press, Las Vegas (2007)

3. Mechkaroska, D., Popovska-Mitrovikj, A., Bakeva, V.: Cryptcodes based on quasigroups in Gaussian channel. Quasigroups Relat. Syst. **24**(2), 249–268 (2016)
4. Mathur, C.N., Narayan, K., Subbalakshmi, K.P.: High diffusion cipher: encryption and error correction in a single cryptographic primitive. In: Zhou, J., Yung, M., Bao, F. (eds.) ACNS 2006. LNCS, vol. 3989, pp. 309–324. Springer, Heidelberg (2006). https://doi.org/10.1007/11767480_21
5. Popovska-Mitrovikj, A., Bakeva, V., Mechkaroska, D.: New decoding algorithm for cryptcodes based on quasigroups for transmission through a low noise channel. In: Trajanov, D., Bakeva, V. (eds.) ICT Innovations 2017. CCIS, vol. 778, pp. 196–204. Springer, Cham (2017). https://doi.org/10.1007/978-3-319-67597-8_19
6. Popovska-Mitrovikj, A., Markovski, S., Bakeva, V.: Increasing the decoding speed of random codes based on quasigroups. In: Markovski, S., Gusev, M. (eds.) ICT Innovations 2012, Web Proceedings, pp. 93–102 (2012). ISSN 1857–7288
7. Popovska-Mitrovikj, A., Markovski, S., Bakeva, V.: 4-Sets-Cut-Decoding algorithms for random codes based on quasigroups. Int. J. Electron. Commun. (AEU) **69**(10), 1417–1428 (2015)
8. Popovska-Mitrovikj, A., Bakeva, V., Markovski, S.: On random error correcting codes based on quasigroups. Quasigroups Relat. Syst. **19**(2), 301–316 (2011)
9. Popovska-Mitrovikj, A., Markovski, S., Bakeva, V.: Performances of error-correcting codes based on quasigroups. In: Davcev, D., Gómez, J.M. (eds.) ICT Innovations 2009, pp. 377–389. Springer, Heidelberg (2010). https://doi.org/10.1007/978-3-642-10781-8_39
10. Popovska-Mitrovikj, A., Markovski, S., Bakeva, V.: Some new results for random codes based on quasigroups. In: 10th Conference on Informatics and Information Technology with International Participants, Bitola, Macedonia, pp. 178–181 (2013)
11. Hwang, T., Rao, T.R.N.: Secret error-correcting codes (SECC). In: Goldwasser, S. (ed.) CRYPTO 1988. LNCS, vol. 403, pp. 540–563. Springer, New York (1990). https://doi.org/10.1007/0-387-34799-2_39
12. Zivic, N., Ruland, C.: Parallel joint channel coding and cryptography. Int. J. Electr. Electron. Eng. **4**(2), 140–144 (2010)

Author Index

Printed in the United States
By Bookmasters